IUTAM Bookseries

Volume 44

The IUTAM Bookseries publishes the refereed proceedings of symposia organised by the International Union of Theoretical and Applied Mechanics (IUTAM).

Every two years the IUTAM General Assembly decides on the list of IUTAM Symposia. The Assembly calls upon the advice of the Symposia panels. Proposals for Symposia are made through the Assembly members, the Adhering Organizations, and the Affiliated Organizations, and are submitted online when a call is launched on the IUTAM website.

The IUTAM Symposia are reserved to invited participants. Those wishing to participate in an IUTAM Symposium are therefore advised to contact the Chairman of the Scientific Committee in due time in advance of the meeting. From 1996 to 2010, Kluwer Academic Publishers, now Springer, was the preferred publisher of the refereed proceedings of the IUTAM Symposia. Proceedings have also been published as special issues of appropriate journals. From 2018, this bookseries is again recommended by IUTAM for publication of Symposia proceedings.

Indexed in Ei Compendex and Scopus.

Kentaro Kato · Ayumu Inasawa ·
Masaharu Matsubara
Editors

Proceedings of the 10th IUTAM Symposium on Laminar-Turbulent Transition

IUTAM Laminar-Turbulent Transition;
September 2–6, 2024, Nagano, Japan

Springer

Editors
Kentaro Kato
Department of Mechanical Systems Engineering
Shinshu University
Nagano, Japan

Ayumu Inasawa
Department of Aeronautics and Astronautics
Tokyo Metropolitan University
Hino, Tokyo, Japan

Masaharu Matsubara
Department of Mechanical Systems Engineering
Shinshu University
Nagano, Japan

ISSN 1875-3507 ISSN 1875-3493 (electronic)
IUTAM Bookseries
ISBN 978-981-96-9828-8 ISBN 978-981-96-9829-5 (eBook)
https://doi.org/10.1007/978-981-96-9829-5

This Springer imprint is published by the registered company Springer Nature Singapore Pte Ltd.
The registered company address is: 152 Beach Road, #21-01/04 Gateway East, Singapore 189721, Singapore

Scientific Committee Appointed by the IUTAM Bureau

Yohann Duguet, University Paris-Saclay, France
Ardeshir Hanifi, KTH Royal Institute of Technology, Sweden
Markus J. Kloker, University of Stuttgart, Germany
Masaharu Matsubara, Shinshu University, Japan
Marcello A. F. Medeiros, University of São Paulo, Brazil
Peter Schmid, King Abdullah University of Science and Technology, Saudi Arabia
Tamer Zaki, Johns Hopkins University, USA
Jerzy M. Floryan, IUTAM Representative, University of Western Ontario, Canada

The picture of the participants

Local National Organising Committee

Ayumu Inasawa, Secretary, Tokyo Metropolitan University
Kentaro Kato, Secretary, Shinshu University
Genta Kawahara, Osaka University
Masaharu Matsubara, Chairman, Shinshu University
Shoji Sakaue, Osaka Metropolitan University
Kosuke Suzuki, Shinshu University
Yuji Tasaka, Hokkaido University
Naoko Tokugawa, Japan Aerospace Exploration Agency

Organization

Program Committee Chairs

P. Henrik Alfredsson, KTH Royal Institute of Technology, FLOW Engineering Mechanics, Stockholm, Sweden
Ayumu Inasawa, Tokyo Metropolitan University, Department of Aeronautics and Astronautics, Hino, Japan
Kentaro Kato, Shinshu University, Department of Mechanical Systems Engineering, Nagano, Japan
Masaharu Matsubara, Shinshu University, Mechanical Systems Engineering, Nagano, Japan

Reviewers

P. Henrik Alfredsson, KTH Royal Institute of Technology, FLOW Engineering Mechanics, Stockholm, Sweden
Hans P. Barth, German Aerospace Center (DLR), Department of High Speed Configurations, Göttingen, Germany
Marco Costantini, German Aerospace Center (DLR), Institute of Aerodynamics and Flow Technology, Göttingen, Germany
Christoph Hader, University of Arizona, Department of Aerospace and Mechanical Engineering, Tucson, USA
Ardeshir Hanifi, KTH Royal Institute of Technology, FLOW Engineering Mechanics, Stockholm, Sweden
Shumpei Hara, Doshisha University, Department of Mechanical Engineering and Science, Kyotanabe, Japan
Masashi Ichimiya, Tokushima University, Department of Mechanical Science, Tokushima, Japan

Ayumu Inasawa, Tokyo Metropolitan University, Department of Aeronautics and Astronautics, Hino, Japan
Markus J. Kloker, University of Stuttgart, Institute of Aerodynamics and Gas Dynamics, Stuttgart, Germany
Masaharu Matsubara, Shinshu University, Mechanical Systems Engineering, Nagano, Japan
Massimo Miozzi, Institute of Marine Engineering, National Research Council (CNR), Rome, Italy
Takuto Ogawa, Tohoku University, Institute of Fluid Science, Sendai, Japan
Daniel Rodriguez, Universidad Politécnica de Madrid, Department of Applied Mathematics, Madrid, Spain
Tristan M. Römer, University of Stuttgart, Institute of Aerodynamics and Gas Dynamics, Stuttgart, Germany
Shoji Sakaue, Osaka Metropolitan University, Department of Aerospace Engineering, Sakai, Japan
Neil Sandham, University of Southampton, Aeronautics and Astronautics, Southampton, UK
Leandro Souza, University of Sao Paulo, Department of Applied Mathematics and Statistics, Sao Carlos, Brazil
Christian Stemmer, Technical University of Munich, Engineering Physics and Computation, Garching b. Munich, Germany
Naoko Tokugawa, Japan Aerospace Exploration Agency (JAXA), Aviation Technology Directorate, Mitaka, Japan
Francesco Tocci, German Aerospace Center (DLR), Department of High Speed Configurations, Göttingen, Germany
Takahiro Tsukahara, Tokyo University of Science, Department of Mechanical and Aerospace Engineering, Noda, Japan
Aiko Yakeno, Tohoku University, Sendai, Japan

Participants

P. Henrik Alfredsson, KTH Royal Institute of Technology, Sweden
Masahito Asai, Tokyo Metropolitan University, Japan
Elliot J. Badcock, Imperial College London, UK
Marina Barahona, Delft University of Technology, The Netherlands
Hans P. Barth, German Aerospace Center (DLR), Germany
Pietro C. Boldini, Delft University of Technology, The Netherlands
Junyoung Byeon, Pusan National University, Republic of Korea
José I. Cardesa, ONERA—Toulouse, France
Jordi Casacuberta, Delft University of Technology, The Netherlands
Yifu Chen, Delft University of Technology, The Netherlands
Giuseppe Chiapparino, Technical University of Munich, Germany
Suhun Cho, Gwangju Institute of Science and Technology, Republic of Korea
Marco Costantini, German Aerospace Center (DLR), Germany
Stuart Craig, University of Arizona, USA
Jeffrey D. Crouch, The Boeing Company, USA
Benjamin Dimond, German Aerospace Center (DLR), Germany
Ming Dong, Chinese Academy of Sciences, China
Yohann Duguet, Université Paris-Saclay, France
Jenna Eppink, NASA Langley Research Center, USA
Hermann F. Fasel, University of Arizona, USA
Jerzy M. Floryan, University of Western Ontario, Canada
Alessandro Franchini, DynFluid, Arts et Métiers ParisTech, France
Jens H. M. Fransson, KTH Royal Institute of Technology, Sweden
Ricardo Frantz, DynFluid, Arts et Métiers ParisTech, France
Song Fu, Tsinghua University, China
Kozo Fujii, Tokyo University of Science, Japan
Stanislaw W. Gepner, Warsaw University of Technology, Poland
Erwin R. Gowree, ISAE-SUPAERO, France
Paul Griffiths, Aston University, UK
Zhengfei Guo, University of Stuttgart, Germany
Mattias K. Gustavsson, Hot Disk AB, Sweden

Juan Guzman-Inigo, City University of London, UK
Christoph Hader, University of Arizona, USA
Mohammed H. Hamede, Brandenburg University of Technology, Germany
Riwan Hammachi, ONERA-Toulouse, France
Ardeshir Hanifi, KTH Royal Institute of Technology, Sweden
Shumpei Hara, Doshisha University, Japan
Stefan Hein, German Aerospace Center (DLR), Germany
Yoshiki Hiruta, Kyoto University, Japan
Tomotaka Ichikawa, Shibaura Institute of Technology, Japan
Masashi Ichimiya, Tokushima University, Japan
Ayumu Inasawa, Tokyo Metropolitan University, Japan
Takahiro Ishida, Japan Aerospace Exploration Agency (JAXA), Japan
Yoshio Ishii, Soka University, Japan
Solkeun Jee, Gwangju Institute of Science and Technology, Republic of Korea
Minjae Jeong, Gwangju Institute of Science and Technology, Republic of Korea
Michael Karp, Technion—Israel Institute of Technology, Israel
Kentaro Kato, Shinshu University, Japan
Genta Kawahara, Osaka University, Japan
Takuya Kawata, Shibaura Institute of Technology, Japan
Rich Kerswell, Cambridge University, UK
Markus J. Kloker, University of Stuttgart, Germany
Kazuki Kohyama, Nippon Telegraph and Telephone Corporation, Japan
Marios Kotsonis, Delft University of Technology, The Netherlands
Adrian Kozluk, Osaka University, Japan
Viswanathan Kumaran, Indian Institute of Science, India
Hyeonjin Lee, Pusan National University, Republic of Korea
Ray-Sing Lin, Gwangju Institute of Science and Technology, Republic of Korea
Jens Lunte, German Aerospace Center (DLR), Germany
David J. Lusher, Japan Aerospace Exploration Agency (JAXA), Japan
Masaharu Matsubara, Shinshu University, Japan
Yuki Matsukawa, Tokyo University of Science, Japan
Marcello A. F. Medeiros, University of Sao Paulo, Brazil
Iliya Milman, Technion—Israel Institute of Technology, Israel
Chloe Mimeau, Conservatoire National des Arts et Métiers, France
Massimo Miozzi, CNR-INM, Italy
Babak Mohammadikalakoo, Delft University of Technology, The Netherlands
Pradeep Moise, Indian Institute of Technology Kanpur, India
Mohammad Moniripiri, KTH Royal Institute of Technology, Sweden
Shingo Motoki, Osaka University, Japan
Shahid Mughal, Imperial College London, UK
Masato Nagata, Kyoto University, Japan
Kosuke Nakagawa, Tokyo University of Science, Japan
Igor Nesteruk, National Academy of Sciences of Ukraine, Ukraine
Martin Oberlack, TU Darmstadt, Germany
Takuto Ogawa, Tohoku University, Japan

Keisuke Ohira, Ryoyu Systems Co., Ltd., Japan
Yuya Ohmichi, Japan Aerospace Exploration Agency (JAXA), Japan
Duncan M. Ohno, Technion—Israel Institute of Technology, Israel
Shinya Okino, Kyoto University, Japan
Pedro Paredes, National Institute of Aerospace, USA
Donghun Park, Pusan National University, Republic of Korea
Victoria Maria Prieto Rueda, ONERA and ISAE-SUPAERO, France
Vishnu Ravindran, Okinawa Institute of Science and Technology, Japan
Nassim Razaaly, Pprime Institute, France
Georgios Rigas, Imperial Collage London, UK
Alberto F. Rius-Vidales, Delft University of Technology, The Netherlands
Jean-Christophe Robinet, DynFluid, Arts et Métiers ParisTech, France
Daniel Rodriguez, Universidad Politecnica de Madrid, Spain
Jonas Roller, Heidelberg University, Germany
Tristan M. Römer, University of Stuttgart, Germany
Shoji Sakaue, Osaka Metropolitan University, Japan
Neil D. Sandham, University of Southampton, UK
Taraneh Sayadi, Conservatorie National des Arts et Métiers, France
Oliver T. Schmidt, University of California San Diego, USA
Atsushi Sekimoto, Okayama University, Japan
Adam E. Skora, University of Arizona, USA
Michael K. Sleeman, California Institute of Technology, USA
Loic Sombaert, DAAA-ONERA, France
Leandro F. Souza, University of Sao Paulo, Brazil
Marlon Sproesser Mathias, University of Sao Paulo, Brazil
Christian Stemmer, Technical University of Munich, Germany
Sharon O. Stephen, University of Sydney, Australia
Caihong Su, Tianjin University, China
Divek Surujhlal, German Aerospace Center (DLR), Germany
Peter S. B. Szabo, Brandenburg University of Technology, Germany
Kunihiko Taira, UCLA, USA
Yuya Taki, Soka University, Japan
Sumit S. Tambe, Indian Institute of Science, India
Nobutaka Taniguchi, The University of Tokyo, Japan
Hideyuki Tanno, Japan Aerospace Exploration Agency (JAXA), Japan
Alexander Theiss, German Aerospace Center (DLR), Germany
Francesco Tocci, German Aerospace Center (DLR), Germany
Naoko Tokugawa, Japan Aerospace Exploration Agency (JAXA), Japan
Takahiro Tsukahara, Tokyo University of Science, Japan
Manas M. Vashishtha, Indian Institute of Space Science and Technology, India
Alexander Wagner, German Aerospace Center (DLR), Germany
Andre Weingartner, KTH Royal Institute of Technology, Sweden
Christoph Wenzel, University of Stuttgart, Germany
Xuesong Wu, Imperial College London, UK
Youcheng Xi, Tsinghua University, China

Aiko Yakeno, Tohoku University, Japan
Shuto Yatsuyanagi, Japan Aerospace Exploration Agency (JAXA), Japan
Imge Yigili, KTH Royal Institute of Technology, Sweden
Zhenyang Yuan, KTH Royal Institute of Technology, Sweden
Zhongyu Zhang, Tianjin University, China

Preface

The tenth IUTAM Symposium on Laminar-Turbulent Transition was held on 2–6 September 2024 at Shinshu University, Nagano, Japan. This series of symposia started in Stuttgart in 1979 and the symposia have been held every five years since then. This was the second time the symposium was held in Japan; the first time was 30 years ago, in 1994 at Tohoku University in Sendai. Six of the participants in Nagano did also participate in Sendai.

The program consisted of invited plenary and keynote talks, as well as contributed oral and poster presentations, in total more than one hundred contributions with almost 135 registered participants representing 18 countries from five continents.

There were eight internationally renowned, invited plenary and keynote speakers.

Rich Kerswell, (Cambridge University, UK), Marios Kotsonis (Delft University of Technology, the Netherlands), Pedro Paredes (National Institute of Aerospace, USA), Taraneh Sayadi (Conservatoire National des Arts et Métiers, France), Henrik Alfredsson (KTH Royal Institute of Technology, Sweden), George Rigas (Imperial College, UK), Daniel Rodriguez (Universidad Politecnica de Madrid, Spain) and Takahiro Tsukahara (Tokyo University of Science, Japan).

In addition a closing talk was given by Prof. Hermann Fasel who made a historical exposé on how computations for stability and transition have evolved over the period since the first symposium in 1979 (at which he attended and also was the co-editor of the proceedings). The title of the presentation was: IUTAM Symposia Laminar-Turbulent Transition: History of employing CFD.

The present volume contains a selected number of reviewed contributions representing the width of the subject areas presented at the symposium. A broad description of the various topics may be listed according to the session titles of the symposium program:

- High-speed flows
- Boundary layer transition
- Cross-flow instability
- Free-stream turbulence
- Roughness

- Separation
- General instability
- Complex flows

The talks and poster presentations had a balanced mixture of theoretical, numerical and experimental approaches, showing clearly that in order to move the area of instability and transition forward all these approaches are needed. That experiments still have a role to play is especially true for boundary layer transition and separation at all speed ranges, since boundary conditions (such as roughness, free-stream turbulence, sound, heat transfer etc.) play an important role for how disturbances enter into and develop in the system.

In addition to the scientific program, Wednesday afternoon was set aside for visits to some different local cultural sites (Matsumoto Castle, Togakushi Chusha and Okusha Shrines) and the participants had the chance to practice tea ceremony, "kendo"—Japanese styled fencing—, "kyudo"—Japanese styled archery—, playing the "koto" (Japanese harp), enjoy a stay at an "onsen" (hot spring) and discover the great use of fluid dynamics in a "wasabi" farm, and feeling circumstances how to establish faith based on ambient nature.

During the symposium banquet on Thursday evening a traditional Japanese style serving with eight different local dishes was enjoyed together with drinks, beer and local assortment of sake.

The symposium was held in a congenial atmosphere with ample time for scientific exchanges after the talks, during the poster sessions, lunches and coffee breaks. We were also lucky that the typhoon named Shanshan approaching Japan the days before the symposium more or less dissipated on September 1, and all participants were able to reach Nagano safely and without delay.

The Local Organising Committee would like to acknowledge the financial and in-kind support received from several sources: Nagano Prefecture, Nagano City, Shinshu University and SURCAS. We also thank the Scientific committee for their work selecting the invited speakers and evaluating the submitted abstracts, the anonymous reviewers for scrutinizing the submitted contributions for this volume as well as the chairpersons for leading the various sessions during the symposium.

On behalf of the Local National Organising Committee

Nagano, Japan — Masaharu Matsubara
Hino, Japan — Ayumu Inasawa
Nagano, Japan — Kentaro Kato
Kosuke Suzuki

Contents

Chapter 1
Turbulent Spot Measurements: Wall-Pressure and Streamwise-Velocity Correlation

Imge Yigili, André Weingärtner, and Jens H. M. Fransson

Abstract Turbulent spots are essential in the late stage of the transition process from laminar to turbulent flow in boundary layers and exhibit complex spatial and temporal characteristics. This study presents a novel experimental approach to investigate the properties of turbulent spots. A unique measurement setup is introduced, integrating wall-mounted microphones to record wall-pressure fluctuations and hot-wire anemometry for time-resolved velocity signals. Our experimental results validate the effectiveness of the technique, demonstrating its ability to grasp the propagation characteristics of turbulent spots with high fidelity. The findings underscore the potential of this approach for advancing our understanding of boundary-layer transition processes and the nature of turbulent spots.

Keywords Turbulent spots · Hot-wire anemometry · Wall-mounted microphones

1.1 Introduction

Turbulent spots are localized regions of turbulence within an otherwise laminar boundary layer and appear in the late stage of most transition scenarios. The spot production rate and their downstream development play a critical role in predicting the extent of the transition zone in boundary layers. They are believed to originate from a secondary instability on the modulated base flow by a primary instability, such as Tollmien-Schlichting waves, unsteady streamwise streaks induced by free-stream turbulence (FST) and velocity deficit/excess behind wall-mounted roughness elements. Turbulent spots were first reported by Emmons [1] as patches of turbulent flow that have similar properties to a fully turbulent boundary layer, and with characteristic spreading parameters that typify the transition region in a boundary layer.

I. Yigili · A. Weingärtner · J. H. M. Fransson (✉)
FLOW, Department of Engineering Mechanics, KTH Royal Institute of Technology, Stockholm, Sweden
e-mail: jensf@kth.se

K. Kato et al. (eds.), *Proceedings of the 10th IUTAM Symposium on Laminar-Turbulent Transition*, IUTAM Bookseries 44, https://doi.org/10.1007/978-981-96-9829-5_1

Previous studies have investigated artificially generated turbulent spots to explore their growth and propagation characteristics, employing techniques such as hot-wire anemometry, particle image velocimetry (PIV) [2], and visualization techniques [3] to capture their dynamic behavior. It has been concluded that several factors including the Reynolds number, pressure gradient, surface roughness, free-stream turbulence, etc. affect the shape and growth characteristics of a turbulent spot. Yet, some characteristics of the spots have been reported to be consistent in many studies. The general shape of a turbulent spot exhibits an arrowhead shape [4], with different leading edge and trailing edge velocities. In the leading edge region of the spot, there is excess velocity over a region of velocity deficit in the wall-normal direction, which is denoted as the *overhang* region [5]. Independent of the scale of the spot, the leading and trailing edge speeds are reported to be around $0.88U_\infty$ [6] and $(0.5–0.62)U_\infty$ [2, 4, 7], respectively. The mean spot propagation speed is measured to be about $0.7U_\infty$ [8] and the half-spreading angle to be 8–11° [7].

In this study, an intermittency estimation technique is employed to track the turbulent spots. To track individual spots, spanwise wall-mounted microphone arrays are utilized to measure the signature of wall-pressure fluctuations at different streamwise positions at the same time. The microphones are mounted under individual pinhole orifices and the turbulent part of the acquired signal is identified using an in-house intermittency detection code [10]. Simultaneous microphone measurements with a hot-wire anemometry probe are performed to capture a turbulent spot in all directions in space.

1.2 Experimental Setup and Data Acquisition

Measurements are performed in the Minimum Turbulence Level (MTL) wind tunnel at the Fluid Physics Laboratory at KTH Royal Institute of Technology. The test section of the MTL is 7 m long with a cross-sectional area of $1.2 \times 0.8\,\text{m}^2$. The maximum speed is 69 m/s and has a relatively low background disturbance level of 0.025%. A 4.5 m long flat plate is mounted and covers the whole witdth of the test section. To maintain a short pressure-gradient region, the plate is designed with an asymmetric leading edge and a trailing edge flap.

The plate is equipped with a total of 1320 wall-mounted microphones, which consist of 101 individual streamwise microphones and 23 spanwise microphone arrays with 53 microphones each. The spanwise arrays are designed such that they are placed following a zigzag pattern to maximize the spatial resolution. The center-to-center distance between the microphones is 4 mm, which results in a spatial resolution of 2 mm in the spanwise direction. The total width of a spanwise array is 104 mm. The microphones are mounted under an individual pinhole with a diameter of 0.4 mm. The arrays are mounted from the bottom of the plate, with a gradually increasing distance in the downstream direction. Four spanwise arrays, i.e. a total of 212 microphones, can be amplified and sampled simultaneously. The sampling frequency is 5 kHz and

Fig. 1.1 Simultaneous measurements with the hot-wire and wall-mounted microphone arrays

an NI cDAQ-9189 equipped with eight NI9205 modules are used for data acquisition. Further details of the setup can be found in [9].

A traversable hot-wire anemometry probe was employed simultaneously with the microphone measurements, to obtain time-resolved velocity signals localized in space (cf. Fig. 1.1). Data was collected at a sampling frequency of 5 kHz. The probe was calibrated in the freestream inside the tunnel against the dynamic pressure obtained from a Prandtl tube connected to a Furness FCO560 differential manometer. Artificial turbulent spots are generated by means of a puff through a hole with a diameter of 0.4 mm on the flat plate. The puff signal is a sinusoidal burst with one period and a frequency of 100 Hz, and is generated using a function generator, supplied to an audio amplifier driving a loudspeaker, which is connected to the hole on the flat plate through a tightly sealed silicon hose.

Two hundred turbulent spots are generated and mean indicator function, $I(t)$, is calculated based on the wall-pressure fluctuations measured with the microphones. The intermittency detection algorithm uses a high-pass filter on the microphone signals, where the cut-off frequency is chosen based on a visual inspection of the microphone signals obtained from all four spanwise arrays. Then, a criterion function, $C(t)$, is computed using the Hilbert transform of the data, over successive small smoothing intervals based on the sampling frequency. A user-set treshold value, C^{th} is chosen as a cut-off, to identify the laminar and turbulent regions of the signal according to

$$I(t) = \begin{cases} 1 & \text{if } C(t) \geq C^{th} \\ 0 & \text{if } C(t) < C^{th} \end{cases}$$

Further details regarding the intermittency detection algorithm can be found in [10].

1.3 Results

1.3.1 Microphone Measurements

Wall-pressure signatures of artificially generated turbulent spots are measured and the phase-average for 200 spots are shown in Fig. 1.2a. Time information is converted to spatial information using a convective velocity of $U_{\text{spot}} = 0.7U_\infty$:

$$x^* = U_{\text{spot}}(t_{\text{mid}} - t) + x_{\text{meas.}} \tag{1.1}$$

where t_{mid} is the average of the leading and trailing edge time steps, and $x_{\text{meas.}}$ is the location of the microphone array. With this setup, it is possible to identify the leading edge and trailing edge of the spots at each measurement location. Hence, the propagation characteristics of the spots can be easily calculated.

In many previous studies, the propagation characteristics were calculated based on the ensemble-average of many spots. With the current setup, due to the capacity of measuring different streamwise positions at the same time, individual spots can be

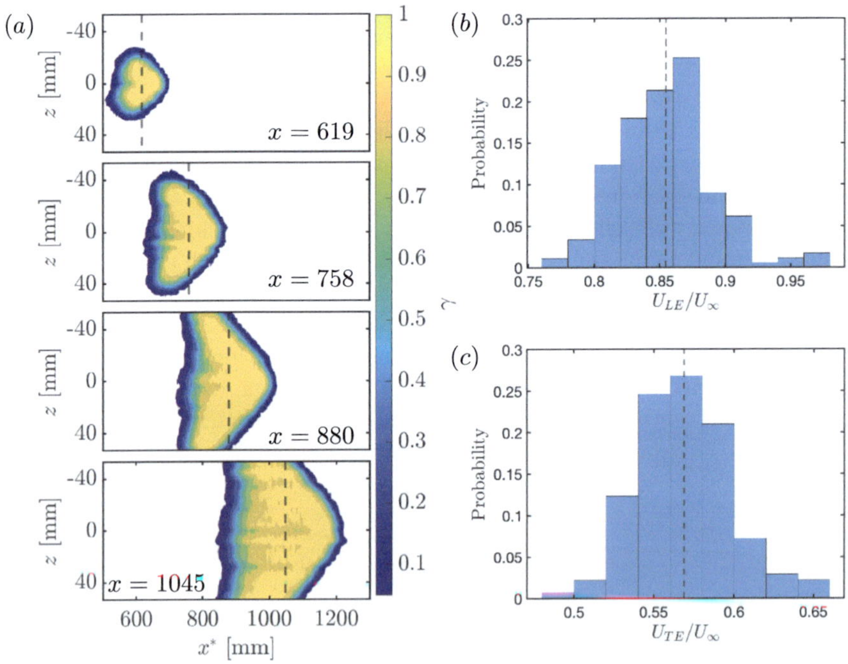

Fig. 1.2 **a** Intermittency factor (γ) averaged over 200 turbulent spots. From the top to the bottom, each figure shows the measurements with one microphone array. Dashed lines show the streamwise locations of the arrays. Histograms of leading edge **b** and trailing edge **c** velocities the artificial turbulent spots. Vertical dashed lines indicate the mean values

measured and the statistics can be investigated. In Fig. 1.2a, the intermittency factor, γ, calculated as the averaged value of $I(t)$ for each time step is plotted.

The leading and trailing edges are detected to be the first and last time steps that are identified as turbulent in the signal by the microphone closest to the center line of the spot. The dashed lines in Fig. 1.2b and c show the mean values for individual spots of the leading and trailing edge speeds. The values are consistent with what is reported in the literature, showing that the new setup is capable of measuring and tracking turbulent spots accurately.

The spanwise edges of the turbulent spots are detected as the first points detected as turbulent, in a spanwise scan. The half-spreading angle (α) is calculated as the slope of the line connecting the two edge points, and the mean value for all spots give 9.9°, which is consistent with the previously reported values. However, as observed in Fig. 1.2a, the spot already becomes larger than the measurement domain in the third array, hence, the spreading angle calculation can only be done using the first two measurements.

1.3.2 Microphone and Hot-Wire Measurements

Simulatenous measurements with the hot-wire and microphones are performed to investigate the correlation between the velocity and wall-pressure signals of a turbulent spot. The hot-wire is traversed in the wall-normal direction while the microphone measures the wall-pressure signature of the spots continuously. The current setup enables the capturing of an individual turbulent spot in the streamwise and spanwise directions, and in the wall-normal direction through ensemble average.

Figure 1.3a shows that the microphones cannot capture the overhang region. Hence, the turbulent part of the wall-pressure signal is somewhat different than the turbulent part of the velocity signal.

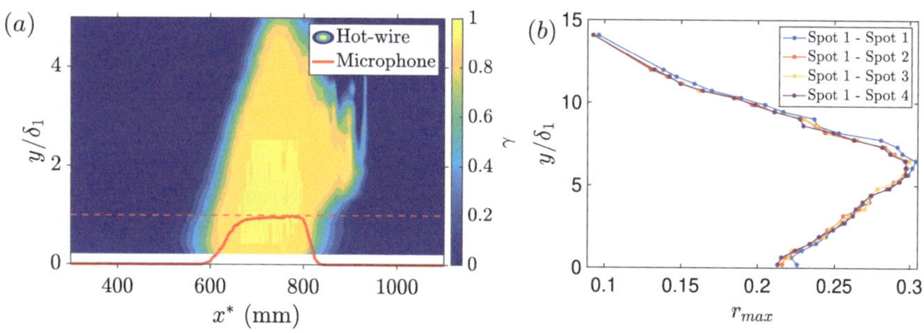

Fig. 1.3 **a** Intermittency factor (γ) measured by the hot-wire in the wall-normal direction. The red signal is obtained with the central microphone. **b** Maximum correlation coefficient ($r_{\max}$) between the velocity and wall-pressure signals, calculated at different wall-normal locations. Different cases represent different spots measured with the hot-wire and the microphones

Figure 1.3b shows the maximum correlation coefficient between the time-resolved velocity signal measured by the hot-wire with the mean flow modulation removed, and the wall-pressure fluctuations measured by the microphones. The microphones have a cut-off frequency of 100 Hz, hence, they only detect the high frequency content. Thus, only the turbulent parts of both signals are taken into account and maximum correlation is found for different time lags. The hot-wire is traversed in the wall-normal direction and there is a peak in the maximum correlation coefficient at around $y/\delta_1 = 5$. However, the maximum correlation value is slightly lower when spot 1 measured with the microphones is correlated to another spot measured with the hot-wire. Thus, it can be concluded that the turbulent fluctuations in the velocity signal are hardly correlated to the turbulent fluctuations in the wall-pressure signal for a turbulent spot. Since the two signals have no physical relation, the correlation maximum is only due to a general similarity of the signals.

1.4 Conclusions

A novel measurement setup is used to acquire time-resolved velocity signals and wall-pressure fluctuations simultaneously with hot-wire anemometry and wall-mounted microphones, respectively. Artificial turbulent spots are generated and their wall-pressure signatures are recorded. The leading and trailing edge velocities are calculated to be $0.85U_\infty$ and $0.56U_\infty$, respectively. The half-spreading angle is calculated to be 9.9°. These values are in agreement with what has been reported previously in the literature, which proves that the setup is capable of tracking turbulent spots accurately.

Simultaneous measurements with the hot-wire and the microphones show that the microphones cannot capture the overhang region of a turbulent spot. The correlation of the velocity signals obtained with the hot-wire and the wall-pressure signal obtained with the microphones change very slightly when calculated for different spots, which suggests that there is a only very weak correlation.

References

1. Emmons, H.W.: The Laminar-Turbulent transition in a boundary layer-Part I. J. Aeronaut. Sci. (Institute of the Aeronautical Sciences) **18**(7), 490–498 (1951). https://doi.org/10.2514/8.2010
2. Schröder, A., Geisler, R., Elsinga, G.E., Scarano, F.: Investigation of a turbulent spot and a tripped turbulent boundary layer flow using time-resolved tomographic PIV. Exp. Fluids **44**, 305–316 (2008). https://doi.org/10.1007/s00348-007-0403-2
3. Matsui, T.: Visualization of turbulent spots in the boundary layer along a flat plate in a water flow. In: Fasel, E.A. (ed.) Laminar-Turbulent Transition, IUTAM Symposium, pp. 289–296 (1980)
4. Wygnanski, I., Sokolov, M., Friedman, D.: On a turbulent 'spot' in a laminar boundary layer. J. Fluid Mech. **78**, 785–819 (1976)

5. Gad-El-Hak, M., Blackwelder, R.F., Riley, J.J.: On the growth of turbulent regions in laminar boundary layers. J. Fluid Mech. **110**, 73–95 (1981)
6. SavaÃƒÂ¡Ã‚Â¹Ã‚Â¡, Ö., Coles, D.: Coherence measurements in synthetic turbulent boundary layers. J. Fluid Mech. **160**, 421–446 (1985)
7. Schubauer, G.B., Klebanoff, P.S.: Contributions on the mechanics of boundary-layer transition. NASA Technical Report (1956). https://ntrs.nasa.gov/citations/19930092285
8. Ching, C.Y., LaGraff, J.E.: Measurement of turbulent spot convection rates in a transitional boundary layer. Exp. Thermal Fluid Sci. **11**, 52–60 (1995)
9. Weingärtner, A.: Experimental studies on laminar boundary layers: instability, transition and control. Ph.D. Thesis. KTH Royal Institute of Technology (2024)
10. Mamidala, S.B., Weingärtner, A., Fransson, J.H.M.: Leading-edge pressure gradient effect on boundary layer receptivity to free-stream turbulence. J. Fluid Mech. **935**(A30) (2022)

Chapter 2
DNS Study of Flow Mechanism Contributing to Drag Reduction Over Distributed Micro Roughness

Takuto Ogawa and Aiko Yakeno

Abstract Direct numerical simulations (DNSs) were performed for flow near the wall surface with a distributed micro roughness (DMR) arrangement with various roughness densities. The Reynolds number was set to $Re = 3,535$ based on the laminar boundary-layer thickness at the inlet boundary (δ_s), with reference to the previous study (Hamada, Yakeno, and Obayashi, *IJHFF*, Vol. 104, 2023). The lengths of the computational domain in the streamwise and spanwise directions are $L_x/\delta_s =$ 56.5 and $L_y/\delta_s = 11.3$, respectively. An artificial disturbance was added at the inlet boundary to induce Tollmien-Schlichting (T-S) waves. The results of DNSs show that the DMR with the appropriate roughness density achieves a skin friction coefficient (C_f) reduction in comparison with the smooth flat plate case. The gradual collapse of the large-scale vortex structure suppresses the abrupt fluctuation in the C_f and increase in turbulent kinetic energy, thus achieving the C_f reduction.

Keywords Transitional flow · Gaussian roughness · Roughness density

2.1 Introduction

Innovative technologies to improve aircraft aerodynamics are needed to reduce environmental impact and achieve sustainable air transportation in the face of future aviation demand growth [1]. In general, skin friction drag is one of the most significant aerodynamic drag form for commercial aircraft. Riblets can reduce friction drag by up to several percent under ideal flow conditions [2]. However, friction drag increases when the deviation between the streamwise direction and the riblet direction exceeds a certain level. In addition, it is difficult to place a riblet on an aircraft because the fuselage and wings are three-dimensionally curved and the wing angle is set back. Tani [3] proposes that hydrodynamically smooth roughness has the effect of reducing friction drag. Note that the hydrodynamically smooth roughness refers to roughness that is lower than the viscous sublayer. In recent years, with the develop-

T. Ogawa (✉) · A. Yakeno
Institute of Fluid Science, Tohoku University, Aoba-ku Sendai, Japan
e-mail: takuto.ogawa.a4@tohoku.ac.jp

K. Kato et al. (eds.), *Proceedings of the 10th IUTAM Symposium on Laminar-Turbulent Transition*, IUTAM Bookseries 44, https://doi.org/10.1007/978-981-96-9829-5_2

ment of numerical computation and measurement techniques, the characteristics of roughnesses on scales close to smooth roughness, has been analyzed. Several studies have been shown that changes in friction drag are greatly affected not only by the height of the roughness but also by its shape [4, 5]. Hamada et al. [6] analyzed the mechanism by which the roughness reduces friction drag by performing DNS for the spanwise uniform sinusoidal roughness surface, the sandy roughness surface based on the actual sandpapers, and the randomly arranged roughness surface based on the Gaussian distribution. The results show that the roughness based on the sandpaper and the Gaussian distribution have a high drag reduction effect. Our research team have named the roughness based on the Gaussian distribution Distributed Micro Roughness (DMR). At present, artificial creation of shapes that exceed the effect of the sandpaper case has not been achieved. And, the effects of the height, diameter and density of each roughness on the flow have not been investigated. In this study, DNSs of the flow over several DMRs are performed using the roughness density as a parameter. The artificial shapes based on the randomly arranged Gaussian distribution, which can simulate sandpaper, are used to quantitatively set the roughness density. The roughness densities are validated in a lower density range than in case of the previous study [6] in order to analyze the influence of each roughness. The effect of the roughness density on the flow is evaluated in terms of the drag reduction.

2.2 Problem Settings

The analysis target is the flow near the wall surface with the smooth surface and the DMR. The Reynolds number is set to $Re = 3,535$ based on the boundary-layer thickness at the inlet boundary (δ_s). Hereafter, all physical quantities are non-dimensionalized by the boundary-layer thickness or the free stream velocity. The lengths in the streamwise, spanwise, and wall-normal direction are set to $L_x = 56.5$, $L_y = 11.3$, and $L_z = 28.3$, respectively. The shape of roughness is defined by a Gaussian function. Therefore, the wall shape of DMR is expressed by the height h from the reference surface, as shown in Eq. 2.1. In this study, four DMRs are discussed as shown in Table 2.1, and the effect of each DMR on the drag reduction is discussed. The parameter σ, which has the dimension of length for each roughness, is set to 1/6 based on δ_s. The roughness diameter ϕ is defined as the range satisfying the six-sigma criterion of the Gaussian distribution and is set to 1. Each roughness is randomly distributed in the range of $2.8 \leq x \leq 42.4$. In the roughness arrangement, α_i in Eq. 2.1 is set to an arbitrary fixed value. The overall height is then compressed so that the height falls within the range. The roughness height is set to $h = [-0.141 : 0.141]$ which is one of the effective values and confirmed in the previous study [6, 7]. The roughness density is expressed as ρ, the number of roughness elements per unit area.

$$h(x, y) = \sum_{i=1}^{N} \alpha_i \exp\left(-\frac{(x - x_i)^2 + (y - y_i)^2}{2\sigma^2} \right) \quad (2.1)$$

2.3 Numerical Methods

The governing equations are the unsteady three-dimensional compressible Navier-Stokes equations. The 6th-order compact difference scheme is used to spatial differencing of the metric, jacobian, convective and viscous terms. The scheme is higher spatial accuracy and resolution than conventional compressibility schemes. The three-step total variation diminishing (TVD) Runge-Kutta method is used for time integration. The low-pass explicit filter is applied at the outlet boundary of the computational domain. Figure 2.1 shows computational grids used in this study. Grid nesting is applied to capture turbulence with high accuracy while reducing the computational cost. The number of grid points in Zone 1 for the entire computational domain and Zone 2 for the area near the DMR are approximately 60 and 80 million, respectively. The Volume Penalization (VP) method is used for represent the complex three-dimensional shape of the DMR [8]. Solid wall condition is applied to smooth surface boundaries located upstream and downstream of the DMR. At the inlet boundary, the time-averaged velocity distributions (u_b, v_b, w_b) are given by the Blasius solution, u_{Blasius}, and the density and pressure are set to the same values as in the free stream. In addition, the wall-normal velocity w_b' is added as an artificial disturbance to induce T-S wave at the inlet boundary. They are described as follows:

Table 2.1 DMR conditions for computational cases

ρ	ϕ	h
1, 4, 9, 25	1	[−0.141:0.141]

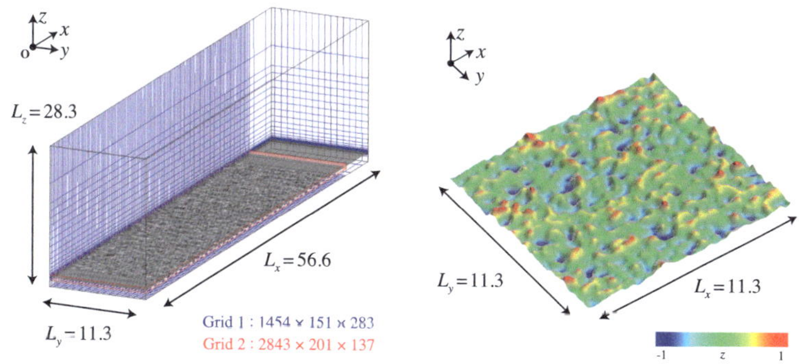

Fig. 2.1 Computational grids and an example of DMR ($\rho = 9, \phi = 1, h = [-1 : 1]$)

$$u_b(z) = u_{\text{Blasius}}(z), \tag{2.2}$$

$$v_b(z) = 0, \tag{2.3}$$

$$w_b(z) = w_{\text{Blasius}}(z) + w_b', \tag{2.4}$$

$$w_b' = 800(w_{\text{Blasius}}(\infty) - w_{\text{Blasius}}(z))\sin(2\pi Ft + 2\pi z). \tag{2.5}$$

The normalized frequency F is set to 0.707 because it is the most unstable frequency at the present Reynolds number $Re = 3,535$. The coefficient of $(w_{\text{Blasius}}(\infty) - w_{\text{Blasius}}(z))$ is set to 800 so that the maximum value reaches the same order as the outflow velocity. At the outlet boundary, the density and pressure are free conditions (Dirichlet boundary conditions). At the far boundary in the wall-normal direction, the pressure and streamwise velocity are fixed to their free flow values, and the wall vertical velocity and spanwise velocity are free conditions. At the spanwise boundaries, the periodic conditions are applied. In-house code LANS3D is used for solving the equations [9]. An initial calcuation is performed until the flow at the trailing edge of the roughness ($x = 42.4$) fully develops into turbulence. Subsequently, an additional calculation up to $t = 100$ is performed to statistically analyze the flow characteristics.

The turbulence contribution to the skin friction drag is calculated by considering the Reynolds shear stress based on FIK identity, as shown in Eq. 2.6 [10]. The Reynolds shear stress effect on the drag coefficient is quantitatively measured by the second term of FIK identity (ΔC_f). Therefore, in the present study, the effect of the roughness density on the turbulence development is evaluated by ΔC_f.

$$C_f = \frac{4(1-\delta)}{Re_\delta} + 2\int_0^1 2(1-z)(-\overline{u'w'})dz - 2\int_0^1 (1-z)^2\left(\overline{I_x} + \frac{\partial \bar{u}}{\partial t}\right)dz, \tag{2.6}$$

$$\text{where} \quad \overline{I_x} = \frac{\partial \overline{uu}}{\partial x} + \frac{\partial \bar{u}\bar{v}}{\partial x} - \frac{1}{Re_\delta}\frac{\partial^2 \bar{u}}{\partial t}. \tag{2.7}$$

2.4 Results and Discussion

Figure 2.2 show the second term of FIK identity (ΔC_f) and the turbulent kinetic energy (TKE) along the vertical direction at each x position for the smooth flat plate (Baseline) and DMR cases, respectively. In each case, fluctuation of ΔC_f due to T-S wave effects are observed near the inlet boundary. In the smooth flat plate case, the fluctuation are observed up to $x \sim 30$, with a maximum peak at $x \sim 25$. Thereafter, ΔC_f decreases moderately as it moves downstream. A similar tendency could be observed for TKE. The position of the maximum peak in TKE coincides with that of the maximum peak in ΔC_f. This position also coincides with the position where the large-scale vortex structure collapses, as described below, and a relationship between the collapse of the vortex structure and the increase in ΔC_f could be inferred. In the DMR cases, the tendency of ΔC_f and TKE are slightly different from those in the smooth flat plate case. In each case, ΔC_f is higher than in the smooth flat plate case

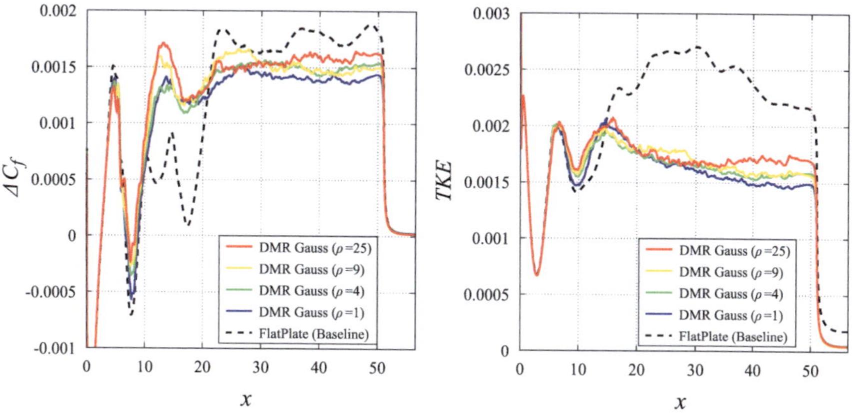

Fig. 2.2 Integral value of ΔC_f (second term of FIK identity, left side) and the turbulent kinetic energy (TKE, right side) for the wall-normal (z) direction along the streamwise (x) direction

at $10 \lesssim x \lesssim 20$. At $x \sim 15$, ΔC_f in the DMR cases is approximately 2 to 3 times higher than that in the smooth flat plate cases. However, the subsequent change is relatively gradual. The DMR cases do not have a high peak as in the smooth flat plate case, and in many cases the value remains low. A similar tendency is observed in TKE.

2.5 Conclusions

DNSs of the flows over the several distributed micro roughnesses (DMRs) with roughness density as a parameter was performed to analyze the mechanism by which DMR achieves the drag reduction. DMRs with appropriate roughness density could suppress rapid fluctuations in ΔC_f by gradually collapsing the large-scale vortex structure generated by the T-S wave. In the future works, we will investigate the effect of other parameters such as height, and we will extract a feature value of DMR to estimate the influence of the DMR on the flow field.

References

1. ICAO, I.C.A.O.: A proposal from ICAO. Technical report (2016). https://aviationbenefits.org/blog/2016/04/a-proposal-from-icao/, 30 Sept 2023
2. Walsh, M.J., Sellers III, W.L., McGinley, C.B.: Riblet drag at flight conditions. J. Aircr. **26**(6) (1989). https://doi.org/10.2514/3.45804
3. Tani, I.: Drag reduction by riblet viewed as roughness problem. Proc. Jpn. Acad. **64**, 21–24 (1988)

4. Nugroho, B., Monty, J.P., Utama, I.K.A.P., Ganapathisubramani, B., Hutchins, N.: Non-k-type behaviour of roughness when in-plane wavelength approaches the boundary layer thickness. J. Fluid Mech. **911** (2021). https://doi.org/10.1017/jfm.2020.875
5. Jelly, T.O., Busse, A.: Reynolds number dependence of Reynolds and dispersive stresses in turbulent channel flow past irregular near-gaussian roughness. Int. J. Heat Fluid Flow **80** (2019). https://doi.org/10.1016/j.ijheatfluidflow.2019.108485
6. Hamada, S., Yakeno, A., Obayashi, S.: Ultra-fine roughness effect on transition delay using direct numerical simulation. In: Proceedings of 12th International Symposium on Turbulence and Shear Flow Phenomena (TSFP 2022) (2022)
7. Hamada, S., Yakeno, A., Obayashi, S.: Drag reduction effect of distributed roughness on the transitional flow state using direct numerical simulation. Int. J. Heat Fluid Flow **104**, 109–230 (2023)
8. Liu, Q., Vasilyev, O.V.: A brinkman penalization method for compressible flows in complex geometries. J. Comput. Phys. **227**(10), 946–966 (2007). https://doi.org/10.1016/j.jcp.2007.07.037
9. Fujii, K.: Developing an accurate and efficient method for compressible flow simulations - example of CFD in aeronautics. In: The Proceedings: Fifth International Conference on Numerical Ship Hydrodynamics (1990)
10. Kametani, Y., Fukagata, K.: Direct numerical simulation of spatially developing turbulent boundary layers with uniform blowing or suction. J. Fluid Mech. **681**, 154–172 (2011). https://doi.org/10.1017/jfm.2011.219

Chapter 3
Transition Control for Separated Flow over Airfoil by Periodic Body Forcing

Takuto Ogawa, Aiko Yakeno, and Kozo Fujii

Abstract We have investigated the effectiveness of the proposed strategy for controlling transitions by periodic body forcing. The control strategy estimates the effective burst frequency (F^+) based on the positional relationship between the transition point in the uncontrolled flow and the plasma actuator. In the case of the uncontrolled flow around an NACA0015 airfoil at $\alpha = 16$ in $Re = 63,000$, the effective F^+ was estimated to be 20 if the plasma actuator has been installed at 5% from the leading edge. The flow control effect of the control strategy was verified by conducting LESs with these conditions. The results showed that the control strategy can control the transition effectively.

Keywords Transition control · Flow control · Vortex · Plasma actuator

3.1 Introduction

Controlling the transition is very important in terms of the drag reduction and the lift increase. We have proposed the strategy to effectively control the transition based on the findings from our previous studies of separation control by using the DBD plasma actuator. In the present study, we have investigated the effectiveness of the proposed strategy for controlling transitions by periodic body forcing. Not only laminar-turbulent transition but also other transitions such as the transition from laminar state to steady fluctuating states are discussed in the present study. Figure 3.1 shows the proposed strategy for controlling transitions. The position of the control device, the spatial extent that the controlled flow structures reach during the dominant period for the flow control (dominant spatial fluctuation), and the position of the transition point are key parameters for the stability of the periodic flow fluctuation. Here, the

T. Ogawa (✉) · A. Yakeno
Institute of Fluid Science, Tohoku University, Aoba-ku Sendai, Japan
e-mail: takuto.ogawa.a4@tohoku.ac.jp

K. Fujii
Department of Information and Computer Technology, Tokyo University of Science, Tokyo, Japan

K. Kato et al. (eds.), *Proceedings of the 10th IUTAM Symposium on Laminar-Turbulent Transition*, IUTAM Bookseries 44, https://doi.org/10.1007/978-981-96-9829-5_3

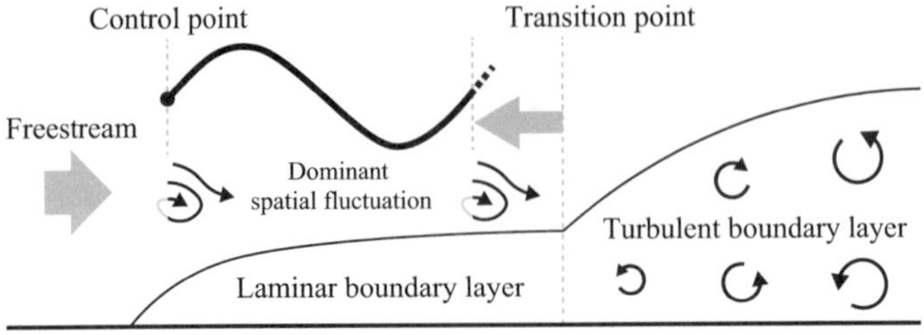

Fig. 3.1 The proposed strategy for controlling transitions proposed by our research team [1]. The spatial extent that the controlled flow structures reach during the dominant period for the flow control (dominant spatial fluctuation) is required to be included in the "non-turbulence" region such as the upstream of the transition point for the high stability of the dominant spatial fluctuation. Otherwise, the stability of the dominant spatial fluctuation becomes worse

dominant spatial fluctuation along the free-stream can be expressed as the ratio of the free-stream velocity to the dominant period of the flow control. The proposed strategy has been clarified in terms of its effectiveness in some conditions in the previous study [1]. In the present study, the generality of the proposed strategy has been investigated in more details.

3.2 Problem Settings

The target flow is the separated flow around a NACA0015 airfoil at $Re = 63,000$. The angle of attack is 16 degrees. The specific heat ratio is set to $\gamma = 1.4$. The Prandtl number is set to $Pr = 0.72$. The condition is same as a free-stream velocity of 10 [m/s] and an airfoil chord length of 0.1 [m] in the wind tunnel experiment [2]. The transition point, non-dimensionalized by the chord length c, is at 0.12 $[x/c]$. The plasma actuator is installed at 0.05 $[x/c]$. According to the proposed strategy, the length of the dominant spatial fluctuation is required to be included in a "non-turbulence" region, such as the upstream of the transition point. Therefore, the length of the dominant spatial fluctuation along the free-stream should be 0.07 $[x/c]$ or less. The burst-mode actuation of the plasma actuator is used for introducing the dominant spatial fluctuation into the flow. The dominant driving frequency F^+ for the burst-mode actuation is non-dimensionalized by the free-stream velocity and the airfoil chord length. The burst ratio for the burst-mode actuation BR is set to 0.2. The lower limit of the effective F^+ estimated from the proposed strategy corresponds to the inverse of the length of the dominant spatial fluctuation along the free-stream (L^+), which is calculated as $F^+ = 1/L^+ = 1/0.07 \approx 14.28$. Four cases are investigated: two cases with $F^+ = 10$ and 20, which are around $F^+ \approx 14.28$, and two cases with $F^+ = 1$ and 5, which are smaller than $F^+ = 10$. Table 3.1 shows the position of the plasma actuator (P), F^+, and corresponding L^+ for four cases. The maximum induced velocity of the plasma actuator is set to a value at which the transition of the flow changes sensitively with the variation of F^+.

Table 3.1 Conditions of the plasma actuator

P (PA position)	F^+	L^+ $(1/F^+)$
0.05	1, 5, 10, 20	1.0, 0.2, 0.1, 0.05

3.3 Numerical Methods

Large eddy simulations (LESs) have been conducted using the in-house code LANS3D [3, 4]. Three-dimensional Navier-Stokes equations are non-dimensionalized by the free-stream density, free-stream velocity, and the chord length of the airfoil, are employed as the governing equations in this study. In the non-dimensional form, the governing equations are represented as follows:

$$\frac{\partial \rho}{\partial t} + \frac{\partial \rho u_k}{\partial x_k} = 0, \tag{3.1}$$

$$\frac{\partial \rho u_i}{\partial t} + \frac{\partial (\rho u_i u_k + p\delta_{ik})}{\partial x_k} = \frac{1}{Re}\frac{\partial \tau_{ik}}{\partial x_k} + D_c S_i(x, y, t), \tag{3.2}$$

$$\frac{\partial e}{\partial t} + \frac{\partial ((e + p) u_k)}{\partial x_k} = \frac{1}{Re}\frac{\partial u_l \tau_{kl}}{\partial x_k} - \frac{1}{(\gamma - 1) Pr Re M_\infty^2}\frac{\partial q_k}{\partial x_k} + D_c S_k u_k, \tag{3.3}$$

$$S_i = Q_c E_i, \tag{3.4}$$

where x_i, u_i, q_i, ρ, p, e, τ_{ij}, δ_{ij}, and t denote the non-dimensional forms of position vector, velocity vector, heat flux vector, density, pressure, energy per unit volume, stress tensor, Kronecker delta, and time, respectively. Non-dimensional parameters Re, M_∞, and Pr denote the Reynolds number, free-stream Mach number, and Prandtl number, respectively and are defined as follows:

$$Re = \frac{\rho_\infty U_\infty c}{\mu_\infty}, \quad M_\infty = \frac{U_\infty}{a_\infty}, \quad Pr = \frac{\mu_\infty C_p}{\kappa_\infty}, \tag{3.5}$$

where μ_∞, ρ_∞, U_∞, c, a_∞, c_p, and κ_∞ denote viscosity, density, velocity, chord length, sound speed, constant pressure specific heat, and heat conduction coefficient, respectively. Here, the subscript ∞ denotes quantity in a free-stream condition. The free-stream Mach number is set to be $M_\infty = 0.2$ for a larger time increment and thus lower costs. In Eqs. 3.2 and 3.3, the last terms on the right-hand side represent the body force and the power added to the unit volume by the body force, respectively. Here, S_i denotes the non-dimensional body force vector, Q_c denotes the electric charge, and E_i denotes the electric-field vector, respectively. The non-dimensional plasma scale parameter (D_c) is defined as follows:

$$D_c = \frac{Q_{c,ref} E_{ref} c}{\rho_\infty U_\infty^2}, \tag{3.6}$$

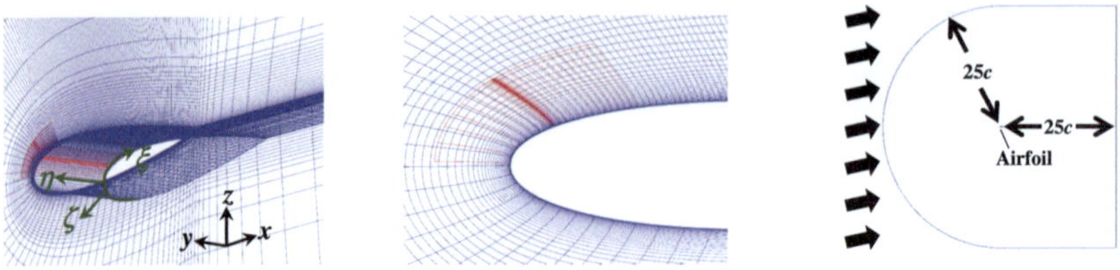

(a) Near the airfoil. (b) Side view. (c) Computational domain and schematic of the inflow.

Fig. 3.2 Computational grids

where the subscript "ref" denotes the reference value. Here, we set $Q_{c,ref}$ and E_{ref} as maximum values of Q_c and E in the simulation results of a Suzen model [5]. The non-dimensionalized plasma scale parameter (D_c), used in this study, is similar to previous studies. The distribution of S_i is determined by the Suzen model. Although the definition of D_c is based on the instantaneous force vector. In the present study, D_c is set to 0.04 in all cases. The maximum induced velocity through the normal actuation with $D_c = 0.04$ in a quiescent field becomes approximately 0.34 times the free-stream velocity [6]. This means that the induced velocity is 3.4[m/s] for the flow condition assumed in the present study when the free-stream velocity is set to $U_\infty = 10$[m/s]. The validity of this model has been demonstrated by the previous studies [7–9]. The zonal method is used to efficiently analyze the localized small-scale flow fluctuation induced by the drive of the plasma actuator. Figure 3.2 shows computational grid. The overall grid is the C-type grid and the distance from the airfoil surface to the outer boundary set to 25 times the chord length. The fine subgrid is placed at the leading edge of the airfoil to analyze the localized small-scale flow fluctuation for the drive of the plasma actuator. The total number of grid points is approximately 18 million and 2 million, respectively.

3.4 Results and Discussion

Figure 3.3 shows the flow characteristics for the proposed strategy ($F^+ = 20$, left side) and the other case ($F^+ = 5$, right side). Noted that the trends do not change significantly for $F^+ = 1, 5$, and 10. Therefore, $F^+ = 5$ is shown as a representative example. The horizontal, vertical, and depth axis represent the free-stream direction, perpendicular direction, and time variation, respectively. The upper surface of the NACA0015 airfoil is illustrated as a gray curve. The iso-surfaces is the time variations of the spanwise- and short-time phase-averaged Q-criterion ($Q = 25$) at each time step. In other words, the filaments of the iso-surfaces could be considered as the advection of coherent vortex structures with an axis in the spanwise direction. In the

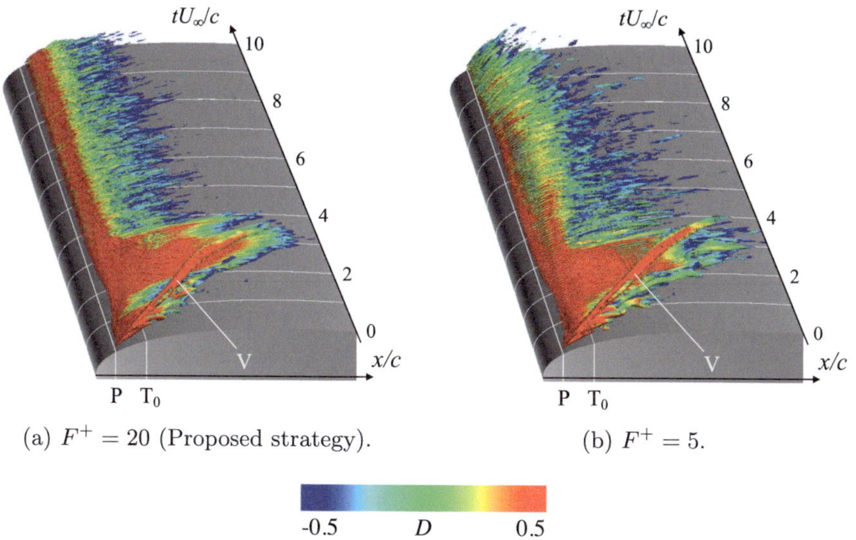

Fig. 3.3 Time variation of the spanwise- and short-time phase-averaged Q-criterion (Q') in each case

short-time phase averaging, three time steps, $T - 1/F^+$, T, and $T + 1/F^+$, are used as the samples for averaging when the current time is defined as $tU_\infty/c = T$. The criterion D for coloring is the ratio of the difference and summation between the spanwise (two-dimensional) component (Q') and the three-dimensional component (Q'') of the Q-criterion ($D = (|Q'| - |Q''|)/(|Q'| + |Q''|)$). Q' dominates when D is positive, and Q'' dominates when D is negative. Point P is the position of the plasma actuator. Point T_0 is the transition point in the no-control case. The transition point is determined from the inflection point of the momentum thickness. In each case, a large-scale vortex structures (V) is shed from near the leading edge when the control is applied. Subsequently, the separated flow attaches, causing the vortex structures to advect closer to the airfoil surface. In these cases, the flow transitions from the separated phase to the attached phase at $0 \lesssim tU_\infty/c \lesssim 4$. The tendency is commonly observed in the burst-mode actuation of the plasma actuator for controlling separated flows around airfoils. In these cases, the trends differs at $tU_\infty/c \gtrsim 4$. In the case of the proposed strategy, the attached phase continues even at $tU_\infty/c \gtrsim 4$. During this period, D is maintained a relatively high and stable value upstream of the transition point. The tendency indicates that coherent vortex structures are stably generated and maintained upstream of the transition point. In contract, in the case of the other case, the flow gradually starts to be separated again from $tU_\infty/c \sim 4$, and a large-scale separation occurs again at $tU_\infty/c = 10$. D gradually decreases upstream of the transition point as the flow shifts to the separation phase. In other words, this indicates that the coherent vortex structures are difficult to generate and maintain upstream of the transition point. The flow in this case alternated between the attached phase and separated phase repeatedly thereafter. The results suggest the effectiveness of the proposed strategy. The coherent vortex structures are stably generated and advect if the dominant spatial fluctuation is included in the upstream of the transition point. In contrast, the vortex structures are easy to collapse if the dominant spatial fluctuation extends downstream of the transition point. It is suggested that the temporal and spatial stability of the dominant spatial fluctuation could be controlled by adjusting the length of the dominant spatial fluctuation with respect to the transition point as a reference.

3.5 Conclusions

The effectiveness of the proposed strategy for controlling transitions by periodic body forcing is investigated. The result of LES shows that the proposed strategy could stably suppresses the flow separation and could maintains 2-dimensional flow fluctuation upstream of the transition point. The tendency supports a certain level of effectiveness of the proposed strategy.

References

1. Ogawa, T., Asada, K., Yakeno, A., Fujii, K.: Revisiting burst drive conditions of DBD plasma actuator for airfoil flow control **2024–0490**, 1 (2024)
2. Sekimoto, S., Nonomura, T., Fujii, K.: Burst-mode frequency effects of dielectric barrier discharge plasma actuator for separation control. AIAA J. **55**(4), 1385–1392 (2017)
3. Fujii, K., Tamura, Y.: Capability of current supercomputers for the computational fluid dynamics. In: Supercomputing '89: Proceedings of the 1989 ACM/IEEE conference on Supercomputing, pp. 71–80. ACM, New York, NY, USA (1989)
4. Fujii, K.: Developing an accurate and efficient method for compressible flow simulations - example of CFD in aeronautics. In: The Proceedings: Fifth International Conference on Numerical Ship Hydrodynamics (1990)
5. Suzen, Y.B., Huang, P.G.: Simulations of flow separation control using plasma actuator. In: AIAA-2006-877 (2006)
6. Sato, M., Okada, K., Aono, H., Asada, K., Yakeno, A., Nonomura, T., Fujii, K.: Les of separated-flow controlled by DBD plasma actuator around NACA 0015 over Reynolds range of $10^4 - 10^6$. In: AIAA 2015-0308, 9–12 Jan 2015
7. Sato, M., Aono, H., Yakeno, A., Nonomura, T., Fujii, K., Okada, K., Asada, K.: Multifactorial effects of operating conditions of dielectric-barrier-discharge plasma actuator on laminar-separated-flow control. AIAA J. **53**(9), 2544–2559 (2015). June
8. Asada, K., Fujii, K.: Computational analysis of unsteady flow-field induced by plasma actuator in burst mode. In: AIAA-2010-5090 (2010)
9. Aono, H., Sekimoto, S., Sato, M., Yakeno, A., Nonomura, T., Fujii, K.: Computational and experimental analysis of flow structures induced by a plasma actuator with burst modulations in quiescent air. Mech. Eng. J. **2**(4), 1–16 (2015)

Chapter 4
Axial Flow Influence on Boundary-Layer Transition on Rotating Slender Cones

K. Kato, K. Yamada, K. Takahara, P. H. Alfredsson, and M. Matsubara

Abstract Flow on a rotating cone with an apex angle of 15 degrees is studied through flake visualizations, both in still fluid and with axial flow. From recorded movies, the reflective light intensity at generating lines is extracted and its development is evaluated in space and time. The results show that increasing axial flow moves transition downstream, whereas increasing rotational speed has the opposite effect. We found that the transition process can be scaled by the von Kármán viscous length scale and a globally defined axial velocity.

Keywords Flow visualization · Three-dimensional boundary layer

4.1 Introduction

Boundary-layer flows on rotating cones in a still fluid and with axial flow have been studied as a canonical model of three-dimensional boundary layers [1, 3], where the geometry is solely defined by the half-apex angle ψ. A laminar boundary layer flow develops near the apex and turns turbulent somewhere along the cone surface. Here, the x-coordinate is defined along the generating line with the origin at the apex. Kobayashi et al. [5] studied the flow on slender rotating cones in axial flow and evaluated it using an axial Reynolds number based on x^* and the velocity U_e^* at the outer edge of the boundary layer (note starred quantities are dimensional). Their results show that by increasing the angular velocity (Ω^*) the transition Reynolds number can reduce dramatically.

For the flow on slender cones, some questions remain. First, what is the proper scale for the flow development? Kobayashi et al. [4] studied the flow in a still fluid using a rotational Reynolds number based on the local radius r^* and azimuthal wall

K. Kato (✉) · K. Yamada · K. Takahara · M. Matsubara
Department of Mechanical Systems Engineering, Shinshu University, Nagano, Japan
e-mail: kentaro_kato@shinshu-u.ac.jp

P. H. Alfredsson
FLOW Engineering Mechanics, KTH Royal Institute of Technology, Stockholm, Sweden
e-mail: hal@kth.se

K. Kato et al. (eds.), *Proceedings of the 10th IUTAM Symposium on Laminar-Turbulent Transition*, IUTAM Bookseries 44, https://doi.org/10.1007/978-981-96-9829-5_4

velocity $r^*\Omega^*$. Recent studies [2, 7] point out that transition does not occur at a unique rotational Reynolds number, but is affected by the angular velocity, and that transition instead occurs at an almost constant Görtler number, clearly indicating the role of centrifugal effects on the transition. Another question relates to the critical flow structure, which appears first and strongly affects the further transition process. Kobayashi et al. [5] reported that the structure becomes ring-like as the axial velocity decreases; meaning that the inclination of the vortex axis with respect to the azimuthal direction $\epsilon \to 0°$, where positive ϵ is defined as the same direction as that of crossflow vortices on a broad cone in still fluid. However, some recent studies suggest that the critical structures have $\epsilon < 0°$ [2, 6]. The cause of the variation of transition Reynolds number and critical flow structures is still not clear. Further experimental evidences are required, and a gap that we hope to, at least partly, fill with this paper.

4.2 Method

Flow visualizations were made in a towing tank with dimension $0.8 \times 9.0 \times 0.6$ m (width × length × depth) as shown in Fig. 4.1. A horizontally placed cone (half-apex angle $\psi = 7.5°$ and a base diameter of 90 mm) was rotated around its axis both in stationary water and during towing in the horizontal direction. The cone was made of polyacetal, it had a smooth surface and its apex was rounded with a diameter of 5 mm. Experiments were made by independently setting the towing speed U^*_∞ of the trolley and the angular velocity Ω^* of the cone.

Pearl-flake pigment (Iriodin111, Merck Ltd.) with a particle diameter of 7–8 μm and a specific gravity of 3.1 g/cm^3 was mixed with water as a tracer of the flow. The flakes were illuminated by a laser sheet (Green Laser Sheet 1000 m/G, Kato Koken

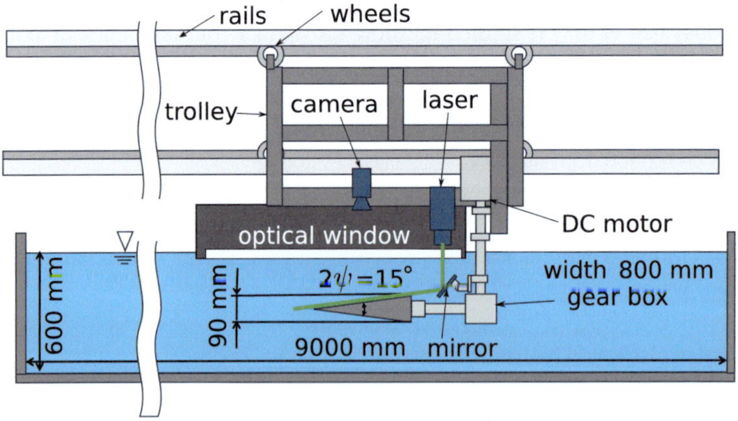

Fig. 4.1 Experimental setup with rotating cone in the towing tank at Shinshu University

Co., Ltd.) with a thickness of 2 mm approximately 1 mm away from a generating line of the cone. The reflective flake patterns were recorded by a camera (FX-30, SONY) fixed on the trolley through an optical window at the water surface.

4.3 Results and Discussion

Typical flake flow visualizations are shown in Fig. 4.2. The left (i) and right (ii) columns show frames of the recorded movies with 30 and 60 rpm, respectively. From the top to the bottom, the axial velocity U^*_∞ increases. In all cases, the cone surface moves from the bottom to the top towards us. Close to the rounded apex, no clear flow pattern is visible indicating that a laminar flow develops. As x^* increases, regular streaky patterns can be seen indicating vortex structures formed due to a primary instability. Further downstream, the flake pattern becomes more chaotic showing that the flow becomes turbulent. Increasing U^*_∞ and decreasing Ω^* seem to shift transition to larger x^*. With U^*_∞, ϵ increases. In still fluid (a), however, primary vortices with various ϵ are observed in time; ϵ is around zero, but sometimes even $\epsilon < 0°$ (opposite to the ones with $U^*_\infty > 0$). This temporal variations of the primary vortex structures probably explain different observations in the literature [2, 5, 6] and can be consistent with a rather gradual peak of spatial growth rate of linear stability analysis on a slender cone in frequency and azimuthal wavenumber space described in Ref. [2].

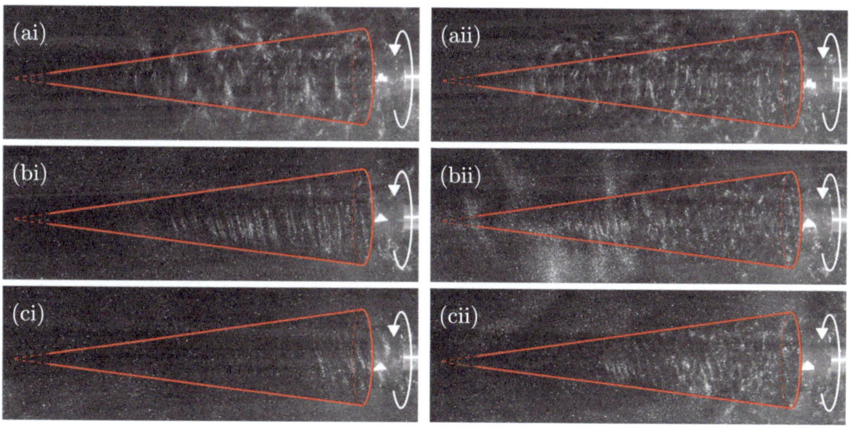

Fig. 4.2 Examples of flake flow visualization in **a** still fluid; and with a towing velocity **b** U^*_∞=0.03 m/s and **c** 0.05 m/s for a cone rotating at (i) 30 rpm and (ii) 60 rpm. The white arrows on the sting downstream the base of the cone show the rotational direction. The red lines show the outer contour of the cone. The apex of the cone is rounded with a diameter of 5 mm (the dashed apex indicates the imaginary apex, which is the origin of x-coordinate)

To characterize the flow in a more quantitative way, the reflective intensity on the generating lines was extracted from the recorded movies and space-time plots were constructed. Figure 4.3a, b shows the space-time plot at a generating line in still fluid and $U_\infty^* = 0.05$ m/s, respectively. The ordinate shows the cone revolution number N (or time normalized by the period of the cone rotation), whereas the abscissa shows the x-coordinate normalized by von Kármán viscous length scale $\delta_\nu^* = \sqrt{\nu^*/\Omega^* \sin\psi}$ (where ν^* is the kinematic viscosity). Figure 4.3c shows the root-mean-square (rms) of the light intensity I_{rms}. To reduce the effect of the background of the recorded movie, the background image was calculated by the temporal minimum of light intensity at each x-location just above the cone surface. The calculated background was subtracted from the original light intensity before calculating the rms of the fluctuating intensity. To smooth the rms curves, rms at several generating lines were calculated and the average of the data was filtered by a moving-average with 10 pixels (corresponding to a spatial resolution of approximately 1 mm).

In still fluid, the emergence of primary flow patterns with various wave lengths can be observed at $x \approx 60$ in Fig. 4.3a, corresponding to the rapid increase of the rms in (c). In the range $60 \lesssim x \lesssim 150$ various primary patterns seem to merge and split, resulting in large-scale structures. Such large-scale structures need longer sampling time to obtain converged statistics, leading here to a rather large variation in the rms. Further downstream, however, the chaotic flow patterns seem to remain until the edge of the cone at x=209 and 295 (Fig. 4.3ai and aii, respectively) where the rms decreases gradually with x.

By increasing the axial velocity, the primary flow patterns seem to be more regular and developed more gradually as shown in Fig. 4.3b and c (see also the corresponding x-locations in Fig. 4.2). The frequency of the primary flow pattern increases with U_∞^*. The dashed lines in (bi, bii) show the convecting velocity $U_\infty^*/\cos\psi$ along the x-direction, which align well with the flake patterns.

Overall, Fig. 4.3c clarifies the development of the flow; increasing U_∞^* and decreasing Ω^* delay transition with respect to x. It is also interesting that the rms shows a rapid growth in the stage where the primary patterns emerge, and collapse nearly on a single curve in the turbulent stage.

Another interesting finding is that the influence of U_∞^* and Ω^* can be reduced by introducing a non-dimensional global axial velocity $U_g = U_\infty^*/\sqrt{\Omega^* \nu^*}$. To demonstrate it, Fig. 4.4ai shows rms as a function of x^* with a nearly constant $U_g \approx 26.5$ and Fig. 4.4bi shows the same data, but as a function of x. This shows that the three cases collapse into a single curve in Fig. 4.4bi, clearly indicating that U_g is an important parameter for the flow development with x. Kobayashi et al. [5] used a similar scaling $S = r^*\Omega^*/U_e^*$. These scalings are essentially the same in the sense of the ratio of circumferential velocity to axial velocity. However, U_g is a global parameter independent on space, can be determined a priori and can be used seamlessly even in still fluid.

Although data with the same U_g show a nice collapse of development of I_{rms} in general, the scaling with U_g may have a singularity at $U_g \approx 0$. Figure 4.4aii shows rms as a function of x^* with $U_g = 0$ and different angular velocities Ω^*, whereas Fig. 4.4bii shows the same data, but as a function of x. Although data with 15 rpm

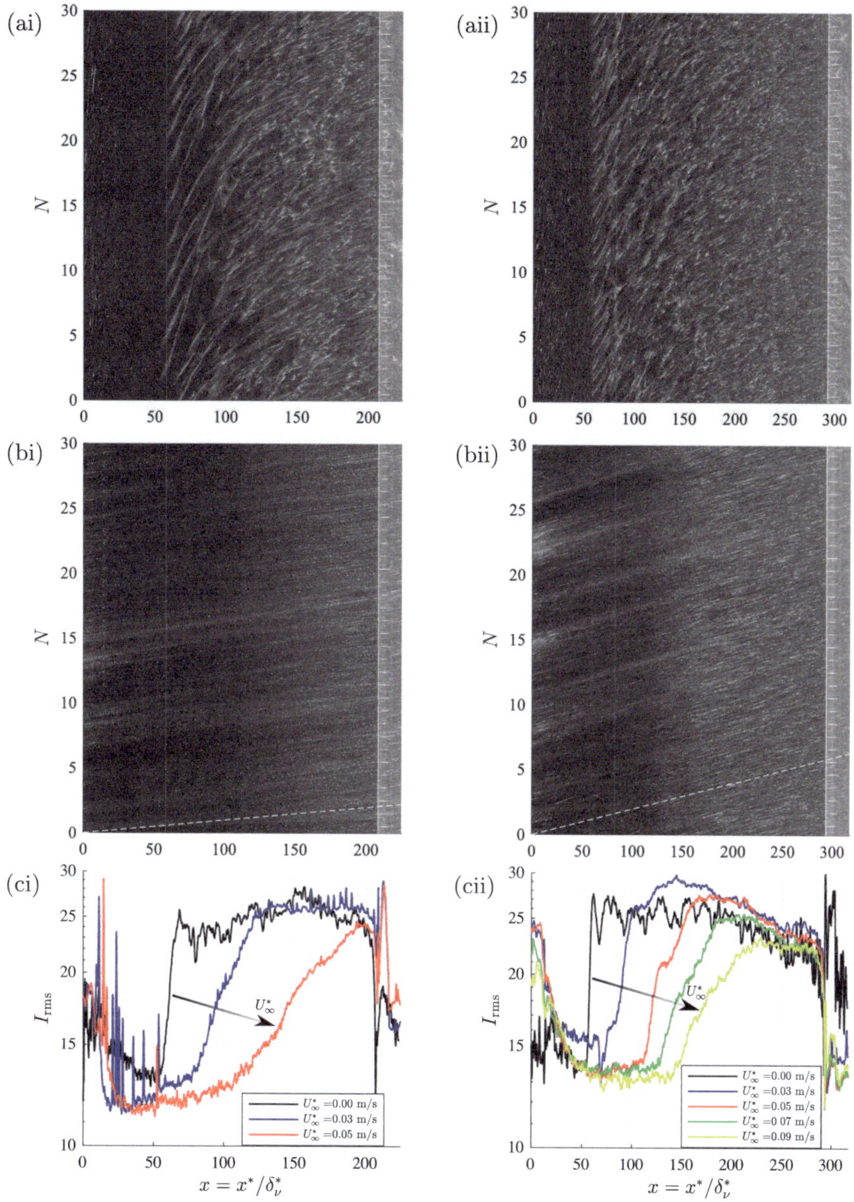

Fig. 4.3 Space-time plots of the flow patterns with axial velocities **ai**, **aii** $U_\infty^* = 0$ m/s and **bi**, **bii** 0.05 m/s. **ci**, **cii** Rms of the light intensity I_{rms} as a function of x with different axial velocities U_∞^*. The left and right columns show data for (i) 30 rpm and (ii) 60 rpm, respectively. The vertical edges at $x = 209$ and 295 in (i) and (ii) indicate the edge of the cone at $r^* = 45$ mm. The dashed lines in **bi**, **bii** show the convecting velocity $U_\infty^*/\cos\psi$ along the x-direction

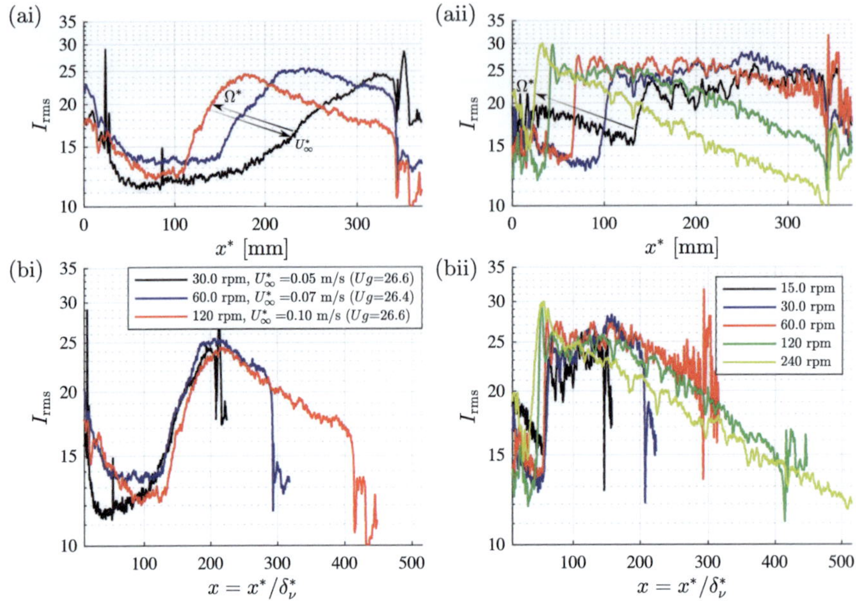

Fig. 4.4 Rms of the light intensity I_{rms} as a function of **a** x^* and **b** x. The left and right columns show data for $U_g = U^*_\infty/\sqrt{\Omega^* \nu^*}$ (i) around 26.5 and (ii) 0 (still fluid), respectively

$\leq \Omega^* \leq 60$ rpm still collapse in Fig. 4.4bii, the increase of I_{rms} shifts systematically to smaller x with Ω^* for $\Omega^* \gtrsim 60$ rpm. A similar trend is also reported by Kato et al. [2]. Further studies are required to understand the cause of this trend.

4.4 Concluding Remarks

The transitional flow on a rotating cone with a half-apex angle of 7.5° in still fluid and axial flow was investigated through flake flow visualizations qualitatively and quantitatively. Space-time plots of the reflective light at a generating line indicate clearly that an axial flow changes the characteristics of the primary flow structures as well as the laminar-turbulent transition (Fig. 4.3a, b). By evaluating the averaged rms of the light intensity on the generating lines the development of the flow can be observed. The results show that a non dimensional global axial velocity $U_g = U^*_\infty/\sqrt{\Omega^* \nu^*}$ controls the flow development with axial flow (Fig. 4.4i); in still fluid ($U_g = 0$), however, transition seems to be promoted by increasing the angular velocity (Fig. 4.4ii).

References

1. Alfredsson, P.H., Kato, K., Lingwood, R.J.: Flows over rotating disks and cones. Ann. Rev. Fluid Mech. **56**, 45–68 (2024). https://doi.org/10.1146/annurev-fluid-121021-043651
2. Kato, K., Segalini, A., Alfredsson, P.H., Lingwood, R.J.: Instability and transition in the boundary layer driven by a rotating slender cone. J. Fluid. Mech. **915**, R4 (2021). https://doi.org/10.1017/jfm.2021.216
3. Kobayashi, R.: Review: Laminar-to-turbulent transition of three-dimensional boundary layers on rotating bodies. ASME. J. Fluids Eng. **116**(2), 200–211 (1994). https://doi.org/10.1115/1.2910255
4. Kobayashi, R., Izumi, H.: Boundary-layer transition on a rotating cone in still fluid. J. Fluid Mech. **127**, 353–364 (1983). https://doi.org/10.1017/S0022112083002761
5. Kobayashi, R., Kohama, Y., Arai, T., Ukaku, M.: The boundary-layer transition on rotating cones in axial flow with free-stream turbulence. JSME Int. J. **30**(261), 423–429 (1987). https://doi.org/10.1299/jsme1987.30.423
6. Kohama, Y.P.: Three-dimensional boundary layer transition study. Curr. Sci. India **79**, 800–807 (2000)
7. Tambe, S., Kato, K., Hussain, Z.: Görtler number-based scaling of boundary-layer transition on rotating cones in axial inflow. J. Fluid. Mech. **987**, R3 (2024). https://doi.org/10.1017/jfm.2024.379

Chapter 5
Numerical Investigations of the Nonlinear Stages of Hypersonic Boundary-Layer Transition for an Ogive Geometry

Christoph Hader and Hermann Fasel

Abstract Numerical investigations were carried out for the boundary layer of an ogive geometry with a blunted nose tip and for the flow conditions of the German Aerospace Center (DLR) hypersonic wind tunnel (H2K) experiments at Mach 5.3. Linear Stability Theory (LST) analysis identified amplified first- and second-mode waves. Various nonlinear transition scenarios were examined using Direct Numerical Simulations (DNS). The simulation results for a "controlled" first mode oblique breakdown showed a transition development in the boundary layer advancing towards the late nonlinear stages and turbulence, with oblique first-mode waves initiating transition. The forcing amplitudes were selected to align the transition onset in simulations with that observed in experiments.

Keywords Hypersonic flow · Boundary-layer transition · DNS

5.1 Introduction

Laminar-turbulent transition in high-speed boundary layers is still a major unresolved topic in fluid dynamics. Transition to turbulence in supersonic/hypersonic boundary layers is associated with significant increases in heat transfer and skin-friction drag, making it a critical focus of research that is relevant for the design and operation of high-speed vehicles. Despite considerable progress in recent years, key aspects of high-speed boundary-layer transition physics remain poorly understood. The so-called "Morkovin roadmap" (see [1]) provides a meaningful classification of the transition process into different "routes" and "stages", which paved the way for systematic theoretical, numerical and experimental investigations. Basically, the transition stages in boundary layers can be classified as linear or nonlinear (see [2]). Substantial progress in the understanding of the linear (primary) instability regime has been made by Mack [3], who pioneered the use of compressible Linear Stability Theory (LST). To this day, transition prediction methods still largely rely on

C. Hader (✉) · H. Fasel
University of Arizona, Tucson, AZ, USA
e-mail: chader@arizona.edu

K. Kato et al. (eds.), *Proceedings of the 10th IUTAM Symposium on Laminar-Turbulent Transition*, IUTAM Bookseries 44, https://doi.org/10.1007/978-981-96-9829-5_5

LST, despite the inherently nonlinear nature of the later stages of transition. Understanding laminar-turbulent transition in low-speed (incompressible) flows is far more advanced than in high-speed flows, primarily due to the drastically increased complexity of the transition physics for high-speed flows. This, in part, is caused by the existence of multiple instability modes already in the linear regime (as found by Mack [3]) and their potential nonlinear interactions in the later stages. Consequently, high-speed transition is highly non-unique and relatively small changes in geometry or flow conditions can lead to markedly different outcomes (e.g. change of dominant transition mechanism(s), shift of transition onset/transition location, etc.). One aspect of the STORT project [4] was the synergistic approach of wind tunnel and flight experiments combined with numerical simulations to investigate boundary-layer transition on a tangent ogive geometry. The present paper provides results for numerical investigations of both the linear and nonlinear transition regimes for the STORT tangent ogive geometry under H2K wind tunnel conditions. The primary objective is to identify the nonlinear mechanisms leading to laminar-turbulent transition in the H2K experiments. To this end, "controlled" transition simulations were conducted for a blunt tangent ogive geometry at $M = 5.3$. For results of a similar investigation for the same geometry at $M = 7.1$ see [5].

5.2 Geometry and Flow Conditions

The STORT ogive geometry of the wind tunnel experiments at the hypersonic wind tunnel H2K at the DLR in Cologne was used for the numerical investigations (see Fig. 5.1). The wind tunnel model is a 50% scale of the forebody of the STORT flight test article [4]. The model of the wind tunnel experiments is a tangent ogive with a base diameter $d_{\text{ogive}} = 0.178$ m. The length of the ideally sharp tangent ogive (without a spherical nose) is $L_{\text{ogive}} = 0.75$ m. For the numerical simulations, the nose of the ogive is modeled as a semi sphere with a radius of $r_{\text{nose}} = 1.25 \times 10^{-3}$ m, which is continuously blended with the tangent ogive geometry. The flow conditions from the H2K experiments at the DLR Cologne, where transition was observed, were

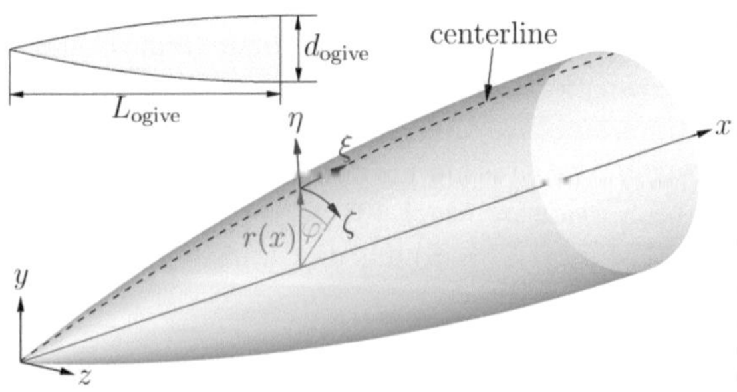

Fig. 5.1 Schematic of the ogive geometry

matched as closely as possible for the numerical investigations [6]. The investigations were carried out for a Mach number of $M = 5.3$ with a unit Reynolds number of $Re_1 = 17.9 \times 10^6\ \mathrm{m}^{-1}$, a stagnation temperature $T_0 = 450$ K, and constant wall temperature $T_{\mathrm{wall}} = 300$ K. Air was modeled as a calorically perfect gas ($\gamma = 1.4$, $Pr = 0.71$, $R_{\mathrm{air}} = 287.15$ J/(kg $\cdot$ K)), with viscosity calculated using Sutherland's law [7].

5.3 Results

The numerical investigations followed a simulation strategy previously used for studying laminar to turbulent transition in high-speed boundary layers (see, e.g., [8, 9]). This approach, illustrated in Fig. 5.2, consists of three steps. First, a precursor base flow, including the ogive nose tip, is obtained from a calculation using CFD++ [10] (Step 1). Once converged, this base flow is re-converged (Step 2) on a subdomain to obtain a "clean" base flow for stability and transition calculations. For this step, an in-house higher-order finite-difference (FD) based Navier-Stokes code [8] is employed on a computational domain starting downstream of the ogive tip, with the precursor solution serving as the initial condition. In the final step, stability and transition calculations are performed using the same FD code (Step 3). The reconverged base flow serves as the initial condition, and disturbances are introduced into the computational domain to initiate transition. The disturbance forcing methods routinely employed for stability and transition simulations can be classified as "controlled" or "natural." In controlled transition simulations, disturbance waves with defined amplitude, frequency, and wavenumber are introduced into the computational domain, whereas in natural transition simulations a broadband disturbance spectrum is forced. Controlled forcing simulations, which have been instrumental in identifying relevant transition mechanisms in previous investigations, were employed for the results discussed in this paper.

To ensure that a reliable base flow was obtained for step 3, the wall-normal profiles of the velocity component tangential to the surface of the ogive at various downstream locations obtained from the precursor calculations (CFD++, Step 1) and the reconverged base flow (FD, Step 2) are compared in Fig. 5.3. The boundary-layer thickness, defined as the wall-normal location where the derivative of the total enthalpy with the

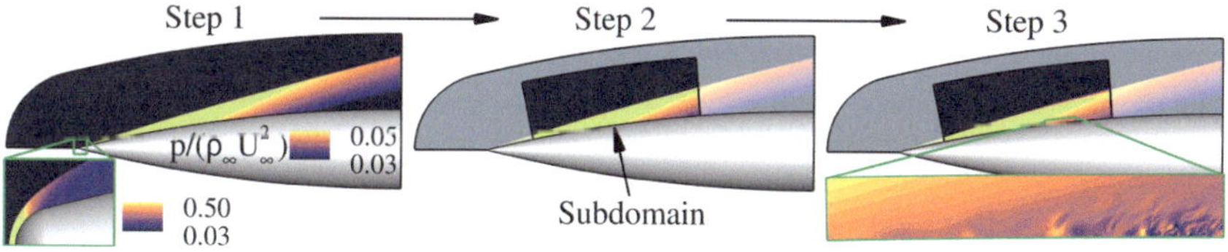

Fig. 5.2 Simulation strategy for numerical transition investigations

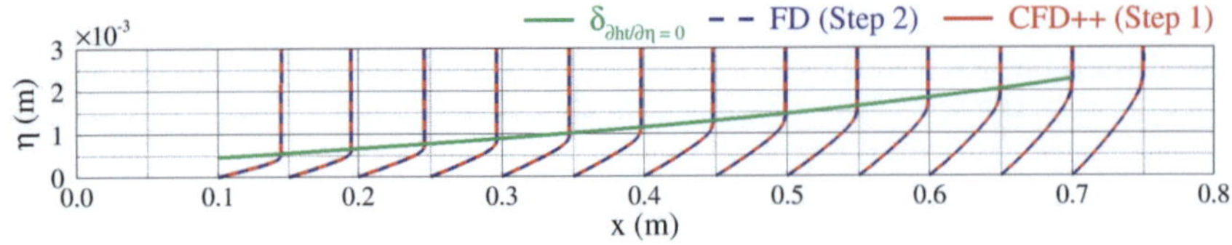

Fig. 5.3 Boundary layer Profiles of the velocity component tangential to the surface of the ogive at various downstream locations obtained from CFD++ (Step 1), and FD (Step 2) calculations

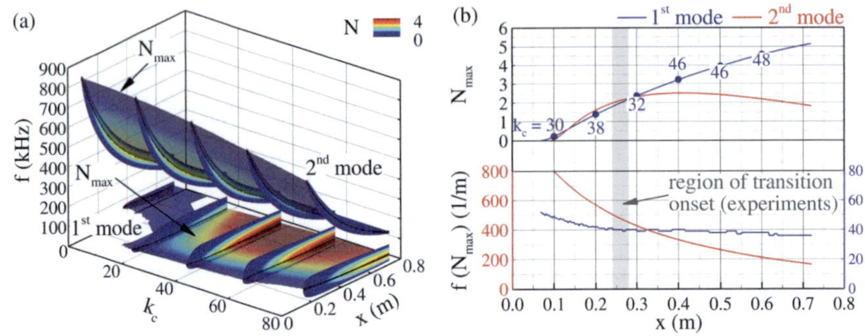

Fig. 5.4 3D N-factor map obtained from LST (**a**), and the maximum N-factors and corresponding frequencies of first and second mode waves (**b**)

respect to the wall-normal coordinate vanishes (see [11]) is also provided in Fig. 5.3. The solutions of the precursor calculation (CFD++) and the base flow reconvergence (FD), are in close agreement (Fig. 5.3), which confirms that a base flow suitable for stability and transition simulations (Step 3) was obtained.

Primary instability investigations were conducted using an in-house LST solver [12], which revealed the presence of both unstable first-mode and second-mode waves (see Fig. 5.4a). The maximum N-factors are displayed in Fig. 5.4a, and indicate that first-mode waves reach higher maximum N-factors than the second-mode waves. The fact that both first and second mode waves are amplified, suggests the possibility of various nonlinear transition scenarios (e.g. first or second mode dominated, nonlinear first mode second mode interactions). To determine the frequencies and wavenumbers relevant for investigating controlled transition scenarios, the LST data were analyzed by extracting the maximum N-factors across all wavenumbers for the first and second modes. The maximum N-factors are shown in Fig. 5.4b, with corresponding frequencies provided at the bottom of the figure with the frequency scale of the first mode on the right hand side. The azimuthal wavenumbers associated with the maximum N-factors for the first mode are marked with circles, along with their k_c values, see Fig. 5.4b. The approximate region of transition onset measured in the H2K wind tunnel experiments [6] are highlighted with a gray shaded area. The maximum N-factors obtained from LST for the first and second modes in the transition onset region are comparable as can be seen in Fig. 5.4b. This suggests that

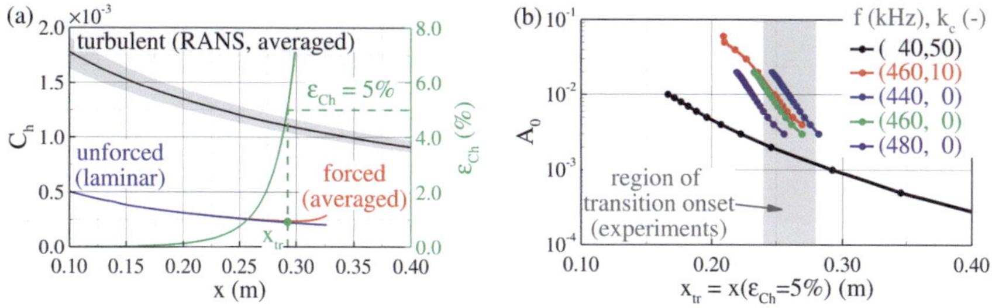

Fig. 5.5 Downstream development of the Stanton number obtained from a base flow and RANS calculations, and from the time- and azimuthally averaged flow from the controlled disturbance calculations (**a**). Transition onset locations for controlled disturbance calculations (**b**)

transition observed in the wind tunnel experiments for this case may be dominated by either the first or second mode, or a nonlinear interaction between them.

In order to investigate the possibility of the various nonlinear transition scenarios leading to boundary-layer transition for the geometry and flow conditions investigated here so-called "transition onset calculations" were carried out. They help to determine if it is possible to obtain transition onset for a particular transition scenario, and what forcing amplitude is required to move the transition onset in the numerical calculations to the approximate transition onset location observed in experiments. To that end, disturbance waves (first or second modes) are introduced into the computational domain through a wall blowing and suction slot. The forcing amplitude is then successively increased until transition onset can be observed. In this paper, transition onset is estimated at the locations where the time- and azimuthally averaged Stanton number departs by 5% from the undisturbed laminar base flow as illustrated in Fig. 5.5a. For reference, the turbulent curve was also obtained from Reynolds Averaged Navier-Stokes (RANS) calculations using CFD++ [10] with three different turbulent models ($k - \epsilon$, Spalart-Allmaras and Menter SST). The average of the three different turbulence models is given as a black solid line in Fig. 5.5a. The gray shaded area indicates the spread of the various turbulence models around their average. Transition onset calculations were carried out for oblique first mode waves, oblique and axisymmetric second mode waves in order to explore the viability of a first or second mode oblique breakdown as well as a fundamental breakdown, respectively (for more details on the nonlinear transition scenarios see for example [9]). The transition onset locations obtained from the onset calculations for different disturbance waves determined from the LST are provided for increasing forcing amplitudes (A_0) in Fig. 5.5b. The transition onset region observed in the experiments is again provided as a gray shaded area. The results in Fig. 5.5b confirm that for all considered cases (first mode oblique, second mode oblique, fundamental breakdown) the transition onset in the simulations can be moved to the downstream region where it was approximately observed in the experiments.

Based on the transition onset calculations, controlled transition DNS were carried out. First, the possibility of a first mode oblique breakdown was investigated. The

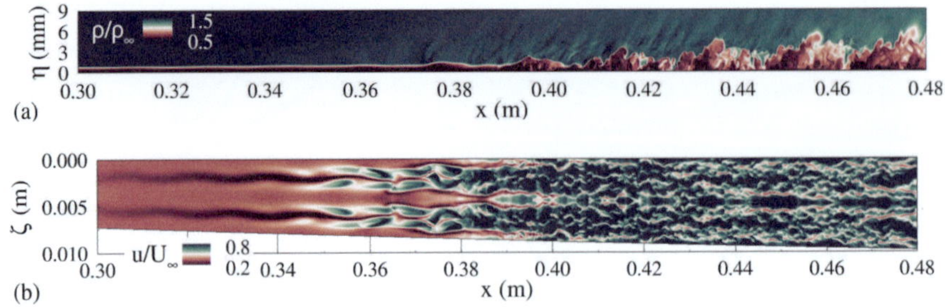

Fig. 5.6 Instantaneous contours of density in a wall-normal plane (**a**), and of the streamwise velocity component in an unrolled $x - \zeta$ plane extracted at a constant distance from the wall (**b**)

disturbance wave parameters (frequency and azimuthal wavenumber) were determined from LST (Fig. 5.4a) and the forcing amplitude was selected based on the transition onset calculations Fig. 5.5b. Contours of instantaneous density contours of this oblique breakdown simulation extracted in a wall normal plane are shown in Fig. 5.6a. These contours show the rapid transition (over a short downstream distance) when initiated by a pair of oblique first mode disturbance waves. In addition, instantaneous contours of the streamwise velocity (u) are provided in an unrolled plane extracted within the transitional boundary layer in Fig. 5.6b. These contours show the development of streamwise streaks which are nonlinearly generated by the self-interaction of the oblique first mode waves followed by a rapid generation of small scales indicating breakdown to turbulence.

The data obtained from the controlled oblique breakdown DNS were averaged in time and in azimuthal direction in order to determine where transition occurs in the simulations. The time-averaged Stanton number on the surface of the ogive is shown in Fig. 5.7a. For the full view of the ogive the data were extended periodically in axial direction for visualization purposes. The time-averaged Stanton number contour is nearly uniform until streamwise streaks appear at $x \approx 0.4$ m. The streak spacing is equivalent to twice the azimuthal wavenumber of the forced oblique disturbance wave to initiate the oblique breakdown. Therefore, the streaks are likely caused by a non-

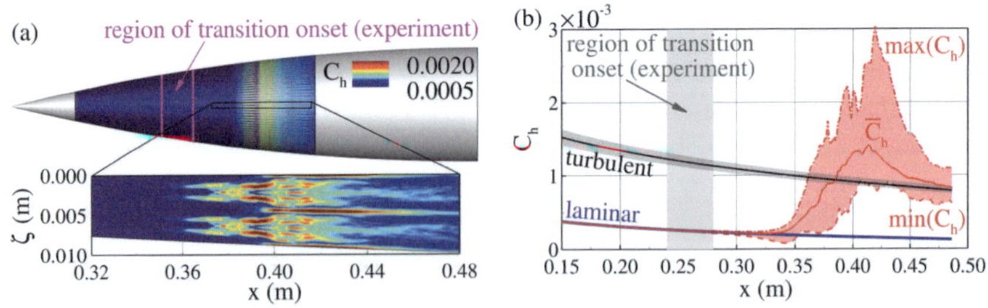

Fig. 5.7 Contours of the time-averaged Stanton number on the surface of the ogive (**a**), and downstream development of the time- and azimuthally averaged Stanton number (**b**)

linear interaction of the oblique disturbance waves (±40, ±50) generating a steady streamwise mode (0, ±100). The development of the time- and azimuthally averaged Stanton number ($\overline{C_h}$) in the downstream direction is plotted in Fig. 5.7b. The red shaded area indicates the minimum and maximum Stanton numbers obtained along the azimuth at each downstream location. The laminar and turbulent (obtained from RANS calculations) curves are provided for reference. The time- and azimuthally averaged Stanton number. After the departure from the laminar curve the Stanton number exhibits a sharp increase towards the turbulent values. After an "overshoot" above the turbulent curve the Stanton number asymptotically approaches the turbulent reference curve. The results here clearly indicate that a first mode oblique breakdown may be a viable nonlinear mechanism that can lead to transition in the experiments at H2K at $M = 5.3$.

5.4 Conclusions

Numerical investigations were carried out for an ogive geometry with a blunt nose tip for the flow conditions of the hypersonic wind tunnel (H2K) of the German Aerospace Center (DLR) at Mach 5.3. Linear Stability Theory (LST) calculations indicate that both first and second mode waves are amplified for the investigated geometry and flow conditions. So-called transition onset calculations confirmed that a first and second mode oblique breakdown as well as a fundamental breakdown could lead to transition onset where observed in the experiments. A controlled transition DNS of a first mode oblique breakdown confirmed that this scenario can lead to transition and thus is a viable candidate for what caused transition in the H2K experiments.

Acknowledgements This work was supported by AFOSR (Air Force Office of Scientific Research) Grant FA9550-21-1-0018, with Dr. Amanda Chou serving as the program manager. Computer time was provided by the Department of Defense (DoD) High Performance Computing Modernization Program (HPCMP) Frontier Project. The authors thank Drs. A. Gülhan and S. Willems (DLR Cologne) for providing experimental data and for the insightful discussions that improved the simulations.

References

1. Morkovin, M.V.: In: Wells, C. (ed.) Viscous Drag Reduction. Springer, US, pp. 1–31 (1969)
2. Stuart, J.T.: J. Fluid Mech. **4**(1), 1 (1958)
3. Mack, L.M.: Jet Propulsion Laboratory (1969)
4. Gülhan, A., Hargarten, D., Zurkaulen, M., Klingenberg, F., Siebe, F., Willems, S., Di Martino, G., Reimer, T.: Acta Astronaut. **211**, 333 (2023)
5. Hader, C., Fasel, H.F.: AIAA 2024-1569 (2024)
6. Willems, S., Gülhan, A.: Personal communication (2023)
7. Sutherland, W.: Phil. Mag. **36**(223), 507 (1893)
8. Laible, A.C., Mayer, C.S.J., Fasel, H.F.: AIAA 2008-4397 (2008)

9. Hader, C., Fasel, H.F.: JFM **869** (2019)
10. Chakravarthy, S., Peroomian, O., Goldberg, U., Palaniswamy, S.: AIAA 1998-5564 (1998)
11. Bertin, J.J.: Hypersonic Aerothermodynamics. AIAA (1994)
12. Haas, A., Hader, C., Fasel, H.F.: AIAA 2021-2848 (2021)

Chapter 6
Instability, Transition and Turbulence on Rotating Disks and Cones—Quo Vadis?

P. Henrik Alfredsson and Kentaro Kato

Abstract The laminar similarity solution of the boundary layer on a rotating disk in still ambient fluid was presented by von Kármán more than 100 years ago. A cross-flow instability was discovered in 1955 through flow visualisation experiments, and that instability has since then been investigated through stability analysis, experiments and numerical simulations. Other studies have investigated if an absolute instability could explain the distinct transition Reynolds number seen in experiments. Later experiments and stability theory have taken on both broad and slender cones to elucidate the influence of the cone apex angle on stability and transition. If you are interested to contribute with research on rotating bodies the present paper can be seen as a guideline—quo vadis.

Keywords Cross-flow instability · Centrifugal instability · Large-scale structures · Meandering

6.1 Prologue

The three-dimensional boundary layers on rotating cones and disks have many interesting features that deserve further study. The geometry is defined by the cone apex angle, see Fig. 6.1 (here the apex angle is given as 2ψ, for the disk $\psi = 90°$, for cones $\psi < 90°$). The flow development is controlled by a Reynolds number ($Re_x = V_w^* x^*/\nu^*$, where $V_w^* = r^*\Omega^*$ is the local wall velocity, $r^* = x^* \sin\psi$ is the distance from the axis of rotation, x^* the distance from the apex along the generating line, Ω^* the angular velocity of the cone/disk, and ν^* the fluid kinematic viscosity, starred quantities are dimensional). In addition the rotating body could be put in

P. H. Alfredsson (✉)
FLOW, Engineering Mechanics, KTH Royal Institute of Technology, Stockholm, Sweden
e-mail: hal@kth.se

K. Kato
Mechanical Systems Engineering, Shinshu University, Nagano, Japan
e-mail: kentaro_kato@shinshu-u.ac.jp

K. Kato et al. (eds.), *Proceedings of the 10th IUTAM Symposium on Laminar-Turbulent Transition*, IUTAM Bookseries 44, https://doi.org/10.1007/978-981-96-9829-5_6

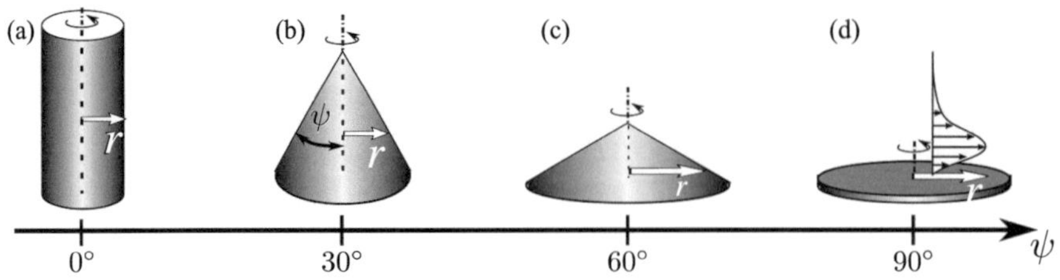

Fig. 6.1 Different types of rotating cones. **a** Cylinder, **b** slender cone, **c** broad cone, **d** disk. The half-apex angle ψ defines the geometry

a co-flowing stream giving an additional flow parameter, but will not be discussed here.

Two main primary instabilities may exist in the boundary layer on rotating cones; cross-flow instability and centrifugal instability. For large apex angles ($45° \lesssim \psi \leq 90°$) on the one hand, the cross-flow instability dominates and is due to an inflectional point in the radial velocity profile (as seen in Fig. 6.1d); on the other hand, for small $\psi \lesssim 45°$ the centrifugal instability seems to dominate. The cross-flow instability is observed as stationary vortices, both in flow visualizations and hot-wire measurements. Travelling disturbances have been observed indirectly (e.g. Ref. [8]) but their origin and structure are not well known. The centrifugal instability on slender cones also results in vortices, similar to Görtler vortices, but they have so far only been observed as travelling. The details of the interchange of these modes around $\psi = 45°$ is not yet clarified. As ψ decreases there is a drastic decrease in both the critical (where the disturbance first is observed) and the transition Reynolds number (where the flow turns turbulent), see Fig. 6.2.

In addition to the primary instabilities, a precursor to the breakdown to turbulence is a secondary instability developing on top of the primary vortices. The developing turbulent boundary layer downstream of transition is also three-dimensional and seems to incorporate some large-scale structures that so far have not been explained.

Here we will not give a review of previous work that started in 1921 with the now famous paper by von Kármán [5], where he among other things derived the exact similarity equations for the laminar boundary layer above a rotating disk. Instead we refer the interested reader to two reviews, one published 30 years ago by Kobayashi [10] and one from 2024 by Alfredsson, Kato & Lingwood [1] that together cover most of the development up to today. The latter includes supplemental material with about 180 references to works published until 2023 (both the paper and supplemental material are open access).

Instead of a review of earlier works we will in the following section directly give the background to some unanswered questions and ideas that may be worthwhile to pursue. We believe that they need to be investigated through experiments, in some cases DNS may be used and theoretical stability analysis should be used wherever possible to validate and clarify the experimental results. Hopefully some aspiring

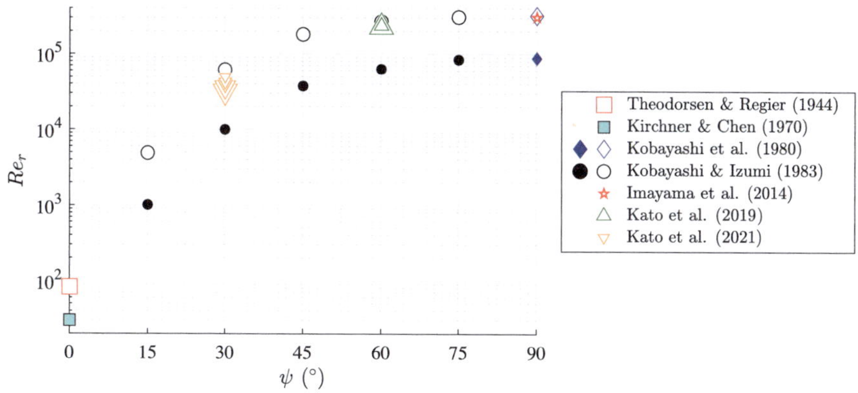

Fig. 6.2 Critical (filled symbols) and transitional (open symbols) Reynolds numbers as function of ψ from various studies reported in the literature, where $Re_r = r^{*2}\Omega^*/\nu^*$. This definition of the Reynolds number is chosen to be able to also include the cylinder ($\psi = 0°$) in the diagram. The size of the triangles at $\psi = 30°$ and $60°$ indicates the rotation rate, the larger the size the higher the rotation rate. As can be seen by increasing the rotation rate for $\psi = 30°$ the transition Reynolds number decreases, whereas for $\psi = 60°$ there is little effect. Figure adopted from Ref. [1].

scientists would find these examples interesting and attack them to further develop our understanding of flows over rotating bodies.

6.2 Quo Vadis—Suggestions for the Future

We divide this section into three subsections; the two first considers instability and transition on rotating disks and broad cones (Sect. 6.2.1) and slender cones (Sect. 6.2.2), respectively. The third subsection (Sect. 6.2.3) deals with the turbulent boundary layer on rotating disks.

6.2.1 Instability and Transition on Rotating Disks and Broad Cones

The rotating disk is the flow case that has been studied most, through experiments, stability analysis and numerical simulations (both DNS and LES). Recent experiments on a broad $\psi = 60°$ cone have shown that it qualitatively resembles the characteristics of transition and instability on the disk but that the most unstable azimuthal wave number (β) for the stationary cross-flow vortices is lower, for a clean surface typically 26 vortices occur upstream transition as compared to 32 for the disk, corresponding to $\beta = 26$ and 32, respectively. For simplicity one can chose the disk as a suitable geometry for the following suggested studies, since broad cones would probably lead to the same conclusions.

6.2.1.1 Wave-Length Selection and Receptivity to Roughness on Broad Cones

When triggering disturbances on the disk small roughness elements, of the order of 5 μm height and 1–2 mm diameter, have been used. With 32 roughness elements evenly distributed in the azimuthal direction at a given radius (usually upstream the critical Reynolds number) the growth of the cross-flow vortices resembles that for the clean disk, although transition occurs at a smaller radius. An interesting idea worth pursuing is to use another number of evenly distributed roughness elements to trigger a less unstable wavenumber, maybe starting from 16 going up to 64, in steps of 8 or less. One would expect slower growth and splitting of vortices if the wave length is larger than the most unstable and merging of vortices if it is smaller. Whether it leads to delayed transition or if it promotes transition is unknown. However, such a study has so far not been done experimentally.

6.2.1.2 Travelling Waves on Broad Cones

In an experiment by Imayama, Alfredsson & Lingwood [4] on a clean disk a smooth, but distinct, hump in the spectra measured by a laboratory-fixed hot wire was discovered around $\omega \sim 40$ (note that for stationary vortices $\omega = \beta$) in the Reynolds number range just prior to transition (see Fig. 6 of that publication). The smooth appearance indicates that it is not a stationary disturbance with respect to the disk but a travelling disturbance. The origin of this hump is unknown, is it a travelling primary mode or due to a secondary instability, or something else?

If the primary cross-flow mode is triggered by roughness elements on the disk surface it may be possible to investigate the origin of this hump in more detail. It would be especially interesting to use two hot wires separated in either the azimuthal or radial direction or maybe both to determine the phase velocity, propagation angle and amplitude distribution of that disturbance.

6.2.1.3 Meandering Cross-Flow Vortices

In experiments it has been observed that after some downstream distance the cross-flow vortices start to meander, i.e. moving back and forth in the radial direction, in a coherent manner over the full disk/broad cone. The amplitude of the meandering increases with x^*. The meandering occurs even when the vortices are triggered by stationary roughness elements. The time scale of the meandering is quite long ($10 \sim 100$ cone revolutions, see e.g. the supplemental movie in Ref. [1]). The reason for the coherent meandering is unknown and would be an interesting research topic.

6.2.1.4 Secondary Instability

So far no experimental studies of the secondary instability related to the cross-flow vortices have been carried out. However, the secondary instability was studied in some detail by Appelquist et al. [2] through DNS. In order to be able to study the secondary instability in an experiment it needs to be triggered when the primary instability has developed and obtained a certain amplitude. How this should be done successfully without disturbing the mean flow is an open question but several possibilities exist; e.g. through a time-dependent disturbance acting through a small hole in the wall, sound waves interacting with a small surface non-uniformity, introduction of a localised heat disturbance through e.g. a pulsating laser or maybe something completely different.

6.2.2 Instability and Transition on Rotating Slender Cones

For cones there exists a transformation that converts the equations for the laminar base flow to that for the disk. The transformation relies on a boundary layer approximation, and for decreasing apex angles it becomes less accurate as shown by Segalini & Camarri [15]. They determined the base flow for two cone apex angles, $\psi = 30°$ and $60°$ and could quantitatively show that the deviation from the similarity solution is mainly observed close to the cone apex and becomes less severe further away from the apex.

The instabilities that are observed experimentally on slender cones are quite different from those on disks and broad cones. These vortices are, at least after some downstream distance, of the Görtler type, and their origin stems from a centrifugal instability. No experiment so far has been able to control the initiation of the vortices and this makes it hard to determine their characteristics since it seems that they are initiated more or less randomly. However, flow visualisation and local linear stability analysis [7] indicate that the dominating instability has a wave number $\beta = -1$ which means that it is inclined in the opposite direction as compared to cross-flow vortices. The small wave number also means that they are almost ringlike as they travel downstream (i.e. in the x-direction) on the cone.

6.2.2.1 Velocity Field, Wave Number and Growth of Vortices

So far there exists no extensive study of the disturbance velocity field related to the instability on rotating slender cones. It seems that hot-wire anemometry in this case may not be the best technique to follow the instability development since the vortices appear randomly. Instead a spatial overview of the disturbances inside the boundary layer obtained by high-resolution PIV would be needed. With time-resolved measurements individual vortices can be followed as they travel downstream and their characteristics determined.

6.2.2.2 Secondary Instability

In the smoke flow visualisation photograph on a $\psi = 30^\circ$ cone by Kohama [13, see Fig. 1c of that paper] a secondary instability developing on the primary vortices can be observed. However, it is not clear whether the instability grows along the primary vortices or if the instability grows as the primary vortices move downstream and the amplitude as well as the Reynolds (or Görtler[1]) number increase. This is clearly an interesting question that should be investigated thoroughly.

6.2.3 Turbulence on the Rotating Disk

There exist some studies where measurements of the mean velocity field of the turbulent boundary layer over rotating disks have been presented. However, there are two conflicting criteria when it comes to the design of an experimental facility, namely to reach high Reynolds numbers where the turbulence is fully developed but still have sufficiently good spatial resolution for accurate measurements. One can show that the viscous length scale ℓ^* is approximately

$$\ell^* \approx \sqrt{\frac{2}{c_{f,\theta}} \frac{x^*}{Re_x}}$$

where $c_{f,\theta}$ is the friction coefficient in the azimuthal direction. To obtain $Re_x = 10^6$ and a viscous length scale of 10 μm we need a disk with at least a diameter of 0.8 m, rotating at about 16 rev/s (40 m/s at the periphery) if the experiments are carried out in air. To set up and balance such a big rotating disk is a non-trivial matter. A large rotating-disk water facility (1.6 m disk diameter) has recently been taken into operation at the University of Melbourne [14] where molecular tagging is used to measure both the azimuthal and radial velocity components in the turbulent boundary layer. However, at this stage (December 2024) only preliminary results are available.

One problem encountered when dealing with turbulent boundary layers is that the friction velocity needs to be determined accurately in order to scale the profiles. In his 1921 paper von Kármán derived integral momentum equations in both the azimuthal and radial directions that may be used to determine the friction at the wall (see Ref. [1] for the equations).

[1] The Görtler number is defined as $G = \frac{V_w^* \theta_V^*}{\nu^*} \sqrt{\frac{\theta_V^*}{r^*}}$ where θ_V^* is the momentum-loss thickness calculated from the azimuthal velocity distribution.

6.2.3.1 Accurate Determination of the Velocity Field

There are so far no sufficiently accurate measurements, especially close to the wall, at high Reynolds numbers of the rotating disk turbulent boundary layer velocity field, both with respect to mean velocities and Reynolds stresses. A complication is the three-dimensional character of the boundary layer, which complicates hot-wire measurements. The azimuthal component can be fairly easily measured, however the radial component would need a V hot-wire configuration to be measured accurately. It may also be possible with high resolution PIV or LDV to carry out such measurements, as well as with the molecular tagging technique used in Ref. [14].

One especially interesting observation from DNS of the turbulent boundary layer on a disk at low Re is that the position of the peak in the radial component does not seem to scale with the standard viscous turbulent scaling [3] but instead with $\sqrt{\nu^*/\Omega^*}$ staying at the same position as in the laminar boundary layer.

6.2.3.2 Large-Scale Structure Identification

In several studies, both flow visualisation and hot-wire measurements, three to four large scale structures have been observed through the transitional and into the turbulent region. The origin of these structures is unclear and their origin and development would be interesting to investigate.

One could not expect that a DNS would be able to reach high Reynolds numbers, especially not if one aims at investigating the large scale structures since that would mean that the full 360° of the disk needs to be simulated and such a simulation would be extremely expensive.

6.3 Epilogue

Axisymmetric rotating bodies (disks and cones) display a number of interesting flow behaviours with respect to instability, transition and turbulence. Some are well described by experiments, simulations and theory ("known knowns") whereas there are others that are identified but are incompletely understood ("known unknowns"). In many papers researchers try to motivate their research on transition and turbulence by using todays "buzzwords" such as "energy savings" and/or "fighting climate change". However, such claims can usually not be substantiated and seem to be more an excuse to carry out the research. Instead we should appreciate that research can be done without direct technical or practical applications and by solving relevant, complex and non-trivial problems, research helps us to fill the space of "known knowns" that at some point in the near or distant future may be useful for an application or an explanation of a phenomenon that we cannot fathom today (“unknown unknowns”)[2].

Acknowledgements This paper is the outcome of an invited keynote presentation by the first author at the 10:th IUTAM Laminar-Turbulent Transition Symposium at Shinshu University, Nagano, Japan 2–6 Sept. 2024. The first author thanks the scientific and organising committees for an excellent Symposium both with respect to its scientific content and social environment. We also want to acknowledge the collaborations with all our co-authors found in the reference list below and with a special thank to Prof. R.J. Lingwood.

References

1. Alfredsson, P.H., Kato, K., Lingwood, R.J.: Flows over rotating disks and cones. Annu. Rev. Fluid Mech. **56**, 45–68 (2024). https://doi.org/10.1146/annurev-fluid-121021-043651
2. Appelquist, E., Schlatter, P., Alfredsson, P.H., Lingwood, R.J.: Transition to turbulence in the rotating-disk boundary-layer flow with stationary vortices. J. Fluid Mech. **836**, 43–71 (2018). https://doi.org/10.1017/jfm.2017.771
3. Appelquist, E., Schlatter, P., Alfredsson, P.H., Lingwood, R.J.: Turbulence in the rotating-disk boundary layer investigated through direct numerical simulations. Eur. J. Mech. B/Fluids **70**, 6–18 (2018). https://doi.org/10.1016/j.euromechflu.2018.01.008
4. Imayama, S., Alfredsson, P.H., Lingwood, R.J.: On the laminar-turbulent transition of the rotating-disk flow: the role of absolute instability. J. Fluid Mech. **745**, 132–63 (2014). https://doi.org/10.1017/jfm.2014.80
5. Kármán, T.V.: Über laminare und turbulente Reibung. ZAMM **1**(4), 238–252 (1921). https://doi.org/10.1002/zamm.19210010401
6. Kato, K., Kawata, T., Alfredsson, P.H., Lingwood, R.J.: Investigation of the structures in the unstable rotating-cone boundary layer. Phys. Rev. Fluids **4**, 053903 (2019). https://doi.org/10.1103/PhysRevFluids.4.053903
7. Kato, K., Segalini, A., Alfredsson, P.H., Lingwood, R.J.: Instability and transition in the boundary layer driven by a rotating slender cone. J. Fluid Mech. **915**, R4 (2021). https://doi.org/10.1017/jfm.2021.216
8. Kato, K., Segalini, A., Alfredsson, P.H., Lingwood, R.J.: Instability and transition on a rotating cone—old problems and new challenges. In: Sherwin, S., Schmid, P., Wu, X. (eds.) 9th IUTAM Laminar-Turbulent Transition Symposium, London 2019 IUTAM Bookseries, vol. 38, pp. 203–213. Springer, Berlin (2022). https://doi.org/10.1007/978-3-030-67902-6_17
9. Kirchner, R.P., Chen, C.F.: Stability of time-dependent rotational Couette flow. Part 1. Experimental investigation. J. Fluid Mech. **40**, 39–47 (1970)
10. Kobayashi, R.: Review: laminar-to-turbulent transition of three-dimensional boundary layers on rotating bodies. ASME. J. Fluids Eng. **116**(2), 200–211 (1994). https://doi.org/10.1115/1.2910255
11. Kobayashi, R., Izumi, H.: Boundary-layer transition on a rotating cone in still fluid. J. Fluid Mech. **127**, 353–64 (1983). https://doi.org/10.1017/S0022112083002761
12. Kobayashi, R., Kohama, Y., Takamadate, C.: Spiral vortices in boundary layer transition regime on a rotating disk. Acta Mech. **35**, 71–82 (1980). https://doi.org/10.1007/BF01190058
13. Kohama, Y.P.: Three-dimensional boundary layer transition study. Curr. Sci. India **79**, 800–807 (2000)

[2] The expressions “known knowns”, “known unknowns” and “unknown unknowns” were brought into public awareness at a press conference by United States Secretary of Defense Donald Rumsfeld in 2002, although had been used in various contexts earlier, for instance for risk assessment, see [17].

14. Mollaei, M., Philip, J., Klewicki, J.C.: An experimental investigation of tangential and radial velocity statistics of a rotating disk boundary layer. In: 13th International Symposium Turbulence Shear Flow Phenomena (TSFP13) Montreal, Canada, June 25–28, (2024). http://www.tsfp-conference.org/proceedings/2023/130.pdf
15. Segalini, A., Camarri, S.: Flow induced by a rotating cone: base flow and convective stability analysis. Phys. Rev. Fluids **4**, 084801 (2019). https://doi.org/10.1103/PhysRevFluids.4.084801
16. Theodorsen, T., Regier, A.: Experiments on drag of revolving disks, cylinders, and streamline rods at high speeds. NACA Technical Report, vol. 793, pp. 367–384. The National Advisory Committee for Aeronautics, Langley Aeronautical Laboratory, Langley Field, VA (1944)
17. https://en.wikipedia.org/wiki/There_are_unknown_unknowns

Chapter 7
Investigation of Turbulent Wedge Intermittency with Time-Resolved Temperature-Sensitive Paint

Benjamin Dimond, Marco Costantini, and Christian Klein

Abstract Turbulent wedges with fully turbulent core and intermittent turbulent wedges were experimentally investigated on a flat plate at high Reynolds numbers and up to high subsonic Mach numbers. With the help of a temperature-sensitive paint capable of resolving fast surface temperature changes at 20 kHz, the time-resolved heat flux distribution and with it the intermittency could be obtained. For turbulent wedges with a fully turbulent core, an intermittent region was observed with a spreading angle similar to that of isolated turbulent spots, suggesting a similar spreading mechanism. Intermittent turbulent wedges yield a similar shape to that of classical turbulent wedges when looking at time-averaged surface temperature and heat-flux. Via the time-resolved measurements, they were found to consist of individual turbulent spots that grow and merge as they propagate downstream.

Keywords Turbulent wedge · Turbulent spot · Time-resolved temperature-sensitive paint · Boundary-layer transition · Intermittency

7.1 Introduction

One special case of laminar-turbulent transition are turbulent wedges-a distinct area of turbulent flow spreading laterally in the downstream direction and forming a triangular shape, hence the term wedge. They are generally caused by surface roughness that leads to premature transition in areas that would otherwise remain mostly laminar ("bypass transition") [1]. Considering this aspect of transition is important in an operational environment as turbulent wedges can be induced by surface imperfections, debris or dead bugs. They also frequently occur in wind-tunnel testing at high Reynolds numbers due to the very high requirement on surface quality that can

Supplementary Information The online version contains supplementary material available at https://doi.org/10.1007/978-981-96-9829-5_7.

B. Dimond (✉) · M. Costantini · C. Klein
Deutsches Zentrum für Luft- und Raumfahrt e.V. (DLR), Göttingen, Germany
e-mail: benjamin.dimond@dlr.de

K. Kato et al. (eds.), *Proceedings of the 10th IUTAM Symposium on Laminar-Turbulent Transition*, IUTAM Bookseries 44, https://doi.org/10.1007/978-981-96-9829-5_7

not always be met [2]. In most cases, turbulent wedges are quasi-stationary in the sense that they consist of a core that has a fully turbulent boundary layer bounded by an intermittent region of alternating laminar and turbulent portions. The intermittency factor γ quantifies the proportion of laminar and turbulent flow, where $\gamma = 0$ denotes a fully laminar and $\gamma = 1$ a fully turbulent boundary layer. Occasionally however, intermittent turbulent wedges arise, that are characterised by an absent turbulent core. They occur for surface roughness heights in between those that would not cause premature transition and those that result in a turbulent wedge with a fully turbulent core [3, 5]. This study investigates both a fully turbulent core wedge and an intermittent turbulent wedge by means of temperature-sensitive paint.

7.2 Experimental Setup

The examined wind-tunnel model was the two-dimensional flat plate *PaLASTra*, which was designed to have a large area of uniform pressure gradient on the surface of interest - the upper side of the model [2]. The experiments were performed in the low-turbulence cryogenic Ludwieg-Tube Göttingen (KRG), at which one can independently set the freestream Mach number M and the chord Reynolds number $Re_c = U_\infty c/\nu_\infty$, with profile chord length c and U_∞ and ν_∞ the freestream velocity and kinematic viscosity, respectively. Due to the working principle of the wind tunnel, a fast expansion of the test gas during a run, the different convective heat flows of laminar and turbulent boundary layers will result in differing surface temperatures. With the help of a fast-response temperature-sensitive paint (iTSP) [4] this different surface temperature development can be recorded. The iTSP used for this investigation is based on Dichlorotris (1, 10-phenanthroline) ruthenium (II) hydrate (98%) as luminophor with a very short luminescent lifetime of less than 1 µs. It was illuminated with 7 high-power LEDs (wavelength 460 nm) installed outside the test-section. The TSP images were than acquired with a Photron MiniAX-200 high-speed camera at 20 kHz. Figure 7.1 shows the top plan view of the wind tunnel model with the iTSP coating (yellow) and the observation area ($55 \times 32\,\text{mm}^2$). The wind tunnel model is further equipped with a central row of pressure taps to determine the pressure distribution and pressure gradient parameter $\lambda_\theta = \frac{\theta^2}{\nu}\frac{\mathrm{d}U_e}{\mathrm{d}x}$ (determined with the boundary layer edge velocity U_e instead of the freestream velocity U_∞); the starboard side of the model was instrumented with hot films, but their data are not considered in the present investigation.

With the measured time series of the surface temperature, the thermal parameters of the TSP layers, and a 1D heat transfer approximation (wall normal direction), the surface heat flux can be determined within the observation area. As discussed in [4], the heat flux values follow a bimodal distribution with distinct peaks corresponding to laminar and turbulent boundary layers. Using Otsu's method, an appropriate threshold can be found that distinguishes the two boundary layer states and enables a direct intermittency determination. The present experimental setup and camera acquisition rate are sufficient to accurately and directly resolve the intermittency values. Further details on the experimental methods and evaluation procedures can be found in [4].

7.3 Results and Discussion

7.3.1 Fully Turbulent Core Wedges

Figure 7.2 shows the intermittency distribution of a turbulent wedge with fully turbulent core, along with the spanwise distribution for 6 streamwise locations. Beside a fully turbulent core, where $\gamma \approx 1$, an intermittent edge area is visible, where the intermittency level steadily decreases in spanwise direction until $\gamma \approx 0$ for a predominantly laminar boundary layer. This intermittent region (between $\gamma = 0.1$ and $\gamma = 0.9$) increases in streamwise direction, resulting in different spreading angles of the wedge for different intermittency levels. The spreading angles were determined to be $9.68(12)^\circ$ at $\gamma = 0.1$ and $8.10(11)^\circ$ at $\gamma = 0.9$ resulting in a difference of about 1.5°.

In a separate investigation, the spreading angle of turbulent spots was found to be $10.0(5)^\circ$ for the same flow conditions [4]. Within the given accuracy, these values are the same as the spreading angle of the turbulent wedge presented here at $\gamma = 0.1$. This correlation illustrates the similarity of growth mechanism of turbulent wedges and turbulent spots and suggests a close link between the two: The wing tips of closely following turbulent spots extend outside the fully turbulent core and comprise the intermittent region while propagating downstream. This assumption is supported by DNS studies by Goldstein et al. [7], who conclude that the spreading mechanism is determined by the outer edges of a turbulent spot and wedge and is independent of the flow structure within. The intermittent region of a turbulent wedge would therefore depend only on frequency and formation location of the wing tips and therefore on

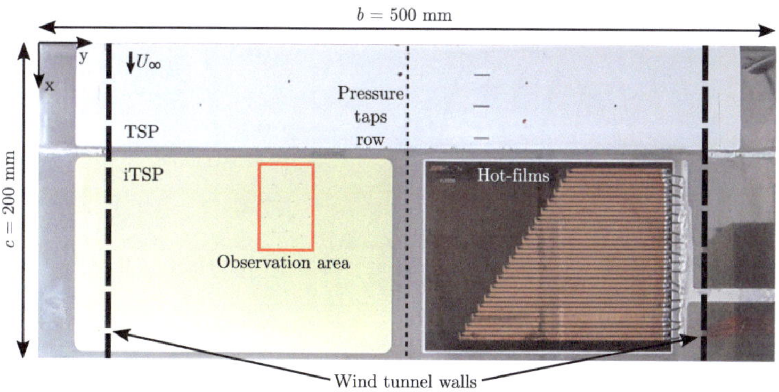

Fig. 7.1 Top plan view of the wind tunnel model (flow from the top)

the flow conditions and the initial phase of the turbulent wedge. This could also explain the larger intermittent region observed by Schubauer and Klebanoff [1] for an incompressible flow (6.4° and 10.6° for the turbulent core and intermittent region, respectively).

7.3.2 Intermittent Turbulent Wedge

The intermittency distribution for an intermittent turbulent wedge is shown in Fig. 7.3. The intermittency value increases in streamwise direction, but never reaches a fully turbulent state ($\gamma = 1$) in the field-of-view. The wedge is presumably caused by the same surface imperfection resulting in the turbulent wedge presented in the previous section, but because of the higher Mach number (and accordingly thicker boundary layer) its impact on the laminar boundary layer is not so strong as to cause a turbulent wedge with fully turbulent core (see Sect. 7.1). The spreading angle is slightly smaller (8.76(6)° for $\gamma = 0.1$) but can be explained by the different flow conditions, as a higher Mach number and more favourable pressure gradient result in a smaller spreading angle [6, 8].

In the supplementary material (Video 1), the time-resolved heat flux distribution for $Re_c = 8 \cdot 10^6$, $M = 0.8$, and $\lambda_\theta = 19.4 \cdot 10^{-3}$ can be seen. It becomes evident, that the intermittent turbulent wedge comprises individual turbulent spots, i.e. patches of higher heat flux levels (dark areas) corresponding to a temporary turbulent boundary layer. Many turbulent spots merge with preceding or following ones as they propagate downstream. Due to the spanwise spot growth the time-averaged results (surface temperature field or intermittency) yields the typical wedge shape.

An example image series of such a turbulent spot can be seen in Fig. 7.4 Figure 7.4a depicts a typical isolated turbulent spot with the common shape of a downstream pointing triangle, growing in both spanwise and streamwise direction as it propagates downstream in the subsequent time steps. A new, independent tur-

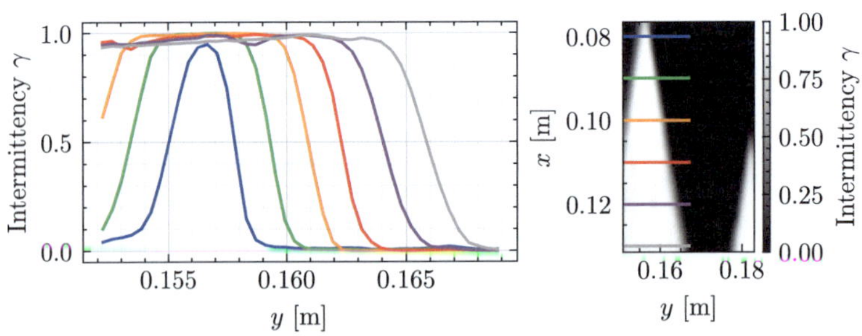

Fig. 7.2 Left: spanwise profiles of intermittency distributions along the turbulent wedge (with fully turbulent core) at the different streamwise locations shown in the right image(flow is from the top), for run conditions $Re_c 7.5 \cdot 10^6$, $M = 0.5$ and $\lambda_\theta = 10.4 \cdot 10^{-3}$

bulent spot develops further upstream, as can be seen for the first time in (c). At image (d) the two spots begin to merge as a third turbulent spot emerges even further upstream in (e).

Figure 7.5 shows three individual examples of turbulent spots recorded at different times. Image (a) was captured shortly after the image series presented in Fig. 7.4. Here, individual, closely following spots form soon after each other and merge in such a way to give the impression of an upstream pointing arrow head. The turbulent spots in (b) and (c) show two similarly sized spots at different streamwise positions, suggesting that the spot formation location also is at different streamwise positions. Throughout the series, turbulent spots were found to form at widespread streamwise positions from outside the observation area until about $x = 0.10$ m.

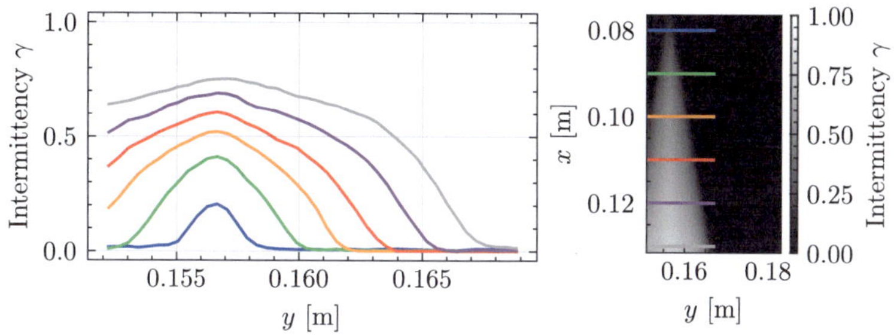

Fig. 7.3 Left: spanwise profiles of intermittency distributions along the (intermittent) turbulent wedge at the different streamwise locations shown in the right image (flow is from the top), for run conditions $Re_c = 8 \cdot 10^6$, $M = 0.8$ and $\lambda_\theta = 19.4 \cdot 10^{-3}$

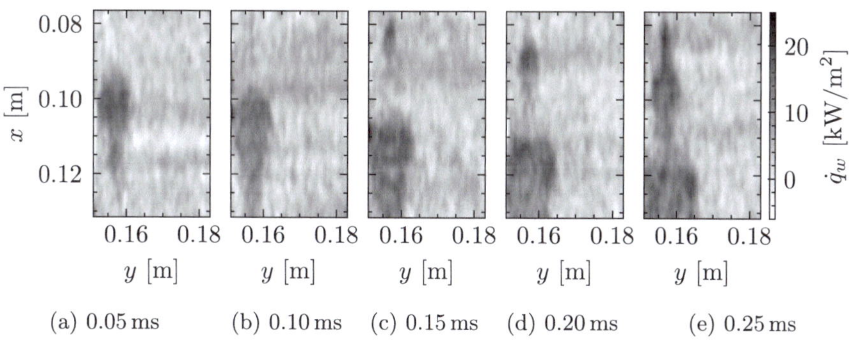

Fig. 7.4 Heat flux distribution series for turbulent spots for $Re_c = 8 \cdot 10^6$, $M = 0.8$, and $\lambda_\theta = 19.4 \cdot 10^{-3}$ after $t_0 = 3.1502$ s. Flow is from the top

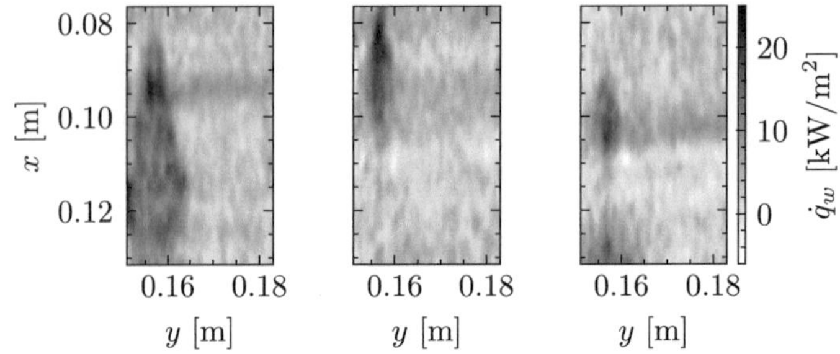

(a) $t = 3.150\,60$ s (b) $t = 3.155\,95$ s (c) $t = 3.156\,55$ s

Fig. 7.5 Heat flux distribution for different turbulent spots with different shapes and streamwise formation locations for $Re_c = 8 \cdot 10^6$, $M = 0.8$, and $\lambda_\theta = 19.4 \cdot 10^{-3}$. Flow is from the top

7.4 Conclusions

With the help of time-resolved temperature-sensitive paint it was possible to measure the surface temperature development at 20 kHz and with it the surface heat flux. This enables the direct determination of intermittency with a high spatial resolution. Two turbulent wedges were examined at $Re_c = 7.5 \cdot 10^6$ and $M = 0.5$ as well as $Re_c = 8 \cdot 10^6$ and $M = 0.8$. The first had a fully turbulent core bounded by an intermittent region spanning 1.5°. The spreading angle of the intermittent region thereby was found to be similar to isolated turbulent spots, suggesting a strong similarity between the respective spreading mechanisms. The second turbulent wedge showed an intermittent behaviour throughout the observation area without a fully turbulent region but with a similar shape and spreading angle. It could be shown for the first time with an imaging technique, that the wedge consists of individual, closely following turbulent spots. The study emphasises the capabilities of iTSP in examining the instationary behaviour of laminar-turbulent transition.

Supplementary Material. Video 1: Intermittent turbulent wedge for $Re_c = 8 \cdot 10^6$, $M = 0.8$, and $\lambda_\theta = 19.4 \cdot 10^{-3}$ resolved at 20 kHz

Disclosure of Interests. The authors have no competing interests to declare that are relevant to the content of this article.

Acknowledgements The authors acknowledge: C. Fuchs, U. Henne, T. Hermann, and T. Kleindienst (DLR) for model setup and support with the experimental setup as well as M. Aschoff, S. Koch, and S. Weiss (DLR) for wind tunnel operation. Furthermore M. Rein and A. Dillmann are thanked for valuable scientific discussion.

References

1. Schubauer, G.B., Klebanoff, P.S.: Contributions on the mechanics of boundary-layer transition. NACA Technical report 1289 (1956)

2. Costantini, M., Hein, S., Henne, U., Klein, C., Koch, S., Schojda, L., Ondrus, V., Schröder, W.: Pressure gradient and non-adiabatic surface effects on boundary-layer transition. AIAA J. **54**(11), 3465–3480 (2016)
3. Clark, J.P., Magari, P.J., Jones, T.V.: On the distribution of the heat transfer coefficient in turbulent and "transitional" wedges. Int. J. Heat Fluid Flow **14**, 217–222 (1993)
4. Dimond, B.: Investigation of boundary layer transition at high Reynolds numbers using time-resolved temperature-sensitive paint. Ph.D. Thesis, Georg-August-Universität Göttingen (2024)
5. Fiala, A., Hillier, R., Estruch-Samper, D.: Roughness-induced turbulent wedges in a hypersonic blunt-body boundary layer. J. Fluid Mech. **754**, 208–231 (2014)
6. Fischer, M.C.: Spreading of a turbulent disturbance. AIAA J. **10**(7), 957–959 (1972)
7. Goldstein, D., Chu, J., Brown, G.: Lateral spreading mechanism of a turbulent spot and a turbulent wedge. Flow Turbul. Combust. **98**, 21–35 (2017)
8. Zhong, S., Chong, T.P., Hodson, H.P.: A comparison of spreading angles of turbulent wedges in velocity and thermal boundary layers. J. Fluids Eng. **125**, 267–274 (2003)

Chapter 8
Boundary-Layer Instability Characteristics of Supersonic and Hypersonic Flows Over Rotating Cones

Alexander Theiss, Jonas Penning, and Stefan Hein

Abstract This study investigates the influence of cone rotation on boundary-layer instabilities in supersonic and hypersonic flows using linear stability theory. In a rotating reference frame, the linearized disturbance equations include terms representing Coriolis and centrifugal forces, whose effects on primary instabilities—such as the first and second modes, crossflow, and centrifugal instabilities—are only partially understood, particularly for compressible flows. The results reveal that rotation destabilizes all instability modes, with the centrifugal instability becoming dominant at higher rotation rates. While centrifugal and metric terms, the latter accounting for surface curvature and conical divergence, exhibit stabilizing effects, Coriolis forces act as a destabilizing mechanism.

Keywords Rotating cone · Hypersonic flow · Linear stability theory

8.1 Introduction

Rotation of an aeronautical object can alter the boundary-layer instability characteristics. This may amplify existing instabilities or promote new mechanisms, such as the centrifugal instability, fundamentally changing the transition process and likely also the laminar-to-turbulent transition locations. While fundamental research on the impact of rotation on boundary-layer instabilities has primarily focused on simple geometries like rotating disks and cones (e.g., see Refs. [1–3]), a comprehensive understanding remains elusive, especially for compressible inflow conditions. Moreover, the effect of the Coriolis and centrifugal force terms in the linearized disturbance equations that appear in a rotating reference frame on the stability of boundary lay-

A. Theiss (✉) · S. Hein
Institute of Aerodynamics and Flow Technology, German Aerospace Center (DLR), Göttingen, Germany
e-mail: alexander.theiss@dlr.de

J. Penning
Institute for the Dynamics of Complex Systems, University of Göttingen, Göttingen, Germany

K. Kato et al. (eds.), *Proceedings of the 10th IUTAM Symposium on Laminar-Turbulent Transition*, IUTAM Bookseries 44, https://doi.org/10.1007/978-981-96-9829-5_8

ers is only partially understood. This study expands the recent work of Song and Dong [3] and Song et al. [4], who have numerically investigated the influence of rotation on the stability characteristics of a boundary layer on 7° half-apex angle cone subjected to supersonic axial flow by considering more half-apex angles and hypersonic axial flow. The main objective of this work is to analyze the impact of rotational effects (the centrifugal and Coriolis force) and metric terms, accounting for surface curvature and conical divergence, on the instability properties of the primary transition mechanisms (i.e., first and second mode waves, crossflow and centrifugal instabilities).

8.2 Methodology

8.2.1 Laminar Basic Flow Computations

The laminar basic flows for the nominally sharp rotating cones (the physical model is depicted in Fig. 8.1) with super- and hypersonic axial inflow at zero angle of attack are computed with the DLR TAU-code [6] on shock-adapted block-structured grids. Grid convergence was checked [7] and the mesh consists of 550 points in x direction and 484 points in z direction. The rotational velocities in the simulation were adjusted to have fixed ratios of circumferential wall velocity to boundary-layer edge velocity (i.e., $\bar{\Omega}$, $\bar{\Omega} \in [0,\ 1.5]$) at $x^\star = 0.4\,\text{m}$.

In total, four selected half-cone angles ($\Psi = 5°, 7°, 10°, 15°$) at two different freestream Mach numbers are investigated. The first Mach number examined is a supersonic case (M3-case) with $Ma_\infty = 3.214$, matching the freestream conditions of Song and Dong [3], and the second Mach number corresponds to a hypersonic case (M6-case) with $Ma_\infty = 6.1$, representing typical low-enthalpy wind tunnel conditions. For both Mach number cases, the freestream temperature is $T^*_\infty = 48.28\,\text{K}$, the pressure is $p^*_\infty = 883.85\,\text{Pa}$, and the isothermal wall temperature is $T^*_w = 300\,\text{K}$.

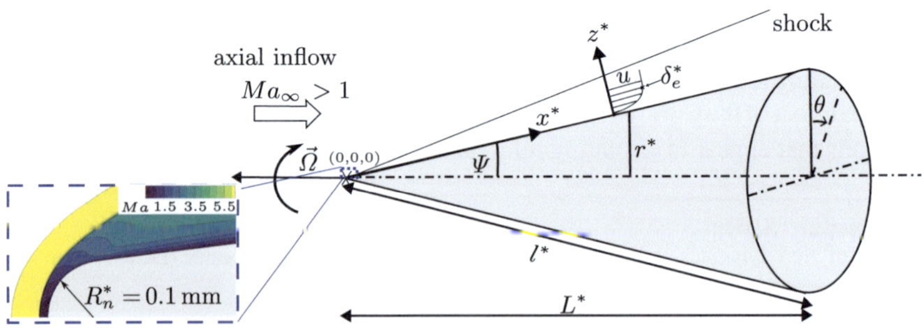

Fig. 8.1 Sketch of the physical model

8.2.2 *Local Linear Stability Theory*

To study the effect of rotation on the primary instability modes at super- and hypersonic speeds, the DLR in-house instability analysis code NOLOT has previously been extended to rotating reference frames [5]. The effect of the rotation vector $\boldsymbol{\Omega}$ is taken into account via terms corresponding to the Coriolis $2\rho\boldsymbol{\Omega} \times \mathbf{U}$ and centrifugal $\rho\boldsymbol{\Omega} \times (\boldsymbol{\Omega} \times \mathbf{r})$ forces ($\mathbf{r}$ is the vector between the rotation vector and the considered point in the boundary layer). This work utilizes the classical quasi-parallel local linear approach, where a flow quantity $\mathbf{q}$ is divided into a steady basic flow $\bar{\mathbf{q}}$ and a disturbance flow $\tilde{\mathbf{q}}$. A modal disturbance approach is made for the perturbed flow that expresses the disturbance as a product of amplitude function $\hat{\mathbf{q}}$ and phase function Θ:

$$\left.\begin{aligned} \tilde{\mathbf{q}}(x, \theta, z, t) &= \hat{\mathbf{q}}(z)\Theta(x, \theta, t) + c.c. \\ \Theta(x, \theta, t) &= \exp\left(i\left[\alpha x + n\theta - \omega t\right]\right) \end{aligned}\right\} \tag{8.1}$$

where t is the time and $c.c.$ denotes the complex conjugate. The streamwise and circumferential wavenumber components are α and n. This work uses the spatial framework; therefore, ω is real and related to the circular frequency of an eigenmode $\tilde{\mathbf{q}}$, while the real part of α, α_r, is related to the physical wavenumber of the disturbance, whereas its imaginary part α_i represents the growth/damping rate in the x direction, $\sigma = -\alpha_i$.

8.3 Basic Flows

The laminar basic flow profiles for the M3 and M6 cases are presented in Fig. 8.2 for selected rotation rates of the 7° cone in a streamline-aligned coordinate system. The velocity profiles are non-dimensionalized using the boundary-layer edge properties of the non-rotating cone. The normalized profiles for the M3 and M6 cases generally align well, with the primary distinction being a more pronounced temperature overshoot in the M6 case due to increased friction. With increasing cone rotation rates, the streamline velocity (U_s) rises as a result of the growing circumferential velocity component at higher $\bar{\Omega}$. Simultaneously, a crossflow velocity profile (V_s) emerges, which likely facilitates rotation-induced instabilities such as crossflow and centrifugal instabilities.

8.4 Results

Figure 8.3 shows the spatial amplification rates of the first and second mode instabilities within the frequency-wavenumber parameter space for both a non-rotating cone ($\bar{\Omega} = 0$) and rotating cone ($\bar{\Omega} = 0.3$) at $x^* = 0.4$ m for the M6-case. Rotation has

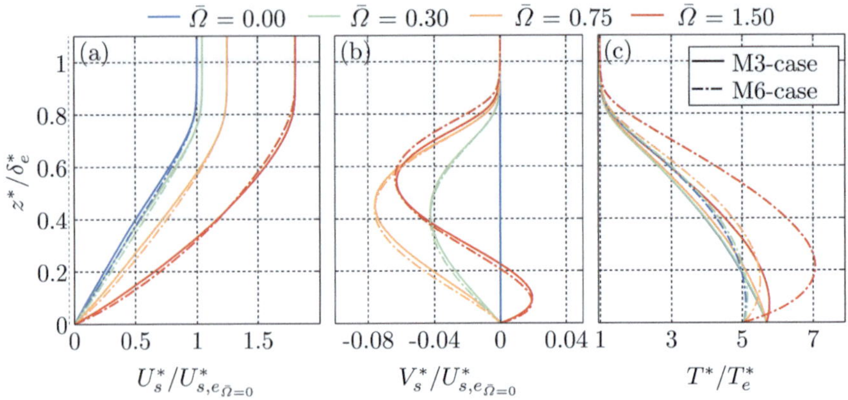

Fig. 8.2 Impact of rotation rate on boundary-layer profiles in streamline coordinate system; $\Psi = 7°$, $x^* = 0.4\,\text{m}$

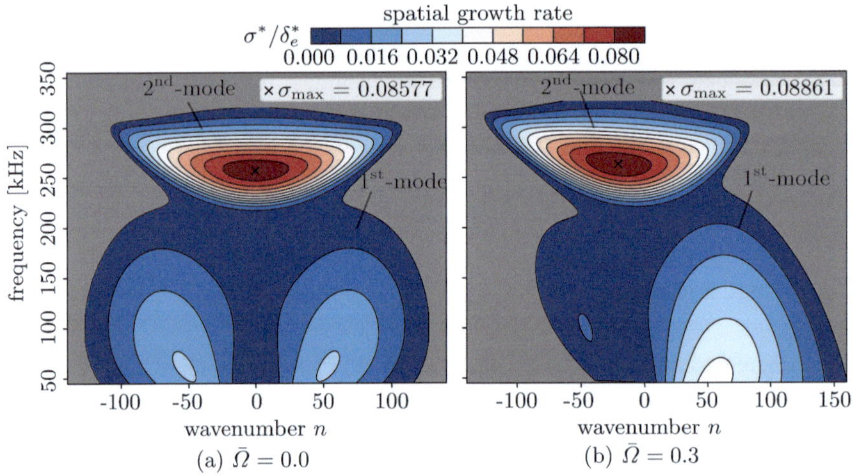

Fig. 8.3 Spatial growth rate of the 1st and 2nd mode instability at $x^* = 0.4\,\text{m}$; M6-case, $\Psi = 7°$

a pronounced destabilizing effect on both instability types. However, the instability characteristics of the first mode are more strongly influenced by rotation. Specifically, the first mode is destabilized for waves propagating in the direction of rotation, while the second mode is destabilized for waves propagating in the opposite direction of rotation. Furthermore, first-mode waves propagating against the direction of rotation are significantly damped, consistent with the results of Song & Dong [3]. In contrast, the damping effect is less pronounced for the second-mode waves propagating in the direction of rotation.

In addition to the first mode (M3 and M6 cases) and the second mode (M6 case only), rotation of the cone causes further rotation-induced instabilities. Figure 8.4a

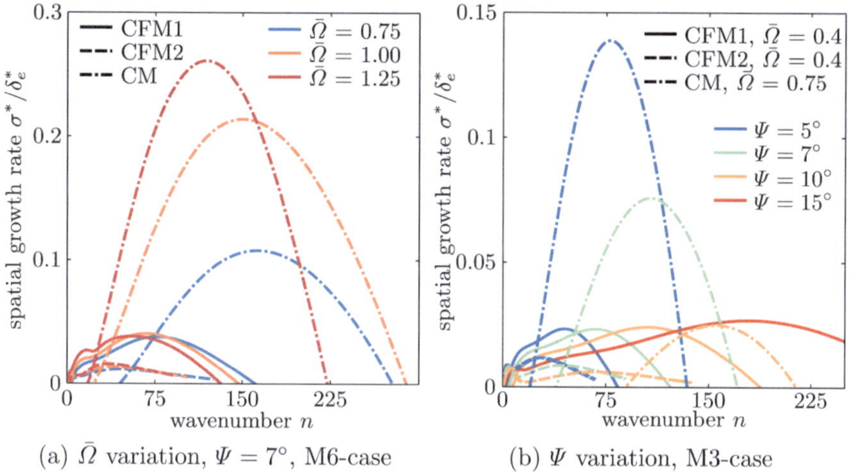

Fig. 8.4 Impact of **a** rotation rate and **b** half-cone angle on the spatial growth rate of rotation-induced stationary ($f = 0\,\text{Hz}$) cross-flow (CFM) and centrifugal (CM) instabilities; $x^* = 0.4\,\text{m}$

depicts the influence of the rotation rate on the spatial growth rate of the stationary crossflow (CFM) and centrifugal instabilities (CM) for the 7° cone at $Ma_\infty = 6.1$. From $\bar{\Omega} \approx 0.5$, the centrifugal instability starts to destabilize at $x^* = 0.4\,\text{m}$ on the 7° cone for both Mach number cases considered. The growth rate of the stationary CM mode increases with $\bar{\Omega}$, and from $\bar{\Omega} \approx 0.75$, the CM mode becomes the most amplified instability type. The maximum growth rate of the crossflow instability also increases with the rotation rate; however, the amplified parameter space shifts with $\bar{\Omega}$, making the effect less apparent for stationary modes with $f = 0$ Hz. Notably, in addition to the most amplified crossflow instability (CFM1), another crossflow instability (CFM2) was identified during this study. To our knowledge, CFM2 has not been previously described in the literature. The CFM2 mode is unstable in a parameter range similar to CFM1 but exhibits significantly lower amplification rates.

The influence of the cone half-angle on the amplification rates of stationary rotation-induced instabilities for the M3 case is presented in Fig. 8.4b. As the cone becomes slimmer, the stationary CM instability is more strongly amplified. However, at $\Psi = 15°$, the stationary CM instability is already damped at the rotation rate considered. For the CFM1 instability, the amplification rate increases with the cone half-angle and spans a broader parameter range. In contrast, the maximum growth rate of the CFM2 instability decreases, following a trend similar to that of the CM instability.

Figure 8.5 depicts the impact of including or neglecting specific terms in the linearized disturbance equations on the amplification rate of the most amplified instability types for the $\Psi = 7°$ cone. In general, the metric and centrifugal terms have a stabilizing effect on the instabilities considered, while the Coriolis force has a

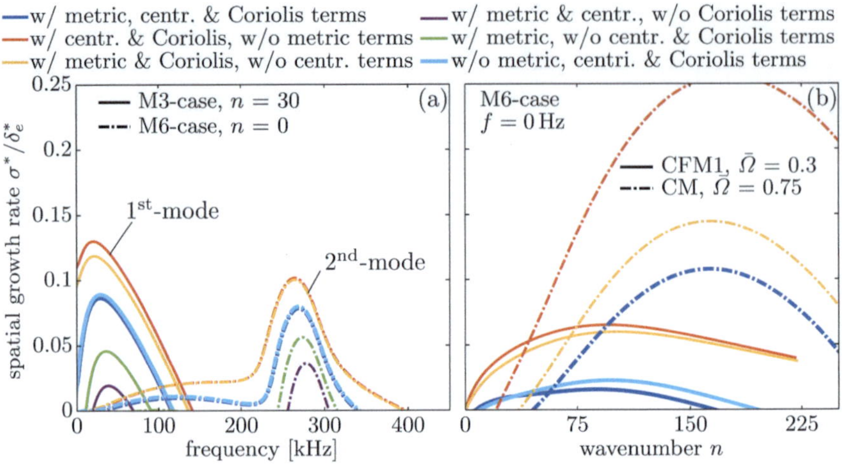

Fig. 8.5 Impact of selected terms in the linearized disturbance equations on the spatial growth rate of **a** the 1st and 2nd mode waves and **b** rotation-induced instabilities; $\Psi = 7°$, $x^* = 0.4\,\text{m}$

destabilizing influence. Interestingly, for the first and second modes, the combined effects of including and neglecting all terms cancel each other out. The centrifugal instability, however, becomes unstable only when the Coriolis terms are included.

8.5 Conclusion

The influence of cone rotation on boundary-layer instabilities in supersonic and hypersonic flows is analyzed through linear stability computations. The effects of centrifugal and Coriolis force terms appearing in the linearized disturbance equations, as well as metric terms accounting for surface curvature and conical divergence, are assessed by comparing computations that include these terms with those that omit them. Results show that the cone rotation has a pronounced destabilizing effect on all instability modes. Specifically, the first mode is destabilized for waves propagating in the rotation direction, while the second mode is destabilized for waves propagating in the opposite direction. Metric terms exhibit a stabilizing influence, counterbalancing the destabilizing effects of the combined rotational terms. A detailed analysis reveals that centrifugal terms contribute to stabilization across all instability modes, whereas Coriolis terms act as the primary destabilizing mechanism.

References

1. Kato, K., Alfredsson, P.H., Lingwood, R.J.: Boundary-layer transition over a rotating broad cone. Phys. Rev. Fluids **4**, 071902 (2019). https://doi.org/10.1103/PhysRevFluids.4.071902
2. Tambe, S., Schrijer, F., Veldhuis, L., Gangoli, Rao A.: Spiral instability modes on rotating cones in high-Reynolds number axial flow. Phys. Fluids **34**, 034109 (2022). https://doi.org/10.1063/5.0083564
3. Song, R., Dong, M.: Linear instability of a supersonic boundary layer over a rotating cone. J. Fluid Mech. **955**, A31 (2023). https://doi.org/10.1017/jfm.2022.1087
4. Song, R., Dong, M., Zhao, L.: Effect of cone rotation on the nonlinear evolution of Mack modes in supersonic boundary layers. J. Fluid Mech. **971**, A4 (2023). https://doi.org/10.1017/jfm.2023.629
5. Dechamps X., Hein S.: Extension of the PSE code NOLOT for transition analysis in rotating reference frames. In: New Results in Numerical and Experimental Fluid Mechanics XI, pp. 179–188 (2018). https://doi.org/10.1007/978-3-319-64519-3_16
6. Schwamborn, D., Gerhold, T., Heinrich, R.: The DLR TAU-code: recent applications in research and industry. In: European Conference on Computational Fluid Dynamics, Delft, Netherland (2006)
7. Penning, J.: Linear instability analyses of supersonic and hypersonic flows over rotating cones, Master‘s thesis, Universität Göttingen (2024)

Chapter 9
Interaction of Crossflow Modes with Forward-Facing Steps: Insights Gained from DNS

F. Tocci, G. Chauvat, A. F. Rius-Vidales, M. Kotsonis, S. Hein, and A. Hanifi

Abstract Surface roughness, such as forward-facing steps (FFS), significantly impacts the laminar-turbulent transition in swept-wing boundary layers. This study employs direct numerical simulations (DNS) to investigate the non-monotonic relationship between FFS height and transition location, complementing recent experimental findings. Two step heights and a baseline clean surface are analyzed under experimentally representative conditions. The simulations confirm a delay in transition for the shallow FFS and premature transition for the higher FFS, reproducing experimental trends while revealing some discrepancies in the abruptness of transition advancement for the larger step. DNS results provide detailed insights into the interaction of crossflow modes with the FFS, particularly near the step, highlighting the spanwise modulation and the formation of reverse flow regions.

Keywords Direct numerical simulation · Crossflow instability · Forward-facing steps

9.1 Introduction

In recent studies, both experiments (e.g. [1, 2]) and numerical simulations (e.g. [3]) have explored how surface roughness, specifically forward-facing steps (FFS), affects laminar-turbulent transition in swept-wing boundary layers with crossflow instabilities (CFI). Casacuberta et al. [3] focused on steady direct numerical simu-

F. Tocci (✉) · S. Hein
Institute of Aerodynamics and Flow Technology, German Aerospace Center (DLR), Göttingen, Germany
e-mail: francesco.tocci@dlr.de

G. Chauvat · A. Hanifi
FLOW, Department of Engineering Mechanics, KTH Royal Institute of Technology, Stockholm, Sweden

A. F. Rius-Vidales · M. Kotsonis
Department of Flow Physics and Technology, Delft University of Technology, Delft, The Netherlands

K. Kato et al. (eds.), *Proceedings of the 10th IUTAM Symposium on Laminar-Turbulent Transition*, IUTAM Bookseries 44, https://doi.org/10.1007/978-981-96-9829-5_9

lation (DNS), in which no transition occurred, to investigate stationary mechanisms near the FFS on a flat-plate configuration with an imposed pressure gradient, where the stationary crossflow (CF) amplitude remained in the linear regime up to the FFS. The experimental observations of Rius-Vidales and Kotsonis [2] revealed a non-monotonic relationship between step height and the location of transition. Notably, an unprecedented delay in transition—compared to scenarios without FFS—occurs when stationary CF vortices interact with a shallow FFS, while interaction with a higher FFS leads to an upstream advancement of the transition front. Motivated by these experimental findings, the present work aims to replicate the setup and conditions described in [2] via DNS. In addition to reproducing the transition trends, this study examines the interaction of CF vortices with the FFS, highlighting the associated formation of reverse flow regions.

9.2 Flow Conditions and Methodology

The experimental setup described in [2] features the M3J swept-wing model with a sweep angle of $\Lambda = 45°$ and a chord length normal to the leading edge of $c_x = 0.9$ m. The model is mounted vertically in the Low Turbulence Tunnel at the Delft University of Technology at an angle of attack AoA $= -3°$. Discrete roughness elements (DREs) near the leading edge are used to trigger a stationary primary CF mode with a spanwise wavelength $\lambda_0 = 7.5$ mm.

The numerical simulations employ a Cartesian coordinate system (x, y, z), where x is oriented perpendicular to the leading edge of the wing (chordwise), z is parallel to the leading edge (spanwise), and y is normal to both x and z (Fig. 9.1). The velocity components in this system are denoted as U, V, W along the x, y, z directions, respectively. For result presentation, a curvilinear reference system (x_τ, y_η, z) is used, where x_τ is the chordwise direction tangent to the wing surface and y_η is normal to the wing surface. In this system, the velocity components are denoted as u_τ, v_η, w. The freestream velocity for the simulations is set to $U_\infty = 27.542\,\text{m s}^{-1}$, corresponding to a chord Reynolds number given by $Re_{c_x} = U_\infty c_x/\nu = 1.68 \times 10^6$. Three-dimensional (3D) DNS are conducted for the clean case (denoted as s0) and FFSs of two heights, s1 ($h = 368\,\mu$m, $Re_{hh} = u_{\tau_h} h/\nu = 205$, where u_{τ_h} represents the u_τ at the FFS height and FFS chordwise location for the clean basic flow) and s2 ($h = 474\,\mu$m, $Re_{hh} = 322$). The computational domain for the 3D DNS spans a single spanwise wavelength λ_0, with periodic boundary conditions applied along the z-direction. The 3D DNS, constrained by the relatively small domain C in Fig. 9.1, relies on precise boundary conditions derived from a preliminary spanwise-invariant (2.5D) RANS simulation covering the entire airfoil shape and wind tunnel walls. Subsequently, a 2.5D DNS is conducted for the domain B in Fig. 9.1. The effect of DREs used to condition the instability modes in the experiments is accurately reproduced via perturbations computed through nonlinear parabolised stability equations, which are applied as inflow conditions for the 3D DNS. Unsteady noise is also added as a random forcing in the boundary layer close to the inflow such that the laminar-

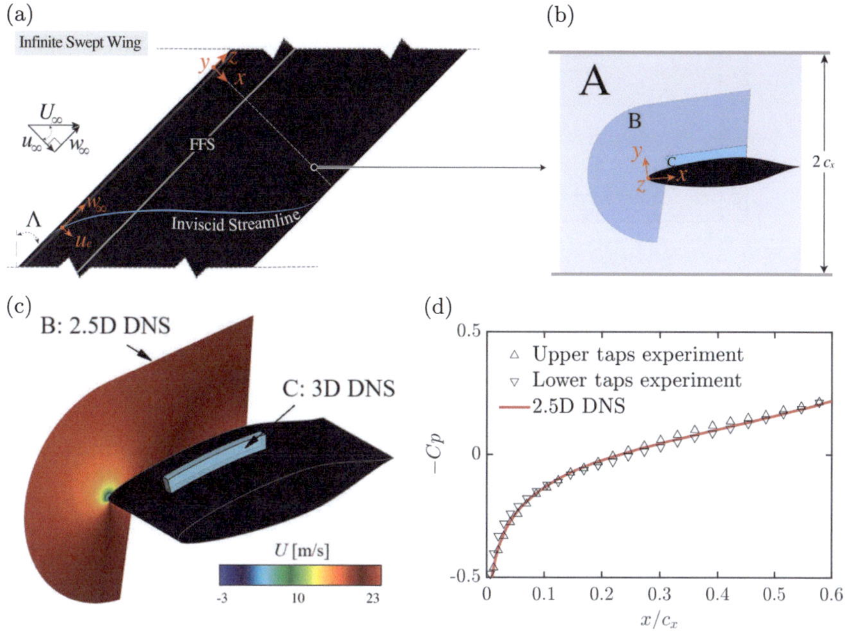

Fig. 9.1 (**a**) Schematic of the M3J infinite swept-wing model. (**b**) Cross-sectional view of a plane normal to the leading edge, showing the wind tunnel walls and computational domains: A for 2.5D RANS, B for 2.5D DNS, and C for 3D DNS. (**c**) Contours of the velocity component U for the clean case from the 2.5D DNS (domain B), with domain C for the 3D DNS highlighted in cyan. (**d**) Comparison of the pressure coefficient C_p between the 2.5D DNS results and experimental data for the clean case

turbulent transition in the simulations approximates that in the experiments for the clean case s0. Further details can be found in [4]. DNS are performed using the spectral element method (SEM) code Nek5000 [5], with a polynomial order $P = 9$.

9.3 Transition Behaviour and CFI Interaction with FFS

Figure 9.2 compares the skin-friction coefficients for s0, s1, and s2, presenting DNS results averaged in time and along the spanwise direction, alongside confidence intervals derived from experimental measurements. The DNS results confirm a delay in transition for FFS s1 and a premature transition for FFS s2, reflecting the non-monotonic relationship between step height and transition location observed in experiments [2]. While the delay associated with s1 shows a relatively good agreement with experimental data, the transition advancement for s2 is noticeably more pronounced in the DNS. This suggests that s2 exhibits supercritical behavior in the simulations, differing from its classification as critical in the experiments. For a detailed analysis

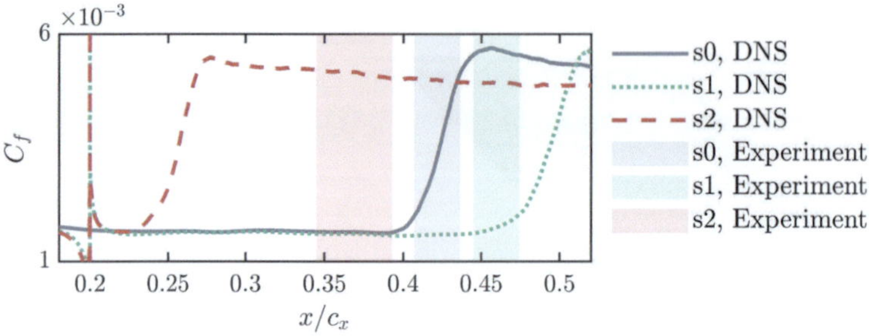

Fig. 9.2 Comparison of skin-friction coefficients for s0, s1, and s2: DNS results (temporally and spanwise averaged) alongside confidence intervals from experimental measurements of the transition location obtained using infrared thermography

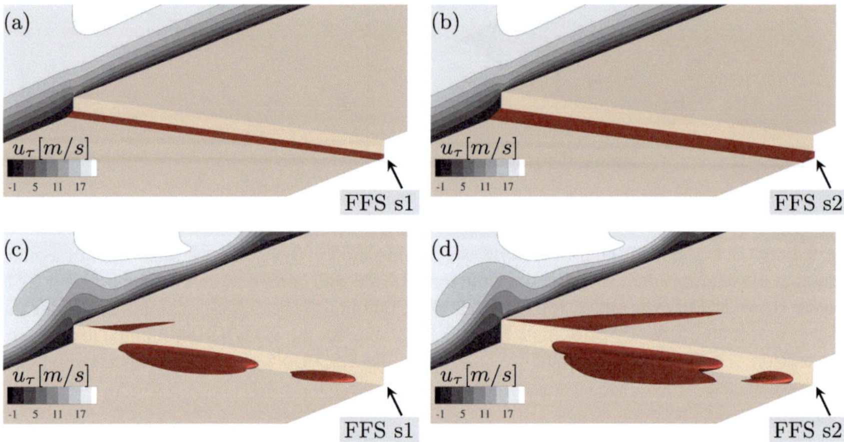

Fig. 9.3 Visualisation of the 2.5D basic flow (**a**, **b**) and the time-averaged 3D DNS (**c**, **d**). (**a**, **c**) correspond to s1 and (**b**, **d**) correspond to s2. The 2.5D basic flows, spanwise-uniform by definition, are extruded for λ_0 to enable comparison. Reverse flow regions are indicated by red isosurfaces of $u_\mathcal{T} = 0$, with grey scale contours of the time-averaged $u_\mathcal{T}$ shown for the plane $x_\mathcal{T} y_\eta$ at $z = -\lambda_z/2$

of the mechanisms underlying the transition behaviour for s0, s1 and s2 the reader is referred to [4].

Figure 9.3 compares the reverse flow regions ($u_\mathcal{T} = 0$) for s1 and s2 in the 2.5D basic flows (Fig. 9.3a, b) and the 3D time-averaged flows (Fig. 9.3c, d) resulting from the interaction of stationary CF vortices with the FFS. In the basic flows, both s1 and s2 exhibit a single reverse flow region upstream of the step, with its size increasing with the step height, resulting in a larger region for s2 compared to s1. Due to the stationary CF vortices in the 3D flows, the reverse flow regions are strongly modulated in the spanwise direction, resulting in isolated zones of reversed flow, both upstream and downstream of the FFS. An analysis of several instantaneous flow fields (not shown) indicates the nearly stationary nature of the spanwise-modulated reverse flow regions. The spanwise modulation of the flow is particularly evident in the 3D fields for s1 and s2, as shown in Fig. 9.4, where the clean case s0 serves as a baseline for comparison. As noted in [1], the modulation of the reverse flow regions is strongly influenced by the amplitude of the stationary CF at the onset. In the present study, the primary CFI reaches the FFS location with a relatively high amplitude, with the maximum CFI spanwise velocity reaching approximately 13% of the local freestream spanwise velocity, leading to a strong spanwise deformation of the shear layer. Compared to the clean configuration, the trajectory of the primary CF vortex (identified using the Q-criterion) is significantly altered in the presence of a FFS, deviating from its original path and interacting with the reverse flow region. At the location of interaction between the CF vortex and the FFS, the reverse flow region observed in the basic flow disappears. However, downstream of the FFS edge, a relatively thin yet elongated reverse flow region forms near the wall. Upstream of the step, the spanwise-modulated reverse flow region exhibits a larger volume compared to its uniform counterpart in the basic flow (Fig. 9.3). Additionally, secondary spanwise-oriented vortical structures emerge within the largest reverse flow region upstream of the step and near the wall downstream of the FFS. For the higher step s2, a secondary vortical structure develops further downstream, closely aligned with the trajectory of the primary CF vortex and located near the wall. The amplitude of the stationary CF vortex in a localized region downstream of the FFS is reduced compared to the clean case, as detailed in [4]. While the higher FFS s2 induces greater overall distortion than s1, the qualitative similarities in the observed flow field distortions suggest that time-averaged flow characteristics alone are inadequate to fully explain the transition trends. This emphasizes the need to consider unsteady mechanisms and secondary instability interactions for a comprehensive understanding of the observed transition behavior.

9.4 Conclusion

This study successfully reproduces the non-monotonic relationship between FFS height and transition location observed in experiments, confirming a delay for the shallow step and premature transition for the higher step. DNS reveal spanwise-modulated reverse flow regions extending both upstream and downstream of the FFS, resulting from the interaction of stationary CF vortices with the FFS. Despite the greater distortion induced by the higher step, the qualitative similarity in time-averaged flow field characteristics across step heights points to the importance of investigating secondary instabilities and unsteady mechanisms for a more comprehensive interpretation of the transition process.

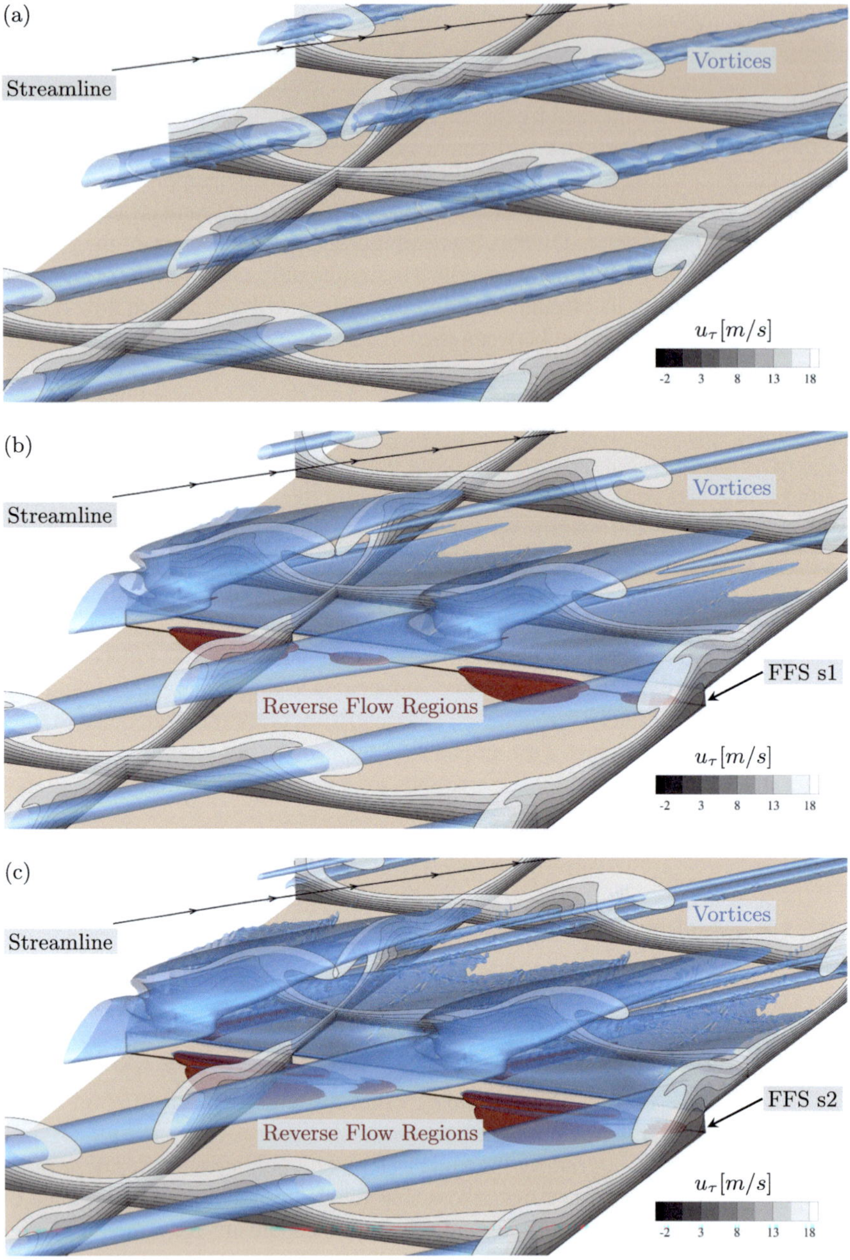

Fig. 9.4 Visualisation of the time-averaged flow field development for s0 (**a**), s1 (**b**), and s2 (**c**). Reverse flow regions are indicated by the red isosurface of $u_\tau = 0$, while vortical structures are represented by a light blue isosurface of the Q-criterion. Grey scale contours of the time-averaged u_τ are shown for multiple slices in the $x_\tau y_\eta$ and $z y_\eta$ planes. The domain is duplicated in the spanwise direction for visualization purposes

Acknowledgements F. Tocci and S. Hein gratefully acknowledge the scientific support and HPC resources provided by the German Aerospace Center (DLR). The HPC system CARO is partially funded by the Ministry of Science and Culture of Lower Saxony and the Federal Ministry for Economic Affairs and Climate Action. F. Tocci and G. Chauvat acknowledge funding from the European Union's Horizon 2020 research and innovation programme through the Marie Skłodowska-Curie grant agreement No. 675008. G. Chauvat has also been funded through the European Research Council under grant agreement 694452-TRANSEP-ERC-2015-AdG.

References

1. Eppink, J.L.: Mechanisms of stationary cross-flow instability growth and breakdown induced by forward-facing steps. J. Fluid Mech. **897**, A15 (2020). https://doi.org/10.1017/jfm.2020.367
2. Rius-Vidales, A.F., Kotsonis, M.: Impact of a forward-facing step on the development of cross-flow instability. J. Fluid Mech. **924**, A34 (2021). https://doi.org/10.1017/jfm.2021.689
3. Casacuberta, J., Hickel, S., Westerbeek, S., Kotsonis, M.: Direct numerical simulation of interaction between a stationary crossflow instability and forward-facing steps. J. Fluid Mech. **943**, A46 (2022). https://doi.org/10.1017/jfm.2022.456
4. Tocci, F., Chauvat, G., Rius-Vidales, A.F., Kotsonis, M., Hein, S., Hanifi, A.: Crossflow instability interaction with forward-facing steps: direct numerical simulation of transition. To be submitted (2025)
5. Fischer, P.F., Lottes, J.W., Kerkemeier, S.G.: Nek5000 Web Page (2008). http://nek5000.mcs.anl.gov

Chapter 10
Numerical Investigations of Stability and Transition for a Blunt Swept Flat Plate at Mach 10

Andrew Hartman, Christoph Hader, and Hermann Fasel

Abstract Numerical investigations were carried out to investigate the linear and nonlinear stability regime with respect to second-mode and crossflow instabilities for a blunt flat infinite swept plate at Mach 10. The infinite swept plate model has a leading edge radius of 5mm and a sweep angle of 75°. Linear Stability Theory (LST) identified multiple instability modes, including second-mode waves, traveling crossflow modes, and steady crossflow modes. The linear instability with respect to steady crossflow modes was further analyzed using Navier-Stokes calculations with controlled low-amplitude disturbance input to ensure linear development. These calculations identified the dominant steady crossflow mode with the highest growth rate. Nonlinear saturation of this mode was then investigated by increasing the forcing amplitude, forming the basis for secondary instability investigations. Secondary instability was examined by introducing a low-amplitude wave packet via short-duration pulse forcing in the presence of the saturated steady crossflow mode.

Keywords Hypersonic flow · Boundary-layer transition · DNS

10.1 Introduction

The so-called Flat-BOLT [1] was found to be suitable surrogate model of the original BOLT geometry in the sense that both the base flow and the stability characteristics (and thus the resulting transition predictions) were very well captured by the Flat-BOLT geometry. Comparison of the base flow and the stability characteristics of a flat plate "Delta Wing" with the analogous Flat-BOLT have shown very good agreement except for the inboard region near the symmetry plane of the respective geometries [2]. These differences are caused by the lower sweep angle of the leading edge in the centerline region for the BOLT and Flat-BOLT, which leads to a lower edge-Mach numbers in the inboard region compared to the Delta Wing. Therefore, it was concluded that the Delta Wing can also be considered as a direct surrogate geometry for

A. Hartman (✉) · C. Hader · H. Fasel
University of Arizona, Tucson, AZ, USA
e-mail: andrewmach1@arizona.edu

K. Kato et al. (eds.), *Proceedings of the 10th IUTAM Symposium on Laminar-Turbulent Transition*, IUTAM Bookseries 44, https://doi.org/10.1007/978-981-96-9829-5_10

investigating stability and transition in the outboard regions of the original BOLT geometry, albeit at somewhat lower fidelity than the Flat-BOLT. The base flow calculations (see Fig. 10.2) show very little downstream variation of the boundary layer thickness in the outboard region of the Delta Wing. This suggests that this outboard region of the Delta Wing can be investigated with an even simpler geometry, namely an infinite blunt swept flat plate (see Fig. 10.2). Therefore for these investigations, an infinite swept plate model was chosen. This model geometry makes it possible to perform Direct Numerical Simulations (DNS) of the nonlinear transition stages for several scenarios for which transition is dominated by (i) crossflow instability (ii) second mode waves and (iii) possible nonlinear interactions between crossflow and second mode instability.

The objective of the present investigations is to shed light on the nonlinear transition physics that are relevant for the outboard region of a Delta Wing and thus for the Flat-BOLT geometry. For our investigations, the flow conditions of the M = 10 transition experiments for a straight cone [3] at the Arnold Engineering Development Complex (AEDC) are used. Results from primary instability investigations using Linear Stability Theory (LST) indicate the existence of both second-mode, and crossflow-mode instabilities for the investigated geometry and flow conditions ($M = 10$ and $Re_1 = 15.3 \times 10^6$ 1/m). LST analysis for crossflow instabilities was carried out with respect for both stationary and traveling crossflow modes. In this paper the steady crossflow instability was investigated. Low amplitude ("linear") DNS calculations were performed to determine which spanwise wavenumber would likely lead to transition furthest upstream. Investigations into the primary and secondary instability of this crossflow mode were then performed. The approach taken to analyze the secondary instability was similar to that in [4], where a low amplitude pulse disturbance wave was introduced in the presence of a saturated steady crossflow mode in order to examine the response to a wide range of frequencies and spanwise wavenumbers that may potentially be amplified.

10.2 Geometry and Flow Conditions

In the present paper, a "Delta Wing" shaped flat plate and an infinite swept plate geometry were considered for stability and transition investigations. The infinite swept plate serves as a simplified model of the Delta Wing. Both geometries feature a sweep angle of 75°, and the leading-edge shape (perpendicular to the leading edge) is a semi-circle with a radius of $r_{\text{nose}} = 5$ mm (see Fig. 10.1a). The dimensions and coordinate systems for the finite Delta wing and the infinite swept wing (outboard region) are shown in Fig. 10.1a (left). The computational domain used for the infinite swept plate investigations is depicted in Fig. 10.1a (right). The infinite swept plate models the outboard region of the Delta wing, where the base flow properties are practically independent of the spanwise location. For calculations involving the infinite swept plate, x' and z' denote the coordinates perpendicular and parallel to the leading edge, respectively (see Fig. 10.1b).

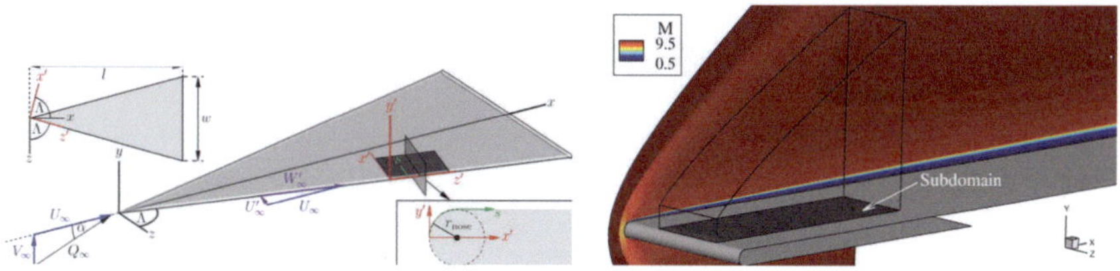

(a) (b)

Fig. 10.1 Schematic of the Delta wing. Dark gray section shows an example of the region on the Delta wing that is modeled by the infinite swept plate geometry (**a**), Computational domain for investigations of the infinite swept blunt flat plate (**b**)

For the numerical investigations the flow conditions from the experiments for a sharp cone [3, 5] were matched as closely as possible. The flow conditions investigated here were $M = 9.86$, $Re_1 = 15.03 \cdot 10^6$ 1/m, $p_0 = 21.8$ MPa, $p_\infty = 563.4$ Pa, $T_0 = 1092$ K, and $T_\infty = 53.4$ K. For the simulations the fluid properties of nitrogen ($\gamma = 1.4$, $Pr = 0.72$, $R_{air} = 296.8$ J/(kgK)) were used because the working fluid in the AEDC T9 experiments was nitrogen [3, 5]. The fluid is considered to be a calorically perfect gas and the viscosity is calculated using Sutherland's law [6].

10.3 Results

Baseflow: To evaluate how well the infinite swept flat plate approximates the base flow of the finite Delta wing, base flow calculations were performed for both geometries under identical flow conditions (see Sect. 10.2). Boundary-layer velocity profiles were extracted along inviscid streamlines starting at the mid-span of the Delta wing and along corresponding streamlines for the infinite swept flat plate. A comparison of the boundary-layer thickness, determined using the enthalpy derivative criterion by [7], and the streamwise velocity profiles obtained from the base flow calculations are shown in Fig. 10.2. The boundary-layer thickness development near the leading edge ($x = 0,$ m) matches closely between the two geometries, with only minor

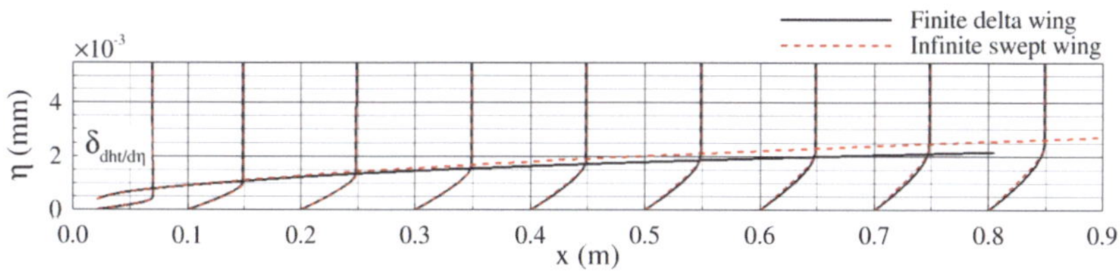

Fig. 10.2 Comparison of the streamwise velocity profiles of the finite delta wing and infinite swept plate

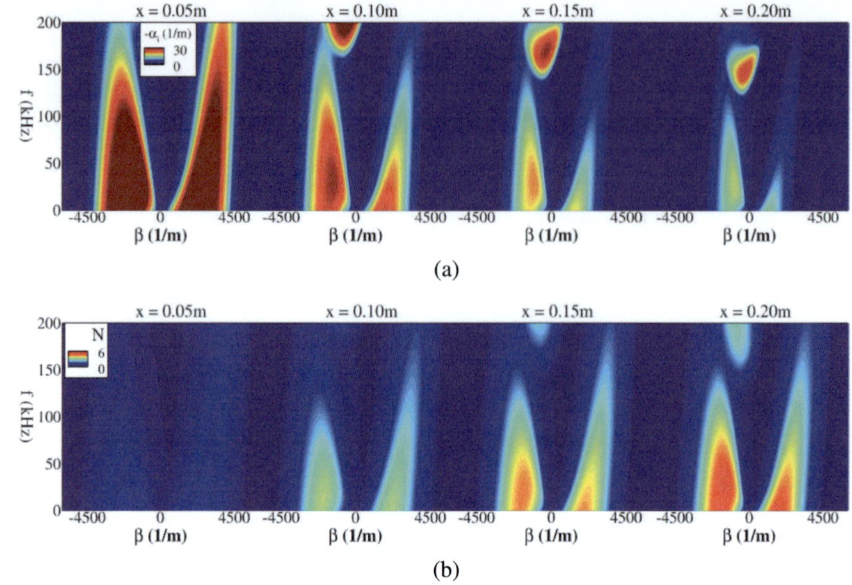

Fig. 10.3 Growth rates (**a**) and n-factors (**b**) obtained from linear stability theory (LST) calculations at several downstream locations

deviations as x increases. Similarly, the streamwise velocity profiles are in good agreement. These findings confirm that the infinite swept plate provides a suitable simplified model for the outboard region of the finite Delta wing.

Linear Stability Theory (Crossflow Instability): LST results for the infinite swept plate geometry at several x locations are shown in Fig. 10.3. Both inboard traveling crossflow modes (negative β) and outboard traveling crossflow modes (positive β) exhibit strong amplification near the leading edge (see Fig. 10.3a). Farther downstream, the outboard traveling crossflow modes show significantly higher growth rates compared to the inboard traveling modes. Additionally, second-mode instabilities are strongly amplified at all downstream locations as seen in Fig. 10.3a. These findings suggest that nonlinear interactions between crossflow and second-mode disturbances are possible and may play a role in transition.

Low Amplitude "Linear" DNS Calculations (Crossflow Instability): Results from Navier-Stokes calculations, where steady crossflow modes were introduced through a wall-blowing and suction slot with a very small forcing amplitude are shown in Fig. 10.5a. Iso-surfaces of the w-velocity (see Fig. 10.5a) confirm that steady streamwise crossflow modes can be successfully generated in the computational domain using this forcing technique. For sufficiently small forcing amplitudes, the Navier-Stokes calculations capture the linear behavior of the disturbance development. This is demonstrated in Fig. 10.4a, which shows close agreement between the amplitude development obtained from LST and that from low-amplitude "controlled" disturbance calculations using the full Navier-Stokes equations. The discrepancies in

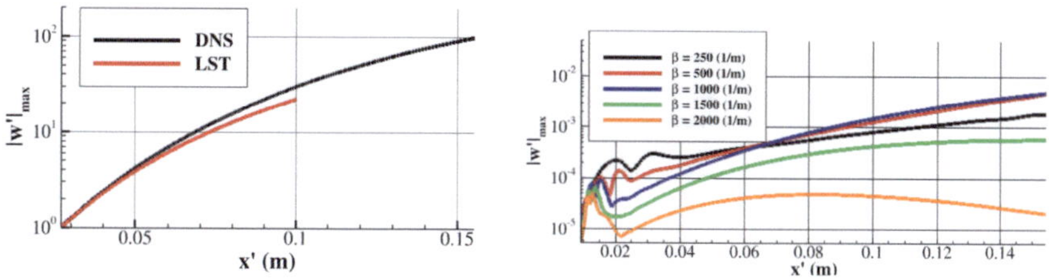

(a) (b)

Fig. 10.4 Amplitude comparison between LST and DNS for f = 0 kHz $\beta = 1000$ (**a**), W-velocity amplitudes versus x' for several steady crossflow wavelengths (**b**)

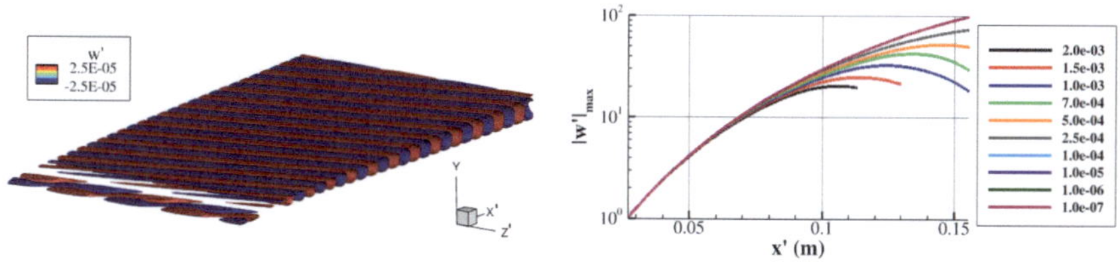

(a) (b)

Fig. 10.5 W-velocity disturbance iso-surface showing primary instability streaks for $\beta = 1000$ (**a**), W-velocity amplitudes for varying forcing amplitudes for f = 0 kHz $\beta = 1000$ (**b**)

the amplitude development may be due to the underlying assumptions of LST, such as the parallel flow assumption, which are not made in the Navier-Stokes calculations.

The amplitude development of steady crossflow modes, based on the w-velocity in the direction perpendicular to the leading edge (x'), is shown in Fig. 10.4b. Consistent with LST results, strong amplification was observed for spanwise wavenumbers in the range $\beta = 500$ to $\beta = 1000$ in the Navier-Stokes calculations.

Primary Crossflow Instability: In preparation for transition simulations to investigate the nonlinear stages and breakdown to turbulence, the nonlinear saturation of the dominant primary instability was investigated. The dominant primary steady crossflow mode, identified as the one with the largest growth rates just downstream of the leading edge, was found to be a steady crossflow mode with a spanwise wavenumber of $\beta = 1000$ (see above). The forcing amplitude required to observe the nonlinear development and ultimately saturation was determined by incrementally increasing the amplitude of the primary wave until departure from the linear amplitude development and finally nonlinear saturation was observed. This progression is shown in Fig. 10.5b, where nonlinear effects first are observed at an amplitude of 2.5×10^{-4}, and nonlinear saturation begins to occur earliest for an amplitude of 2.0×10^{-3}.

Secondary Instability of Steady Crossflow Mode: After determining a suitable forcing amplitude for the nonlinear saturation of the primary instability, the secondary instability regime was investigated. A low-amplitude wave packet was introduced

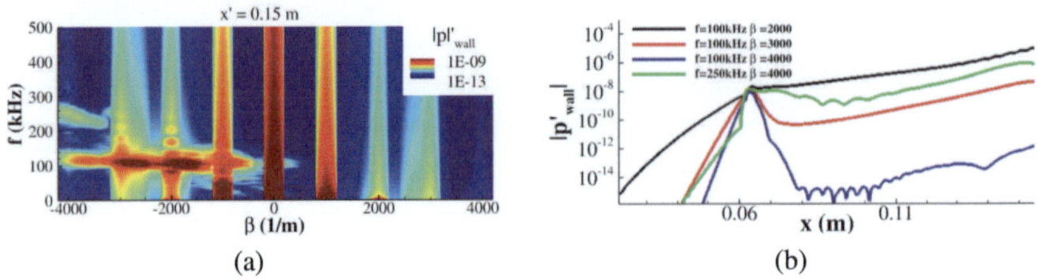

Fig. 10.6 Wall pressure disturbance amplitude spectra at a downstream location of $x' = 0.15$ (**a**), Wall pressure disturbance amplitudes for several frequencies and spanwise wavenumbers that were continuously forced as a part of the secondary instability study

using short-duration pulse forcing slightly upstream of the location where the primary wave saturates nonlinearly. This introduced a wave packet into a modified base flow where the primary instability vortices had already developed and saturated. This approach, which is based on the investigations of low-speed crossflow-dominated transition by [4], enables the examination of the response to a wide range of frequencies and spanwise wavenumbers that may be susceptible to secondary instability. The results of the secondary instability calculation are shown in Fig. 10.6a. Strong secondary amplification was observed around 100 kHz for $\beta = 2000$ and $\beta = 3000$, with an additional amplification region identified near $f = 250$ kHz for $\beta = 4000$.

Additional calculations were performed for the frequencies and wavenumbers identified as relevant in the secondary instability investigation. For these calculations, the disturbance waves were introduced into the computational domain by continuously forcing through a wall-blowing and suction slot. This method allows for a "controlled" disturbance input by controlling the frequency and wavenumber of the forced disturbance wave or waves, in the spirit of Schubauer & Skramstad [8], to investigate the linear and nonlinear development of select disturbances. The results of these calculations are presented in Fig. 10.6b. The highest amplitude was obtained for $f = 100$ kHz and $\beta = 2000$, while the highest growth rate was observed for $f = 250$ kHz and $\beta = 4000$ for the investigated secondary disturbance waves. Which of these secondary disturbances may play a dominant role in the transition process remains uncertain and is the subject of future investigations. The results indicate that the secondary growth rates do not change significantly with varying spanwise wavenumber. This behavior contrasts with other secondary instability studies in high-speed boundary layers, where large secondary growth rates were typically observed within a narrow spanwise or azimuthal wavenumber range.

10.4 Conclusion

The stability characteristics of a blunt flat plate with an infinite sweep were investigated. The baseflow properties for an infinite swept plate were found to closely match those of a Delta Wing shape flat plate. Linear stability theory results indicated a strong crossflow instability (traveling and stationary modes) as well as the presence of second mode instabilities. Low amplitude "linear" DNS calculations were performed to determine which spanwise wavenumbers for the stationary crossflow mode were most amplified. It was determined that $\beta = 1000$ had the highest growth rates and would thus be a good candidate for further nonlinear breakdown investigations. The nonlinear saturation of the primary instability was then determined and a secondary instability was investigated using both a pulse forcing and controlled forcing approach.

This secondary instability analysis showed that for $f = 100\,\text{kHz}$, $\beta = 2000$, and $f = 250\,\text{kHz}$, $\beta = 4000$, significant secondary amplification was observed for the investigated flow conditions and geometry. It was also found that the secondary growth rates do not change significantly within the investigated range of spanwise wavenumbers. The presented work identified the dominant primary instability and multiple secondary instability modes, which may play a dominant role in the transition process.

In future work, the forcing amplitude of the secondary instability will be successively increased in order to initiate nonlinear boundary-layer transition. Additionally, "natural" transition simulations will be performed using random inflow disturbances to simulate the transition process in a wind tunnel environment without biasing towards any specific nonlinear mechanism.

Acknowledgements This work was supported by ONR Grant N00014-22-1-2138 with Dr. Eric Marineau serving as the program manager. Computer time was provided by the Department of Defense (DOD) High Performance Computing Modernization Program (HPCMP). The views and conclusions contained herein are those of the authors and should not be interpreted as necessarily representing the official policies or endorsements, either expressed or implied, of the Office of Naval Research or the U. S. Government.

References

1. Wernz, S.: AIAA 2023-3442 (2023)
2. Hartman, A., Fasel, H., Wernz, S., Haas, A., Hader, C.: AIAA 2023-3443 (2023)
3. Marineau, E.C., Moraru, G.C., Lewis, D.R., Norris, J.D., Lafferty, J.F., Wagnild, R.M., Smith, J.A.: In: 19th AIAA International Space Planes and Hypersonic Systems and Technologies Conference. AIAA 2014-3108 (2014)
4. Wasserman, P., Kloker, M.: J. Fluid Mech **456**, 49 (2002)
5. Marineau, E., Moraru, G., Lewis, D., Norris, J., Lafferty, J., Johnson, H.: In: 53rd AIAA Aerospace Sciences Meeting. AIAA 2015-1737 (2015)
6. Sutherland, W.: The London, Edinburgh, and Dublin Philosophical Magazine and Journal of Science, vol. 36, issue 223, p. 507 (1893)
7. Bertin, J.J.: Hypersonic Aerothermodynamics. AIAA (1994)
8. Schubauer, G., Skramstad, H.: NASA Report No. 909 (1948)

Chapter 11
Numerical and Experimental Investigations of Transition in a Laminar Separation Bubble

David Borgmann, Shirzad Hosseinverdi, Jesse Little, and Hermann Fasel

Abstract The interaction of separation and transition in a laminar separation bubble (LSB) was investigated in a synergistic fashion using wind tunnel and free-flight experiments, direct numerical simulations (DNS), and linear theory. Initial DNS resulted in a larger separated region compared to the wind tunnel experiments. However, the addition of random disturbances to the local upstream boundary layer in DNS, to account for freestream turbulence in wind tunnel experiments, resulted in remarkable agreement in topology and frequency content of the LSB. Scientific free-flight experiments showed LSB characteristics matching those from the wind tunnel experiments for the same Reynolds number and angle of attack. In all cases, linear stability theory provided accurate estimates for the unstable frequency content.

Keywords Boundary layer separation · Stability · Transition

11.1 Introduction

Unmanned Aerial Vehicles (UAVs) with military, commercial, and recreational applications operate at flight velocities resulting in moderate Reynolds numbers $O(10^4$–$10^6)$, based on the mean aerodynamic chord. At such Reynolds numbers and the typically low freestream turbulence (FST) environment encountered in flight, the laminar boundary layer, especially for laminar airfoils, can separate when subjected to strong adverse pressure gradients (APG). Downstream of the location of laminar separation, hydrodynamic instability of the inviscid shear layer causes rapid formation of intense spanwise vortical structures [2, 13, 14]. Subsequent three-dimensional

D. Borgmann (✉) · S. Hosseinverdi · H. Fasel
Department of Aerospace and Mechanical Engineering, The University of Arizona, Tucson, AZ, United States
e-mail: davidborgmann@arizona.edu

J. Little
Department of Mechanical and Aerospace Engineering, The Ohio State University, Columbus, OH, United States

K. Kato et al. (eds.), *Proceedings of the 10th IUTAM Symposium on Laminar-Turbulent Transition*, IUTAM Bookseries 44, https://doi.org/10.1007/978-981-96-9829-5_11

deformation of these structures [11–13] leads to breakdown to turbulence, further enhancing mixing and leading to turbulent reattachment. In a time-averaged sense, this causes formation of a closed recirculation region, called a Laminar Separation Bubble (LSB, Gaster [3], Horton [5]). The transition process in the LSB is highly susceptible to a variety of disturbances, i.e. FST and/or disturbances upstream of the LSB due to, for example, surface roughness. In addition, the shear layer caused by the LSB is hydrodynamically unstable via an inviscid instability mechanism. A detailed understanding of the LSB dynamics is required for design of effective flow control strategies to mitigate the negative effects of LSBs and in particular to prevent "bubble bursting" which can lead to sudden stall of an airplane wing. Difficulties in direct comparisons between experiments [1, 2, 12] and numerical simulations [6, 9, 13, 14] arise, in part, from variations of experimental setups used in wind tunnels to generate LSBs (e.g., displacement body [1, 3], contoured ceiling [12]); and flow conditions (e.g., FST level and spectrum, Reynolds number). Numerical and experimental studies have indicated a strong dependence of the bubble topology to variations in Reynolds number, FST, and angle of attack. FST, in particular, strongly influences the transition in both experiments and simulations [1, 6, 8–10], even at low levels of turbulence intensity (Tu $\leq 0.5\%$), thus affecting the reattachment location due to the enhanced mixing in the subsequent transitioned flow. This paper provides an overview of a synergistic research approach undertaken at the University of Arizona for the study of LSBs. The approach combines wind tunnel and free-flight experiments, highly-accurate DNS, and stability analysis. Sample results are presented below.

11.2 Model Geometries

A modified NACA $64_3 - 618$ airfoil is used for all investigations (Fig. 11.1). The wind tunnel model has a chord length of $c = 1$ ft (304.8 mm) and the span is 4 ft. It includes 60 static pressure taps at three spanwise locations distributed along the chord of the airfoil covering the suction and pressure side.

An in-house developed high-fidelity Navier-Stokes code was employed, which was tailored for hydrodynamic stability and DNS of incompressible flows for curvilinear orthogonal grids (see Hosseinverdi and Fasel [7]). In this code, the three-dimensional, unsteady incompressible Navier-Stokes equations in vorticity-velocity formulation are solved. An in-house developed Poisson grid generator was employed for obtaining O-grids with high orthogonality and smoothness for the modified NACA $64_3 - 618$ airfoil geometry (Fig. 11.1). To emulate the effects of FST of the experiments in the simulations, low-amplitude random disturbances were introduced into the boundary layer upstream of separation (see [4]).

For flight experiments, a radio controlled (R-C) flight model was employed with swept wings (35°) using the same airfoil and chord length (perpendicular to the leading edge) as in the wind tunnel experiments and the numerical simulations. The

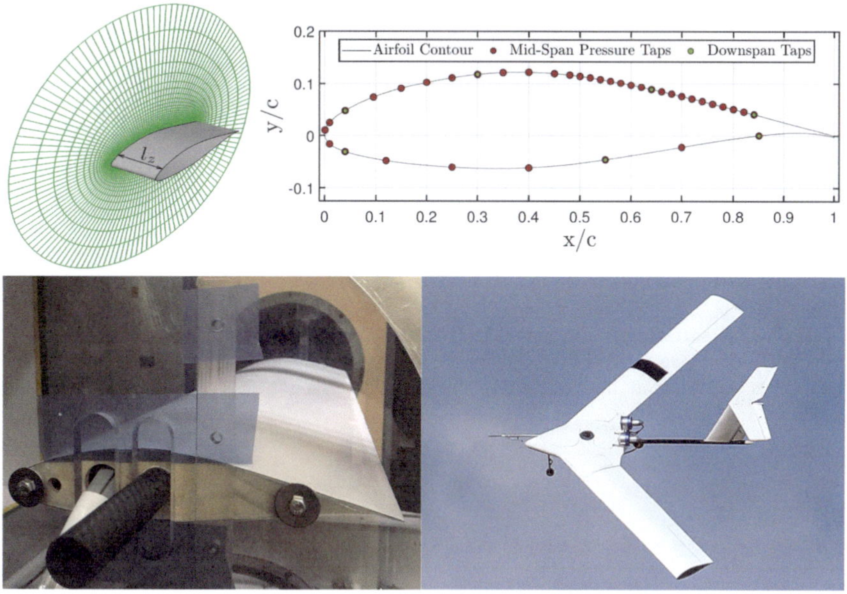

Fig. 11.1 Synergistic approach to LSB investigations. DNS grid (top left); 2D contour of the modified NACA 64_3618 with pressure tap locations (top right); Airfoil model in wind tunnel (bottom left) and R-C model with 35° swept wings

wing section was instrumented such that the C_p distribution could be measured in flight with the same resolution (same pressure tap spacing) as in the wind tunnel experiments (Fig. 11.1, top right).

11.3 Results

Flow separation and self-sustained transition was observed in DNS on the suction and pressure side of the airfoil (suction side in Fig. 11.2a). Figure 11.2b shows a larger pressure plateau (larger mean bubble) for DNS compared to the experiment. Random disturbance input ($O(10^{-7}–10^{-8})$) was introduced to the upstream boundary layer (case DNS-r, Fig. 11.2), in order to model the low disturbance levels present in the wind tunnel. With this addition, the mean bubble topology agrees very well with the experiment. The spectral content in the LSB from the experiment is in good agreement with the results from stability calculations (PSE based on the DNS-r baseflow) and the DNS with random disturbances (Fig. 11.2c). Without the random disturbances, the DNS lead to a larger bubble and a lower dominant frequency range.

Figure 11.3 shows C_p development from flight experiments for two different conditions compared with wind tunnel experiments at the same Reynolds number and angle of attack ($\alpha = 3.3°$ and $-0.5°$). Flight data are for quasi-steady conditions. The

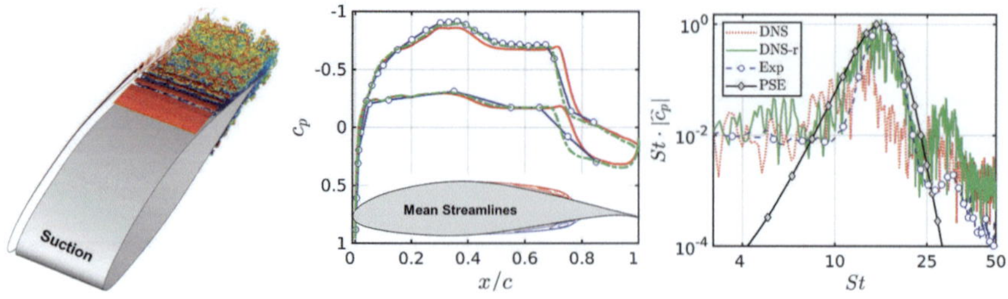

(a) DNS without FST (b) C_p comparison (c) Shear layer spectra

Fig. 11.2 **a** Iso-surfaces of γ_2 structures and contours of spanwise vorticity in the x-y plane from DNS. **b** C_p development and **c** spectra in the LSB shear layer for the DNS and wind tunnel experiments on the airfoil. Inset in **b** shows the LSB on the suction and pressure side of the airfoil from DNS

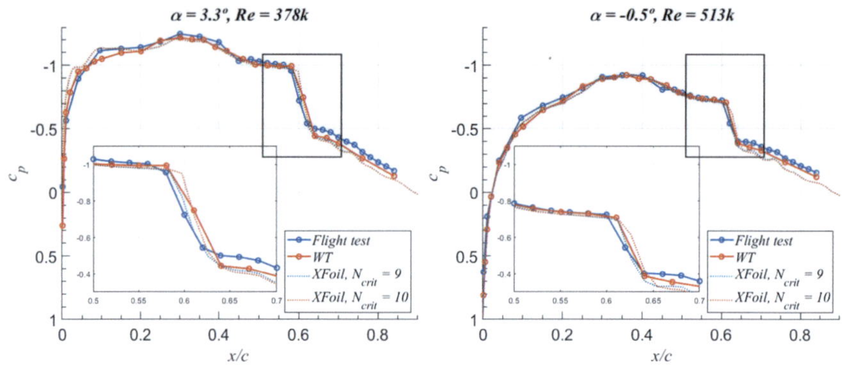

Fig. 11.3 Comparison of mean C_p development for flight tests and wind tunnel experiments; XFoil data plotted for reference

comparison in Fig. 11.3 demonstrates that flight data are in remarkable agreement with the wind tunnel experiments considering the difficulty in accurately maintaining flight speed (and angle of attack) in the flight experiments. However, calculations (XFoil) for $N_{crit} = 9$ and $N_{crit} = 10$ indicate that the FST levels in flight were likely above those in the wind tunnel.

Infrared thermography (IRT) was carried out in both the flight and wind tunnel experiments and comparisons are shown in Fig. 11.4 for the same chord Reynolds number and angle of attack. Characteristic LSB locations in the IRT data are extracted following Spalart and Strelets [15], with the strongest decrease in the heat transfer coefficient (maximum of surface temperature gradient) at separation. Transition conversely is estimated at the strongest increase in heat transfer and the maximum heat transfer coincides with the mean reattachment location. Characteristic locations indicate the presence of a two-dimensional bubble in all cases, even with the 35° sweep angle in the flight tests. Analogous to the surface pressure in Fig. 11.3, the transition

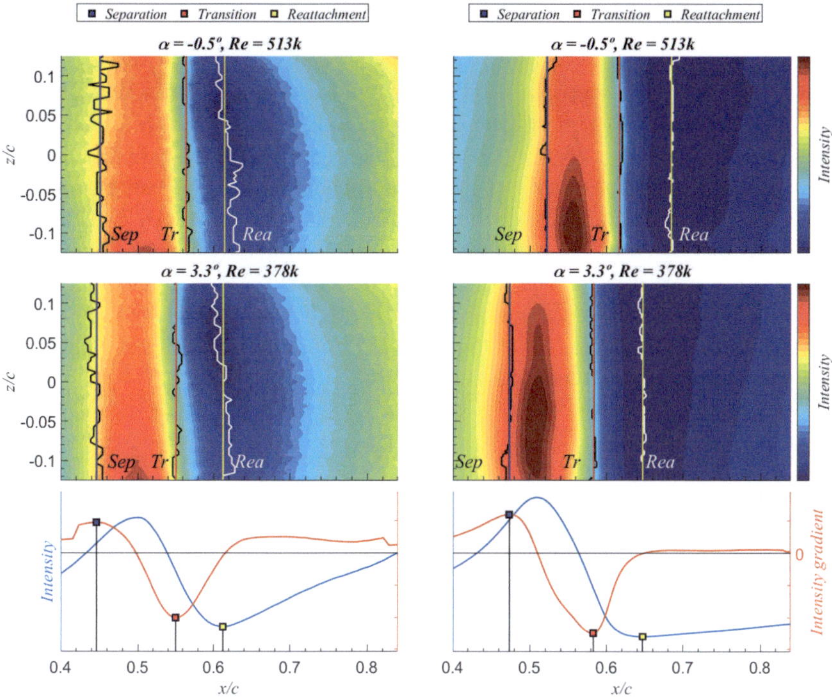

(a) Flight Test (b) Wind Tunnel Experiment

Fig. 11.4 IRT heat maps for flight tests and wind tunnel experiments. Spanwise averaged intensity and intensity gradient shown for $\alpha = 3.3°$ indicating the characteristic locations in the LSBs

and reattachment locations in the wind tunnel experiments were found to be downstream of those for the flight experiments. This is consistent with higher FST in flight. In both wind tunnel and flight tests, the characteristic locations shift downstream with a decrease in angle of attack, despite the differences in Reynolds number.

The IRT measurements in the wind tunnel were carried out in controlled manner with an external heat lamp to increase the surface temperature and to allow for a temperature equilibrium before each measurement. In contrast, the flight tests rely on radiative heating from the sun, leading to uneven surface temperature. In addition, the infrared camera was mounted to the vertical tail of the plane, leading to a significant viewing angle towards the airfoil surface. The small differences in temperature observed in the flight experiments, and changes in flow conditions due to, e.g., wind gust and higher turbulence, reduce the duration at which the desired measured conditions are affecting the surface heating. This reduces the available time for the surface temperature distribution to reach an equilibrium state for each case and may partially explain the smaller downstream shift of the LSB locations in Fig. 11.4 compared to the results from wind tunnel experiments.

11.4 Conclusion

A combined approach based on wind tunnel and free-flight experiments, DNS and linear theory was used to investigate LSBs on a modified NACA $64_3 - 618$ airfoil. DNS results exhibited a mean separated region larger than in the experiments. The earlier onset of transition in the experiment compared to the DNS was attributed to the FST in the wind tunnel. Introduction of low-levels of FST in the DNS accelerated transition, and decreased the mean separated region, which then agreed remarkably well with the experiments. The flight experiments with a R-C flight model using swept wings with the same airfoil and chord length (perpendicular to the leading edge), as in the wind tunnel and DNS, showed impressive agreement with results from wind tunnel experiments and simulations.

References

1. Borgmann, D., Hosseinverdi, S., Little, J.C., Fasel, H.F.: Investigation of laminar separation bubbles using experiments, theory and DNS. In: AIAA Aviation 2021 Forum, p. 2898 (2021). https://doi.org/10.2514/6.2021-2898
2. Diwan, S.S., Ramesh, O.: On the origin of the inflectional instability of a laminar separation bubble. J. Fluid Mech. **629**, 263–298 (2009). https://doi.org/10.1017/S002211200900634X
3. Gaster, M.: The structure and behaviour of laminar separation bubbles. Aeronaut. Res. Counc. Rep. Memo. (3595), 1–31 (1967)
4. Grille Guerra, A., Hosseinverdi, S., Little, J.C., Fasel, H.F.: Unsteady behavior of a laminar separation bubble subjected to wing structural motion. In: AIAA SciTech 2022 Forum, p. 2331 (2022). https://doi.org/10.2514/6.2022-2331
5. Horton, H.P.: Laminar separation bubbles in two and three dimensional incompressible flow. PhD thesis, Queen Mary University of London (1968)
6. Hosseinverdi, S., Fasel, H.F.: Numerical investigation of laminar–turbulent transition in laminar separation bubbles: the effect of free-stream turbulence. J. Fluid Mech. **858**, 714–759 (2019). https://doi.org/10.1017/jfm.2018.809
7. Hosseinverdi, S., Fasel, H.F.: High-order accurate incompressible Navier-Stokes solver based on vorticity-velocity formulation for orthogonal curvilinear grids. In: AIAA Aviation 2021 Forum, p. 2741 (2021). https://doi.org/10.2514/6.2021-2741
8. Istvan, M.S., Yarusevych, S.: Effects of free-stream turbulence intensity on transition in a laminar separation bubble formed over an airfoil. Exp. Fluids **59**(3), 1–21 (2018). https://doi.org/10.1007/s00348-018-2511-6
9. Jacobs, R.G., Durbin, P.A.: Simulations of bypass transition. J. Fluid Mech. **428**, 185–212 (2001). https://doi.org/10.1017/S0022112000002469
10. Klebanoff, P.: Effect of free-stream turbulence on a laminar boundary layer. Bull. Am. Phys. Soc. **16** (1971)
11. Marxen, O., Lang, M., Rist, U.: Vortex formation and vortex breakup in a laminar separation bubble. J. Fluid Mech. **728**, 58–90 (2013). https://doi.org/10.1017/jfm.2013.222
12. Michelis, T., Yarusevych, S., Kotsonis, M.: On the origin of spanwise vortex deformations in laminar separation bubbles. J. Fluid Mech. **841**, 81–108 (2018). https://doi.org/10.1017/jfm.2018.91

13. Postl, D., Balzer, W., Fasel, H.F.: Control of laminar separation using pulsed vortex generator jets: direct numerical simulations. J. Fluid Mech. **676**, 81–109 (2011). https://doi.org/10.1017/jfm.2011.34
14. Rist, U., Maucher, U.: Direct numerical simulation of 2-d and 3-d instability waves in a laminar separation bubble. In: Agard Conference Proceedings, AGARD, pp. 34–34 (1994)
15. Spalart, P.R., Strelets, M.K.: Mechanisms of transition and heat transfer in a separation bubble. J. Fluid Mech. **403**, 329–349 (2000). https://doi.org/10.1017/S0022112099007077

Chapter 12
Experimental Study on the Mechanism of Lateral Growth of Developed Turbulent Wedge

Ayumu Inasawa and Masahito Asai

Abstract The mechanism of lateral contamination in developed turbulent wedge was examined experimentally. Turbulent wedge was generated by a cluster of cylinder roughness elements, and high-resolution hot-wire rakes and PIV were employed to capture and identify turbulent structures developing in the laminar-turbulent interface of the developed turbulent wedge. The results showed that oblique fin-like turbulent structures with oblique (swept-back) angle of about 75° developed in the region away from the wall. Occurrence of fin structures were rather periodic and their frequency gradually decreased downstream, suggesting that a certain frequency selection mechanism exists there. The fin structures caused development and breakdown of near-wall low-speed streaks in the ambient laminar region. Thus, we concluded that development of fin structures was responsible for lateral growth of developed turbulent wedge.

Keywords Developed turbulent wedge · Lateral contamination · Fin structures · Hot-wire rake

12.1 Introduction

Lateral growth of turbulent spot and wedge is one of the most important routes in laminar-turbulent transition in boundary layers. When a boundary layer is subjected to high free-stream turbulence, the transition to turbulence is caused by transient growth followed by the occurrence of turbulent spots [1]. On the other hand, when an isolated roughness/protrusion on a surface whose Reynolds number (based on roughness height and velocity there) is above the critical value [2], it can directly cause a turbulent wedge, contaminating ambient laminar region. Since detailed experimental results on development of turbulent spot and wedge were reported by Schubauer and Klebanoff [3], lots of experimental and numerical studies have been conducted to

A. Inasawa (✉) · M. Asai
Department of Aeronautics and Astronautics, Tokyo Metropolitan University, Tokyo, Japan
e-mail: ainasawa@tmu.ac.jp

K. Kato et al. (eds.), *Proceedings of the 10th IUTAM Symposium on Laminar-Turbulent Transition*, IUTAM Bookseries 44, https://doi.org/10.1007/978-981-96-9829-5_12

clarify the mechanism(s) of their lateral growth. For turbulent spot, breakdown of oblique waves appearing at both tips of the spot was found to be a key mechanism for the lateral growth [4]. For turbulent wedge, on the other hand, successive generation (regeneration) of streamwise vortices/low-speed streaks was observed as an important key event [5–8] although the observations were mainly done in the early stage of lateral growth. Importantly, both turbulent spot and wedge have the same half-spreading-angle of about 6.4° for a fully turbulent region and 10.6° for a region including an intermittent region [3] despite the fact that the turbulent structures in the laminar-turbulent interface region were not the same at least in the early stage of their developments. This suggests that a certain contamination mechanism different from that in the early stage might set in there for turbulent wedge. In the present experimental study, we revisited this problem, especially focusing on the growth mechanism of "developed" turbulent wedge.

12.2 Experimental Setup and Procedure

The experiment was conducted in a low turbulence wind tunnel of open jet type, with a cross-section of $400 \times 400\,\text{mm}^2$. The boundary-layer plate made of blass had a sharp leading edge and was 1000 mm long, 395 mm wide, and 5 mm thick (Fig. 12.1). A small cluster of circular cylinders whose diameter and height were both 3 mm were glued on the boundary-layer plate, aligned in a trapezoidal shape, at a location 150 mm downstream from the leading edge ($x = 150$ mm). As for the coordinate system, x was the streamwise distance measured from the leading edge, y the normal-to-wall distance, and z the spanwise distance measured from the mid-span of the cluster. Lateral (spanwise) spreading of developed turbulent wedge was visualized using a

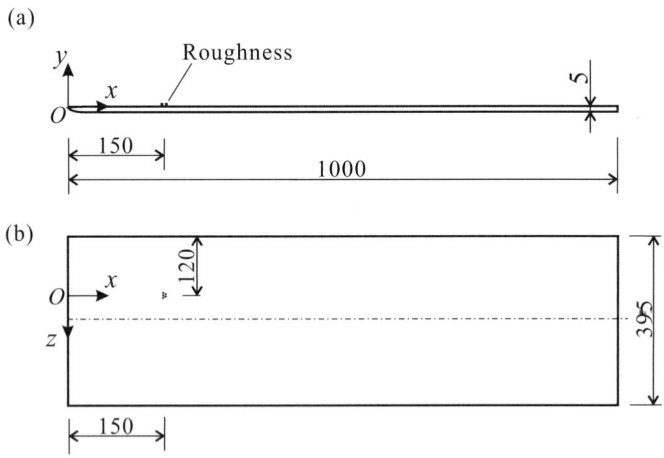

Fig. 12.1 Boundary-layer plate. **a** side view, **b** top view. Dimensions in mm

(a) (b)

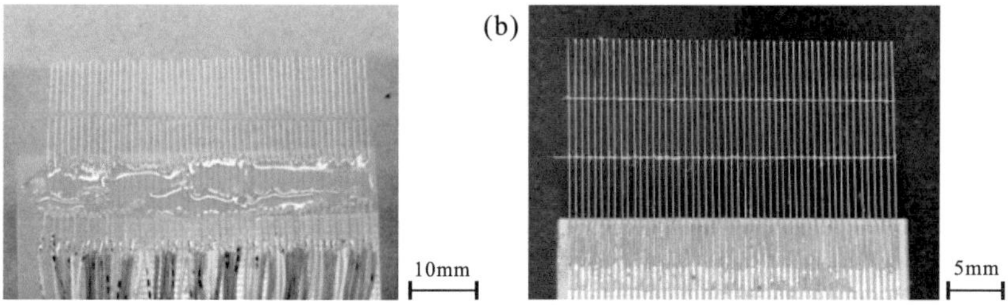

Fig. 12.2 Photographs of two hot-wire rakes installed in the boundary layer. **a** Rake A having 24 sensors with 2-mm-interval, **b** rake B having 48 sensors with 0.6-mm-interval

PIV and a hot-wire rake with 24 sensors (rake A with CTA operation) or 48 sensors (rake B with CCA operation). The spanwise intervals of rake A and rake B were 2 and 0.6 mm, respectively (see Fig. 12.2a, b).

The free-stream velocity was fixed at $U_\infty = 10$ m/s, at which the x-Reynolds number Re_x was 5.4×10^5 at the most downstream observation location, $x = 800$ mm. The half-spreading angle α obtained by following up the z location at which the RMS value of streamwise velocity fluctuations took a maximum in the laminar-turbulence interface region became constant with $\alpha \approx 7°$ for $x \geq 600$ mm, which was close to that reported previously [3]. The boundary layer thicknesses at $x = 800$ mm were 10 mm at the spanwise center of turbulent wedge ($z = 0$) and 5 mm in the ambient laminar boundary layer (Blasius flow), respectively.

12.3 Results

Figure 12.3 shows instantaneous velocity fields in the laminar-turbulent interface in terms of streamwise-velocity contours in the (x, z) plane (parallel to the wall) near the wall ($y \approx 1$ mm) and away from the wall ($y \approx 3$ mm). Note that figures (a) and (b) were taken at different instants. We see the appearance of a large-scale turbulent packet away from the wall, followed by many low-speed streaks in the vicinity of the surface, not unlike a turbulent spot. The large-scale turbulent packet away from the wall was an oblique structure whose swept-back angle was about 75° (see Fig. 12.3b). Such a structure has not been demonstrated in past studies on turbulent wedges. Hereafter, we call this turbulent packet "fin structure".

The intermittent occurrence of fin structures at the interface is more clearly visualized by time-series picture of instantaneous streamwise velocity measured using a hot-wire rake. Figure 12.4 compares flow patterns away from the surface ($y \approx 3$ mm) at $x = 500$ and 800 mm. The protrusion of the fin structure was observed rather periodically in time. Furthermore, the frequency of protrusion decreased gradually with increasing x, that is, $f_m \approx 40$ Hz at $x = 500$ mm and $f_m \approx 30$ Hz at $x = 800$ mm. These suggest that a certain frequency selection mechanism exists at around the laminar-turbulent interface region of developed turbulent wedge.

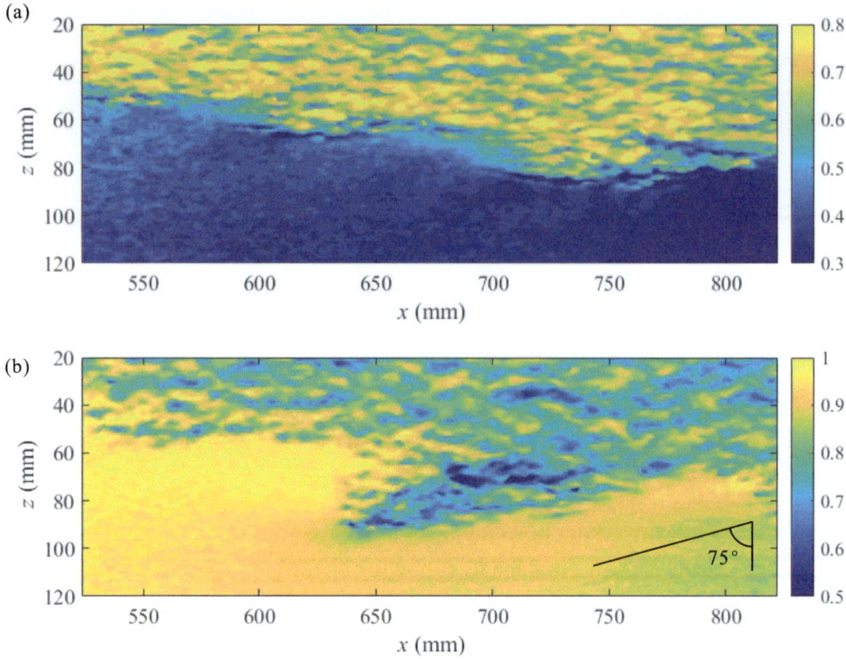

Fig. 12.3 Contour maps of streamwise velocity $(U + u)/U_\infty$ in the laminar-turbulent interface region, measured by PIV. **a** $y \approx 1$ mm, **b** $y \approx 3$ mm. Note that figures **a** and **b** were taken at different instants

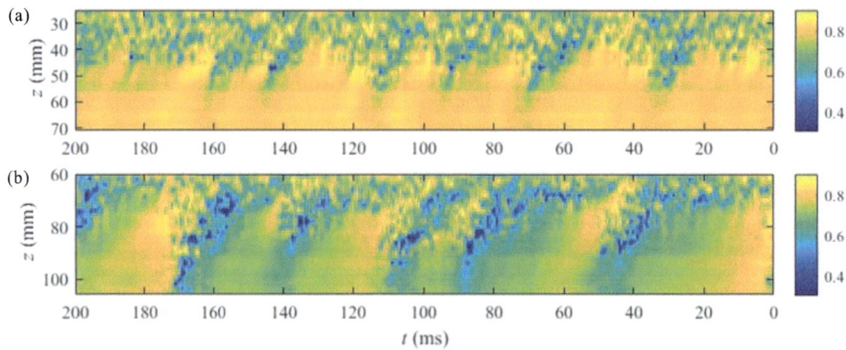

Fig. 12.4 Instantaneous turbulent structure in the laminar-turbulent interface region at $y \approx 3$ mm, measured by hot-wire rake (rake A). **a** $x = 500$ mm, **b** $x = 800$ mm. Color bar denotes the value of $(U + u)/U_\infty$. Time t is taken from right to left

In order to further understand development of wall turbulence structures in the interface region, the response of near-wall flow to large-scale fin structures developing in the region away from the wall was examined using two hot-wire rakes. Rake A (spanwise interval of 2 mm) and Rake B (spanwise interval of 0.6 mm) were

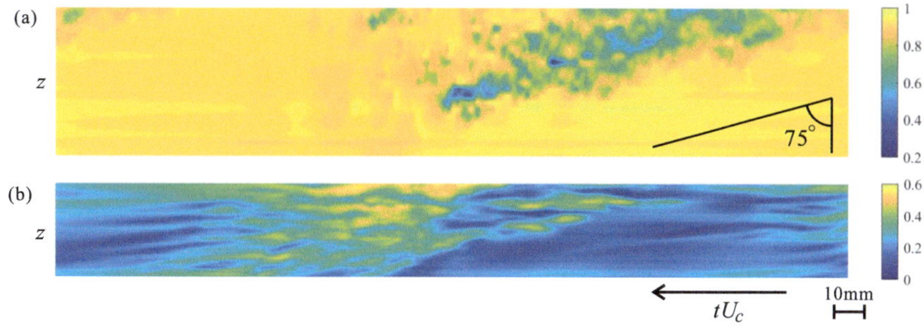

Fig. 12.5 Simultaneous streamwise velocity $(U + u)/U_\infty$ of laminar-turbulent interface of developed turbulent wedge by two hot-wire rakes at $x = 800$ mm. Instantaneous velocity fields **a** and **b** were measured by rake A ($y \approx 5$ mm) and rake B ($y \approx 0.5$ mm), respectively. Note that the horizontal axis (time t from right to left) was converted to streamwise distance tU_c with convection velocity $U_c = 0.6U_\infty$. The spanwise ranges in **a** and **b** are $z = 60$–106 mm and $z = 78$–106 mm, respectively

fixed at $y \approx 5$ mm and $y \approx 0.5$ mm (about 1/4 the displacement thickness of Blasius flow), respectively, at $x = 800$ mm. Figure 12.5 depicts instantaneous flow structures measured simultaneously. Here, the horizontal axis (time t) was converted to the streamwise distance tU_c with convection velocity $U_c = 0.6U_\infty$ which was estimated from a movie of smoke-wire flow visualization. An oblique fin structure whose swept-back angle was the same as that in Fig. 12.3b was reproduced in Fig. 12.5a. The fin structure extending to the otherwise laminar region caused a cluster of low-speed streaks to develop near the wall in the ambient laminar region, as shown in Fig. 12.5b. They were no doubt induced by the fin structure passing away from the wall, just like bypass transition caused by high-intensity turbulence [9].

Thus, we could understand that the lateral growth of developed turbulent wedge having the constant spreading-angle was governed by the development of fin turbulent-structures in the laminar-turbulent interface region.

12.4 Conclusions

The mechanism of lateral growth in developed turbulent wedge was examined experimentally. High-resolution hot-wire rakes and PIV were employed to capture and identify the key coherent turbulent structures responsible for lateral growth of developed turbulent wedge. The results showed that lateral growth of developed turbulent wedge was predominantly governed by oblique fin turbulent structures developing away from the wall. Occurrence of the fin structures was in a rather regular manner, and its frequency gradually changed (decreased) downstream, suggesting that some frequency selection mechanism exists at around the laminar-turbulent interface region. It was also demonstrated that the fin structures caused growth and breakdown of near-wall low-speed streaks in the ambient otherwise laminar region, just like the bypass transition under high-intensity freestream turbulence. These phenomena lead to lateral growth of developed turbulent wedge.

References

1. Matsubara, M., Alfredsson, P.H.: Disturbance growth in boundary layers subjected to free stream turbulence. J. Fluid Mech. **430**, 149–168 (2001). https://doi.org/10.1017/S0022112000002810
2. Morkovin, M.V.: On roughness-induced transition: facts, views and speculations. In: Hussaini, M.Y., Voigt, R.G. (eds.) Proceedings of Instability and Transition, vol. 1, pp. 281–295 (1990). https://doi.org/10.1007/978-1-4612-3430-2_34
3. Schubauer, G.B., Klebanoff, P.S.: Contributions on the mechanics of boundary-layer transition. NACA. TN 1289 (1956)
4. Henningson, D., Spalart, P., Kim, J.: Numerical simulations of turbulent spots in plane Poiseuille and boundary layer flow. Phys. Fluids **10**, 2914–2917 (1987). https://doi.org/10.1063/1.866067
5. Mochizuki, M.: Smoke observation on boundary layer transition caused by a spherical roughness element. J. Phys. Soc. Japan **10**, 995–1008 (1961). https://doi.org/10.1143/JPSJ.16.995
6. Asai, M., Nishioka, M.: Development of wall turbulence structure in transitonal flows. In: Tatsumi, T., Watanabw, E., Kambe, T. (eds.) Theoretical and Applied Mechanics 1996, pp. 121–138. Elsevier (1997)
7. Kuester, M.S., White, E.B.: Structure of turbulent wedges created by isolated surface roughness. Exp. Fluids **57**, 47 (2016). https://doi.org/10.1007/s00348-016-2140-x
8. Goldstein, D., Chu, J., Brown, G.: Lateral spreading mechanism of a turbulent spot and a turbulent wedge. Flow Turbul. Combust. **98**, 21–35 (2017). https://doi.org/10.1007/s10494-016-9748-z
9. Asai, M., Konishi, Y., Oizumi, Y., Nishioka, M.: Growth and breakdown of low-speed streaks leading to wall turbulence. J. Fluid Mech. **563**, 371–386 (2007). https://doi.org/10.1017/S002211200700688X

Chapter 13
Experimental Study of the Instability Amplitude Effect on Critical Step Heights in Crossflow-Dominated Transition

Hans Peter Barth and Stefan Hein

Abstract Reliable tolerance criteria for step heights are necessary for laminar designs. Existing criteria for crossflow-dominated laminar-turbulent transition neglect the impact of crossflow instability amplitude at the step location. This study in an incompressible flow quantifies this effect by varying the instability amplitude through a systematic variation of the diameter of discrete roughness elements used to excite stationary crossflow instabilities. It is shown that critical step height Reynolds numbers decrease approximately linearly with increasing instability amplitude.

Keywords Crossflow · Laminar · Step · Tolerance · Amplitude

13.1 Introduction

Accurate tolerance criteria for two-dimensional surface irregularities such as steps and gaps are important for future laminar wing designs. Research on step height tolerances is more advanced for two-dimensional than for three-dimensional boundary layers. At DLR, the SPECTRA (Swept flat PlatE Crossflow TRAnsition) configuration [1] is used to study crossflow-dominated laminar-turbulent transition. Recently, the configuration has been improved with respect to its experimental capabilities. A new motorized hot-wire positioning system has been developed and a global transition-detection system based on temperature-sensitive paint (TSP) has been integrated [2].

Eppink [3] investigated the impact of forward-facing steps on crossflow-dominated laminar-turbulent transition and demonstrated an effect of the amplitude of stationary crossflow instabilities (CFI) on the criticality of a fixed step height. However, current tolerance criteria for forward-facing steps in crossflow-dominated transition (e.g. the

H. P. Barth (✉) · S. Hein
Department of High-Speed Configurations, German Aerospace Center, Institute of Aerodynamics and Flow Technology, Göttingen, Germany
e-mail: peter.barth@dlr.de

K. Kato et al. (eds.), *Proceedings of the 10th IUTAM Symposium on Laminar-Turbulent Transition*, IUTAM Bookseries 44, https://doi.org/10.1007/978-981-96-9829-5_13

Tufts-Reed criterion [4]) do not account for this effect. In order to study this effect in more detail in this investigation, the CFI amplitude is systematically varied in parallel with a systematic variation of forward-facing step heights in a low-speed facility.

13.2 Experimental Methods

13.2.1 The SPECTRA Configuration

The SPECTRA configuration consists of a model plate, above which a displacement body is mounted in order to imprint a pressure distribution on the plate's upper side. The displacement body used in this investigation (SPECTRA-A) imprints a nearly constant negative pressure gradient along the chordwise coordinate x_c. This together with other design choices (see [1]) enables the isolated investigation of crossflow-dominated laminar-turbulent transition due to the stabilization of other primary instabilities. The configuration closely follows the design principles used by Bippes et al. in their experiments (see [5]).

The model plate consists of two segments, the leading-edge segment and the main segment. At the junction between the two segments, a step (either forward- or backward-facing) can be introduced by shifting the leading-edge segment vertically relative to the main segment. This is done using a system of four stepper motors in conjunction with four draw-wire sensors for step-height control. This system is inspired by the step-height control system developed by Duncan [6] and will be described in more detail in a forthcoming publication. A cross-section of the configuration is shown in Fig. 13.1a. Two coordinate systems are used: A model-oriented coordinate system with the subscript "c" and a coordinate system oriented along the curved boundary-layer edge streamline with the subscript "s". The wall-normal distance is z_c. Both coordinate systems are shown in Fig. 13.1b.

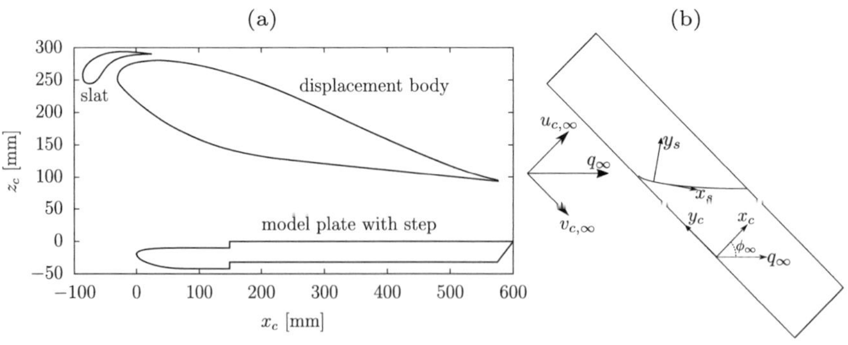

Fig. 13.1 **a** Cross-section of the SPECTRA-A configuration with exaggerated forward-facing step at $x_c = 150$ mm. **b** Top-view sketch of the model plate

13.2.2 Wind Tunnel

The experiments are performed in the DLR-1MG wind tunnel. The wind tunnel is a low-speed facility of Göttingen type and the turbulence level is reported as $Tu = 0.15\%$ [5]. The nozzle has a width of 1 m and a height of 0.74 m. The configuration is fixed in the open test section at a sweep angle of $\phi_\infty = -45.0°$. The freestream unit Reynolds number is $Re'_\infty = 1.90 \cdot 10^6\,\text{m}^{-1}$.

13.2.3 Forcing of Stationary CFI

An array of spanwise-equidistant cylindrical discrete roughness elements (DRE) of constant height h and diameter D is applied to the model surface at a constant x_c location to form a single DRE array configuration. The spanwise spacing between individual DREs corresponds to the spanwise wavelength of the selected forcing mode. Its spanwise wavenumber is $\beta = 500\,\text{m}^{-1}$ and its spanwise wavelength is $\lambda = 12.57$ mm. For the isolated variation of the CFI amplitude, six DRE array configurations with different diameters between $1.18\,\text{mm} \leq D \leq 5.30\,\text{mm}$ are applied at $x_c/c = 0.07$. Two additional DRE array configurations are investigated: In order to achieve an even higher initial CFI amplitude than in the DRE diameter variation, an array with a DRE diameter of 4.42 mm is placed further upstream at $x_c/c = 0.06$ because of the increased receptivity. For a lower initial CFI amplitude, a DRE array with the smallest DRE diameter (1.18 mm) is applied further downstream at $x_c/c = 0.1$ because of the decreased receptivity. The DRE arrays are fabricated using a CNC-controlled foil cutter. The height of the applied DREs is measured using a 3D optical profilometer (KEYENCE VR-5000) as $h = 31(2)\,\mu\text{m}$.

13.2.4 Hot-Wire Anemometry

A probe positioning system is attached to the trailing edge of the model plate. Using a combination of stepper motors and piezo actuators, it allows for very precise positioning of two hot-wire V-probes in the boundary layer on the upper side of the model plate in a wide range of x_c, y_c and z_c coordinates. In V-probes, a perpendicular arrangement of two wires is used to measure the two velocity components in the plane common to the two wires which is the x_c, y_c-plane in this case. Through a rotation by the local sweep angle measured at the upper boundary-layer edge, they are transformed into the streamline-oriented coordinate system introduced in Sect. 13.2.1, resulting in the velocity components u_s and v_s. The probes are mounted at a spanwise distance to each other of 60 mm. Planar y_c-z-scanning measurements are performed at a constant chordwise location x_c. They cover a spanwise distance of 60 mm for each probe and include the boundary-layer edge at the top of the z_c range. The spanwise step size is 1 mm and the wall-normal step size is varied as a function of the wall-normal distance between $1\,\mu\text{m} < \Delta z < 160\,\mu\text{m}$.

13.2.5 Temperature-Sensitive Paint

Temperature-sensitive paint (TSP) measurements are used to detect the laminar-turbulent transition front on the plate's upper side. The plate contains a pocket in which a layer of carbon nanotube heating and a layer of TSP are integrated. LEDs are integrated into the displacement body to excite the TSP. The light emission from the TSP is captured by four miniature cameras that are also integrated into the displacement body. This setup and the measurement technique are described in more detail in [2].

13.3 Results

13.3.1 Characterization of the DRE Array Configurations

The initial CFI amplitude, and consequently the amplitude at the step location, is varied by applying different DRE array configurations to the model as described in Sect. 13.2.3. Otherwise, the freestream conditions, in particular the Reynolds number, remain the same for all investigated configurations.

The amplitude a of the excited CFI is quantified by planar hot-wire measurements at chordwise locations $x_{c,meas}$ in the linear regime of its evolution for each DRE array configuration with the step height set to $k = 0\,\mu\text{m}$. For this, the amplitude $\hat{u}_s(\beta)$ of the spanwise periodicity in the measured velocity component $u_s(y_c, z_c)$ is determined using Fourier decomposition. The amplitude $a(x_{c,meas})$ corresponds to the wall-normal maximum of the evaluated amplitudes $\hat{u}_s$ at the spanwise wavenumber of the artificial excitation $\beta = 500\,\text{m}^{-1}$. N-factors from linear stability analysis are used to extrapolate this amplitude to the chordwise excitation location $x_{c,0}/c = 0.07$: $a_0 := a(x_{c,0}) = a(x_{c,meas}) exp(-\Delta N)$ with $\Delta N = N(x_{c,meas}) - N(x_{c,0})$. For comparability, the amplitudes for the DRE array configurations placed at $x_c/c = 0.06$ and $x_c/c = 0.10$ are also extrapolated to the same chordwise location $x_{c,0}/c = 0.07$. The "initial" amplitudes a_0 are shown in Fig. 13.2 and the maximum and minimum values of this investigation are labeled as $a_{0,max}$ and $a_{0,min}$.

13.3.2 Amplitude Effect on Step Criticality

For each DRE array configuration, the spanwise-averaged transition location is derived from the TSP measurements. The ratio $r_{tr} = (x_{tr} - x_k)/(x_{tr,0} - x_k)$ is used to evaluate the criticality of a given step height k. It compares the distance between the spanwise-averaged transition location in the presence of a step (x_{tr}) and the step location x_k with the same distance in the reference case without a step. The ratio is

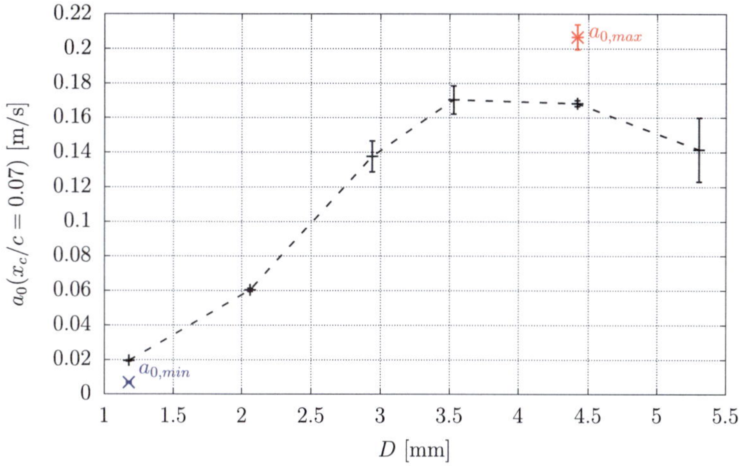

Fig. 13.2 CFI amplitudes derived from extrapolating the measured amplitude data to $x_c/c = 0.07$ using linear stability analysis

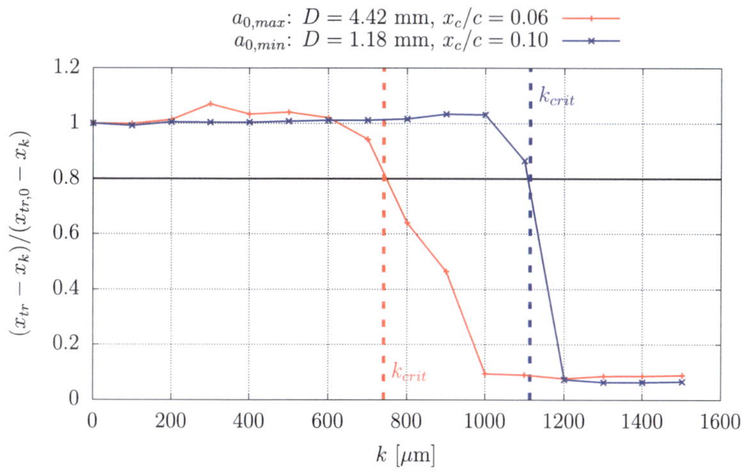

Fig. 13.3 Definition of the critical step height for two exemplary DRE array configurations

close to 1 for small step heights and tends to 0 for step heights at which the presence of the step strongly affects the transition, an observation which is in principal agreement with previous research (e.g. [3, 6]). The step height k at which r_{tr} falls below the arbitrarily chosen threshold value of 0.8 is defined as "critical" and is denoted by k_{crit}. This is shown in Fig. 13.3 for the two DRE array configurations which result in the lowest and highest initial CFI amplitude.

A critical step height is evaluated for each DRE array configuration and the resulting critical step height Reynolds numbers $Re_{k,crit} = u_{c,e}k_{crit}/\nu$ are shown in Fig. 13.4 with $u_{c,e}$ denoting the chordwise boundary-layer edge velocity at the step

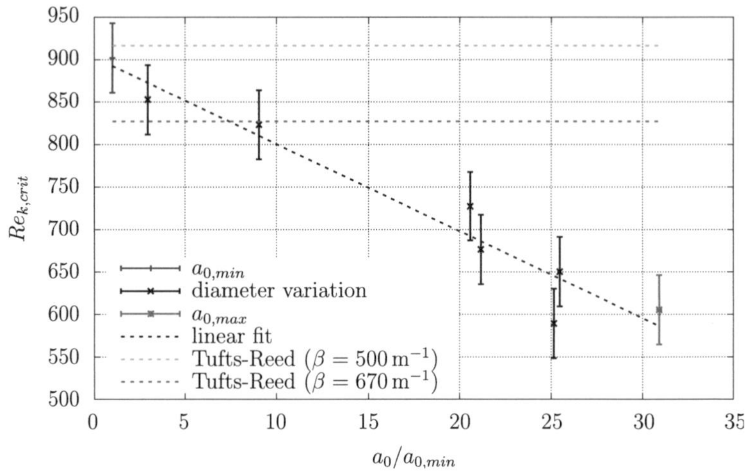

Fig. 13.4 Critical step height Reynolds numbers for different initial CFI amplitudes

location. In addition, the critical step height predicted by the Tufts-Reed criterion (see [4]) for the constant freestream conditions is expressed as a critical step height Reynolds number both for the forced CFI mode with $\beta = 500\,\mathrm{m}^{-1}$ and the CFI mode with the highest N-Factor at the step location with $\beta = 670\,\mathrm{m}^{-1}$. The abscissa shows the initial CFI amplitude referenced to the minimum value $a_{0,min}$.

The critical step height Reynolds number decreases nearly linearly with the initial CFI amplitude. While the Tufts-Reed criterion provides a good estimate of the critical step height Reynolds number for small amplitudes, the experimentally observed critical step height Reynolds numbers deviate significantly from the predicted values for higher CFI amplitudes.

13.4 Conclusion

The initial amplitude of stationary CFI was systematically varied in parallel with a forward-facing step height variation without changing other flow conditions by applying eight different DRE array configurations to the model close to the leading edge. Under the assumption of linear CFI evolution, this would result in an equivalent variation of the CFI amplitude at the step location. Even taking into account non-linear effects upstream of the step, a strong correlation between the initial CFI amplitude and the CFI amplitude at the step can still be assumed. The observation of a nearly linear decrease of the critical step height Reynolds number with increasing initial CFI amplitude confirms and quantifies the effect first observed by Eppink [3]. The direct implication is that the surface roughness quality of the model affects the tolerance for forward-facing steps in crossflow-dominated laminar-turbulent transition. Furthermore, many other factors, e.g. the receptivity and stability conditions, affect the amplitude with which the CFI reaches the step. Many step tolerance experiments include, for example, a variation of the freestream Reynolds number which affects the receptivity and CFI amplification and therefore the CFI amplitude at the step. The isolated amplitude variation in this investigation by a factor of $a_{0,max}/a_{0,min} \approx 30$ is equivalent to a difference in linear integral amplification of $\Delta N = ln(30) = 3.4$. The same difference in expected CFI amplitude at the step location could be achieved in this setup, for example, by shifting the chordwise step location by approximately 20% of the chord length.

This research demonstrates that crossflow instability amplitude is an important parameter to be taken into account in the search for improved tolerance criteria for forward-facing step effects on laminar-turbulent transition.

Acknowledgements The authors would like to acknowledge funding from the DLR-internal project *LamTA*.

References

1. Barth, H.P., Hein, S., Rosemann, H.: Redesigned swept flat-plate experiment for crossflow-induced transition studies. In: Notes on Numerical Fluid Mechanics and Multidisciplinary Design. Springer, pp. 155–165 (2017). https://doi.org/10.1007/978-3-319-64519-3_14
2. Lemarechal, J., Dimond, B.D., Barth, H.P., Hilfer, M., Klein, C.: Miniaturization and model-integration of the optical measurement system for temperature-sensitive paint investigations. Sensors **23**(7075) (2023). https://doi.org/10.3390/s23167075
3. Eppink, J.: The effect of forward-facing steps on stationary crossflow instability growth and breakdown. In: 2018 AIAA Aerospace Sciences Meeting, Kissimmee, AIAA 2018-0817 (2018). https://doi.org/10.2514/6.2018-0817
4. Tufts, M., Reed, H., Crawford, B., Duncan, G., Saric, W.: Computational investigation of step excrescence sensitivity in a swept-wing boundary layer. In: 45th AIAA Fluid Dynamics Conference, Dallas, AIAA 2015-2775 (2015). https://doi.org/10.2514/6.2015-2775
5. Bippes, H.: Basic experiments on transition in three-dimensional boundary layers dominated by crossflow instability. Prog. Aerosp. Sci. **35**(4), 363–412 (1999). https://doi.org/10.1016/s0376-0421(99)00002-0
6. Duncan, G., Crawford, B., Tufts, M., Saric, W., Reed, H.: Effects of step excrescences on a swept wing in a low-disturbance wind tunnel. In: 52nd Aerospace Sciences Meeting, National Harbor, AIAA 2014-0910 (2014). https://doi.org/10.2514/6.2014-0910

Chapter 14
Flow State Transition Process and Oblique Pattern of Taylor–Couette–Poiseuille Flow at a Medium Cylinder Ratio

Yuki Matsukawa, Ryo Araki, and Takahiro Tsukahara

Abstract We have performed direct numerical simulations of counter-rotating Taylor–Couette–Poiseuille flows with a radius ratio of $\eta = 0.5$ to investigate the scenario of turbulent transition phenomena. Near the inner cylinder, streaks aligned with the wall shear stress coexisted with a large-scale oblique mode in a different direction. The flow fields were decomposed using filter operations to identify their characteristic length scales. The low-pass filtered fluctuating velocity field reveals large-scale secondary flows, resembling the localized turbulent patterns observed during subcritical transitions in wall-bounded uni-directional flows. However, the large-scale oblique mode differs from the conventional localized turbulent patterns because the laminar and turbulent regions are not separated. This large-scale oblique mode can be interpreted as a helical version of the mode structure observed at low radius ratios ($\eta \leq 0.5$, Tanaka et al., Int. J. Adv. Eng. Sci. Appl. Math., 2018).

Keywords Direct numerical simulation · Wall-bounded shear flow · Subcritical transition

14.1 Introduction

Wall-bounded shear flows can undergo sudden laminar–turbulent transitions even in the subcritical transition regime (below the critical Reynolds number for linear stability), as the base flow exhibits nonlinear instability to finite-amplitude disturbances [1]. Localized turbulent regions may form characteristic large-scale patterns in this regime, particularly near the global stability critical Reynolds number [2]. For example, turbulent puffs appear in pipe flows [3, 4] and turbulent spots, bands,

Y. Matsukawa (✉) · R. Araki · T. Tsukahara
Tokyo University of Science, Noda-shi, Chiba, Japan
e-mail: 7523701@ed.tus.ac.jp

T. Tsukahara
e-mail: tsuka@rs.tus.ac.jp

K. Kato et al. (eds.), *Proceedings of the 10th IUTAM Symposium on Laminar-Turbulent Transition*, IUTAM Bookseries 44, https://doi.org/10.1007/978-981-96-9829-5_14

and stripes appear in planar flows [5, 6]. In addition, large-scale secondary flows in a direction different from the mainstream occur around the localized turbulence [7].

In annular Poiseuille flow (APF), driven by an axial pressure gradient between concentric pipes, different localized turbulent patterns emerge depending on the radius ratio, $\eta = r_{\mathrm{in}}/r_{\mathrm{out}}$. At high radius ratios, where the inner-to-outer radius ratio is close to 1 ($\eta \to 1$), the geometry becomes similar to planar flows, and turbulent stripes appear. In contrast, at low radius ratios close to 0 ($\eta \to 0$), the geometry becomes similar to pipe flows, and turbulent puffs occur [8]. In a counter-rotating Taylor–Couette flow (TCF) with a high radius ratio, the localized turbulence pattern takes the form of spiral turbulence [9]. On the other hand, in a TCF with a medium or low radius ratio ($\eta \lesssim 0.5$), a mode structure that extends in the axial direction appeared [10].

The Taylor–Couette–Poiseuille flow (TCPF), the subject of this study, is an annular pipe flow that combines the APF and TCF. The flow is driven by a combination of azimuthal wall rotation and axial pressure gradient, resulting in a combined shear flow. A counter-rotating TCPF shows a subcritical transition and generates localized ring-shaped turbulence at $\eta = 0.883$ [11]. In addition, a TCPF with a stationary outer cylinder exhibits switching between supercritical and subcritical transitions [12]. Since the flow structure of combined shear flows such as TCPF is complicated, many such unique phenomena occur. In this study, we performed direct numerical simulations (DNSs) for a TCPF with a medium radius ratio of $\eta = 0.5$ and elucidated a part of the flow state transition process.

14.2 Methodology and Conditions

In this study, the working fluid is an incompressible Newtonian fluid, and we performed DNSs of its TCPF shown in Fig. 14.1. Here, r_{in}, r_{out}, Ω_{in}, and Ω_{out} represent the inner and outer cylinder radii and the angular velocities of the inner and outer cylinders, respectively. The inner and outer cylinders' surface velocities are $u_{\mathrm{in}} = r_{\mathrm{in}}\Omega_{\mathrm{in}}$ and $u_{\mathrm{out}} = r_{\mathrm{out}}\Omega_{\mathrm{out}}$, respectively. Based on these, we define three Reynolds numbers of

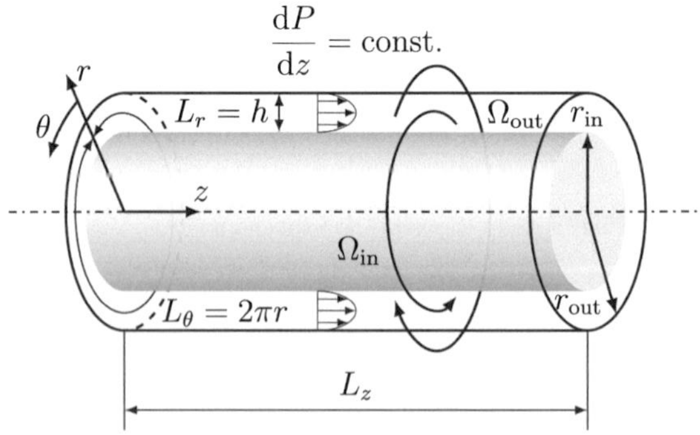

Fig. 14.1 Configuration of Taylor–Couette–Poiseuille flow (TCPF)

$$Re_{\text{in}} = \frac{u_{\text{in}} h}{\nu}, \quad Re_{\text{out}} = \frac{u_{\text{out}} h}{\nu}, \quad Re = \frac{u_0 h}{\nu} \tag{14.1}$$

using the kinematic viscosity ν, the gap width $h = r_{\text{out}} - r_{\text{in}}$ and the relative velocity $u_0 = u_{\text{in}} - u_{\text{out}}$.

The governing equations are the continuity and the Navier–Stokes equations for an incompressible three-dimensional cylindrical coordinate system, expressed as

$$\frac{\partial u_z^*}{\partial z^*} + \frac{1}{r^*}\frac{\partial}{\partial r^*}(r^* u_r^*) + \frac{1}{r^*}\frac{\partial u_\theta^*}{\partial \theta} = 0, \tag{14.2}$$

$$\frac{\partial u_z^*}{\partial t^*} + (\mathbf{u}^* \cdot \nabla^*) u_z^* = -\frac{\partial p^*}{\partial z^*} + \frac{1}{Re}\nabla^{*2} u_z^* + \frac{1}{Re} F(P), \tag{14.3}$$

$$\frac{\partial u_r^*}{\partial t^*} + (\mathbf{u}^* \cdot \nabla^*) u_r^* - \frac{u_\theta^{*2}}{r^*} = -\frac{\partial p^*}{\partial r^*} + \frac{1}{Re}(\nabla^{*2} u_r^* - \frac{u_r^*}{r^{*2}} - \frac{2}{r^{*2}}\frac{\partial u_\theta^*}{\partial \theta}), \tag{14.4}$$

$$\frac{\partial u_\theta^*}{\partial t^*} + (\mathbf{u}^* \cdot \nabla^*) u_\theta^* + \frac{u_r^* u_\theta^*}{r^*} = -\frac{1}{r^*}\frac{\partial p^*}{\partial \theta} + \frac{1}{Re}(\nabla^{*2} u_\theta^* - \frac{u_\theta^*}{r^{*2}} + \frac{2}{r^{*2}}\frac{\partial u_r^*}{\partial \theta}). \tag{14.5}$$

The external force $F(P)$ in Eq. (14.3) is a pressure gradient function that represents the magnitude of the constant pressure-gradient driving force applied in the axial direction of the cylinder, defined by

$$F(P) = -Re\frac{\mathrm{d}P^*}{\mathrm{d}z^*}. \tag{14.6}$$

Here, the superscript $(\;)^*$ denotes nondimensionalization by h and u_0. In this study, we chose Re_{in} and $F(P)$ as the control parameters. Table 14.1 shows the computational condition in this study. The axial friction Reynolds number, $Re_{\tau,z} = u_{\tau,z} h / 2\nu$, is a weighted average value at the inner and outer cylinder walls. In addition, we define the axial bulk Reynolds number as $Re_{\text{m},z} = u_{\text{m},z} h / \nu$.

14.3 Flow State Transition

Figure 14.2 shows the visualization of the fluctuation velocity $u_r' \equiv u_r - \bar{u}_r$ in the z–θ planes at different y_{in}^*. Here, y_{in}^* is the wall-normal height from the inner cylinder, normaized by the gap width. The overbar represents an ensemble-averaged value. A

Table 14.1 Physical and numerical conditions of a TCPF: $Re_{\tau,z}$, the obtained friction Reynolds number in the axial direction; $Re_{\text{m},z}$, the obtained bulk Reynolds number in the axial direction; L_i, the domain length; N_i, the number of grids

η	Re_{in}	Re_{out}	$F(P)$	$Re_{\tau,z}$	$Re_{\text{m},z}$	$L_z \times L_r \times L_\theta$	$N_z \times N_r \times N_\theta$
0.5	600	−2000	5	40.3	898	$51.2h \times h \times 2\pi r$	$1024 \times 64 \times 1024$

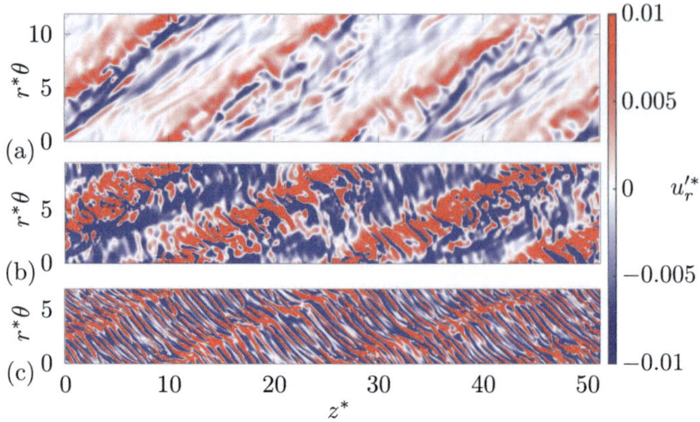

Fig. 14.2 Instantaneous flow field of wall-normal fluctuation velocity $u_r'^*$ in z–θ plane: **a** near the outer wall ($y_{\text{in}}^* \approx 0.9$), **b** the gap center ($y_{\text{in}}^* \approx 0.5$) and **c** near the inner wall ($y_{\text{in}}^* \approx 0.1$)

large-scale oblique mode can be observed throughout the entire gap, spanning from the outer cylinder side in Panel (a) to the inner cylinder side in Panel (c). The stripe is weaker near the outer cylinder than near the inner cylinder. In addition to the large-scale oblique mode, small-scale streaks occur in different directions on the inner cylinder side. These streaks are particularly prominent on the inner cylinder, and appear absent on the outer cylinder. Here, we extract these structures near the inner cylinder by applying filter operations. The low- and high-pass filtered velocity fields near the inner cylinder are shown in Fig. 14.3a, b, respectively. The filtering threshold is $(k_z, k_\theta) = (10, 5)$. In Panel (a), the extracted flow structure appears alongside the low-pass filtered axial and azimuthal velocity vectors. The vector field shows a flow from the blue region (negative u_r') to the red region (positive u_r'), indicating the existence of a large-scale secondary flow. This large-scale secondary flow is similar to the localized turbulence patterns seen during subcritical transitions [7]. However, while the conventional localized turbulence patterns separate clearly the laminar and turbulent regions, this is not the case in Figs. 14.2c and 14.3a. A previous study on low-η TCF [10] have shown that a mode structure that extends in the axial direction appears for $y_{\text{in}}^* \leq 0.5$. Therefore, it is possible that the flow structure seen in this study is an helical version of this mode structure because a TCPF is an annular pipe flow. In addition, we observed that the high-wavenumber streaky structures in Fig. 14.3b is aligned with the direction of the wall shear stress near the inner cylinder. This phenomenon also occurs under other TCPF conditions, where the streaks in the same direction as the wall shear stress coexist with large-scale patterns (ring-shaped turbulence) in a different direction [11].

We quantitatively determine the scale of the flow structure using the pre-multiplied energy spectrum (PMES) [13]. Figure 14.4 presents the PMES for (a) the axial wavenumber and (b) the azimuthal wavenumber near the inner cylinder. In Panel (a), the PMES of $k_z E_{u_r u_r}(k_z)$ shows a peak at $\lambda_z/h \approx 0.78$. The azimuthal counter part, $k_\theta E_{u_r u_r}(k_\theta)$, also shows a peak at $\lambda_\theta/h \approx 0.69$ in Panel (b). The high-wavenumber,

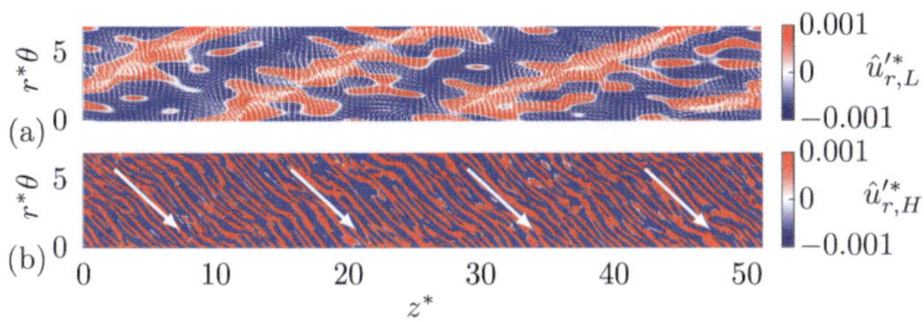

Fig. 14.3 Instantaneous flow field of the wall-normal fluctuation velocity $u_r'^*$ in z–θ plane near the inner cylinder wall ($y^*_{\rm in} = 0.1$): **a** the low-wavenumber structure and **b** the high-wavenumber structure. The white arrows indicate the vectors of the velocity with a low-pass filter ($\hat{u}'^*_{z,L}$, $\hat{u}'^*_{\theta,L}$) in Panel (**a**) and the direction of the wall shear stress in Panel (**b**)

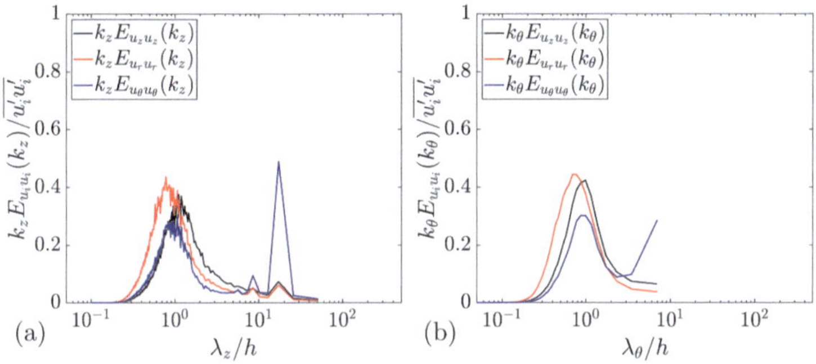

Fig. 14.4 Pre-multiplied energy spectra: **a** $k_z E_{u_i u_i}(k_z)$ of u_i' as a function of λ_z and **b** $k_\theta E_{u_i u_i}(k_\theta)$ of u_i' as a function of λ_θ. Both panels show the results near the inner cylinder wall ($y^*_{\rm in} = 0.1$)

or short-wavelength, structures characterized by these peaks correspond to quasi-streamwise streaks dominating the flow near the inner cylinder. In addition, the blue line indicating the PMES of the azimuthal velocity u_θ shows a peak at $\lambda_z/h \approx 17$, as shown in Panel (a). This peak is consistent with a large-scale oblique mode that occurs from the inner cylinder side to the outer cylinder side, as visualized in Fig. 14.2.

14.4 Conclusion

We performed DNSs for a TCPF with a radius ratio of $\eta = 0.5$ to investigate a subcritical transition in a wall-bounded, combined shear flow. The tested condition was a parameter set of $Re_{\rm in} = 600$, $Re_{\rm out} = -2000$, $F(P) = 5$, and we observed the large-scale oblique mode developed throughout the entire gap. However, near the inner cylinder, small-scale streaks elongated in a different direction than the oblique mode structure were developed. We separated these structures by low- (high-) pass filtering operations. We found that the large-scale secondary flow occurs around the oblique mode and that the streaks are aligned with the direction of the wall shear stress. We also found that the oblique mode is a different flow phenomenon from the conventional localized turbulent pattern of subcritical transition, and we argued that it can be associated with the structures observed in the same system with smaller radius ratios. Furthermore, we employed PMES to quantitatively evaluate the scales of these two structures.

Acknowledgements This work was supported by a Grant-in-Aid for the Japan Society for the Promotion of Science (JSPS) Fellowship 19H02071 and 24KJ2023. Direct numerical simulations were performed on SX-series supercomputers at the Cybermedia Center of Osaka University and the Cyberscience Center of Tohoku University.

References

1. Manneville, P.: Transition to turbulence in wall-bounded flows: where do we stand? Mech. Eng. Rev. **3**(2), 15–00684 (2016)
2. Tuckerman, L.S., Chantry, M., Barkley, D.: Patterns in wall-bounded shear flows. Annu. Rev. Fluid Mech. **52**(1), 343–367 (2019)
3. Reynolds, O.: An experimental investigation of the circumstances which determine whether the motion of water shall be direct or sinuous, and of the law of resistance in parallel channels. Philos. Trans. R. Soc. **174**, 935–982 (1883)
4. Wygnanski, I.J., Champagne, F.H.: On transition in a pipe. Part 1. The origin of puffs and slugs and the flow in a turbulent slug. J. Fluid Mech. **59**(2), 281–335 (1973)
5. Prigent, A., Grégoire, G., Chaté, H., Dauchot, O., van Saarloos, W.: Large-scale finite-wavelength modulation within turbulent shear flows. Phys. Rev. Lett. **89**(1), 145011–145014 (2002)
6. Tsukahara, T., Seki, Y., Kawamura, H., Tochio, D.: DNS of turbulent channel flow at very low Reynolds numbers. In: Proceedings of 4th International Symposium on Turbulence and Shear Flow Phenomena, pp. 935–940 (2005)
7. Duguet, Y., Schlatter, P.: Oblique laminar–turbulent interfaces in plane shear flows. Phys. Rev. Lett. **110**(3) (2013)
8. Ishida, T., Duguet, Y., Tsukahara, T.: Transitional structures in annular Poiseuille flow depending on radius ratio. J. Fluid Mech. **794**, R2 (2016)
9. Coles, D.: Transition in circular Couette flow. J. Fluid Mech. **21**(3), 385–425 (1965)
10. Tanaka, R., Kawata, T., Tsukahara, T.: DNS of Taylor–Couette flow between counter-rotating cylinders at small radius ratio. Int. J. Adv. Eng. Sci. Appl. Math. **10**(2), 159–170 (2018)
11. Matsukawa, Y., Tsukahara, T.: Subcritical transition of Taylor–Couette–Poiseuille flow at high radius ratio. Phys. Fluids **34**(7), 074109 (2022)
12. Matsukawa, Y., Tsukahara, T.: Switching between supercritical and subcritical turbulent transitions in inner cylinder rotating Taylor–Couette–Poiseuille flow. Int. J. Heat Fluid Flow **112**, 109667 (2025)
13. Jiménez, J.: The largest scales of turbulent wall flows. Center for Turbulence Research Annual Research Briefs, pp. 137–154 (1998)

Chapter 15
New Perspectives on Instability of Laminar Separation Bubbles: Have We Been Missing Something Important?

Daniel Rodríguez

Abstract Several decades of research on laminar separation bubbles (LSBs) have amassed evidence of an scenario dominated by the convective amplification of incoming disturbances by the inflectional instability. However, some LSBs present dynamics suggesting that additional, self-excited instability mechanisms are also present. This paper reviews some theoretical/numerical results on self-excited instability, and revisits recent low-Tu wind tunnel experiments which provide evidence of the global instabilities. It is concluded that self-excited mechanisms might have been present and overlooked in many previous studies; some considerations are suggested for future research to allow for their identification.

Keywords Laminar separation bubbles · Global instability

15.1 Introduction

Laminar separation bubbles (LSBs) are a ubiquitous phenomenon in aerodynamic flows, associated with detrimental effects including stall, emergence of sudden pitching and rolling moments and cyclic forces on aerodynamic surfaces and output losses in power turbines. LSBs are particularly relevant in wings of high maneuverability aircraft, sailors and low-pressure turbines, and are gaining a renewed interest with the advent of Micro Aerial Vehicles and Urban Air Mobility.

The appearance of LSBs requires that a laminar boundary layer encounters a strong enough adverse pressure gradient (APG); the boundary layer then separates from the wall, with the separating streamline forming an angle which depends on the APG magnitude and the local boundary layer thickness. The ensuing separated shear layer is highly unstable, leading to a relatively fast transition to turbulence. The turbulent mixing then leads to the reattachment of the mean flow and the formation of a closed recirculation bubble. Within the region enclosed by the time-averaged separation

D. Rodríguez (✉)
School of Aeronautics, Universidad Politécnica de Madrid (ETSIAE-UPM), Madrid, Spain
e-mail: daniel.rodriguez@upm.es

K. Kato et al. (eds.), *Proceedings of the 10th IUTAM Symposium on Laminar-Turbulent Transition*, IUTAM Bookseries 44, https://doi.org/10.1007/978-981-96-9829-5_15

streamline, fluid moves upstream with low velocity compared to the free-stream, which entails a nearly constant pressure value and the consequent modification of the aerodynamic forces and moments. The details of the laminar-turbulent transition and their impact on the mean LSB properties, the aerodynamic performance and its dynamics, are ultimately governed by the hydrodynamic instability mechanisms at play. In addition, a deep understanding of these instabilities can be exploited towards the devise of flow control strategies that promote the growth of flow structures favoring a fast and steady reattachment.

Several decades of research on flow instability and transition on LSBs consolidated the following picture [1–3]. The separated shear layer is unstable to multiple linear and local instability modes, including Tollmien-Schlichting waves (arising from the viscous stresses at the wall), Kelvin-Helmholtz waves (owing to inflectional instability) and Görtler vortices (owing to the concave curvature of the separating streamlines). Hence, the laminar shear layer amplifies virtually any incoming flow disturbance as it is convected downstream. This amplifier scenario is also extended to nonmodal instabilities including streamwise streaks and Orr's mechanism, which act as precursors of the modal ones [4]. However, for quiet inflow scenarios (i.e. in which the amplitude of incoming disturbances is low), experimental evidence demonstrates the dominance of the inflectional instability. Resulting from this instability, spanwise-dominant vortices are formed and shed, and their subsequent secondary instability leads to their three-dimensional (3-D) break-up into turbulence.

The literature on LSBs also displays a broad variety in terms of their size, appearance and frequency of their dominant dynamics. For their characterization, a reduced number of dimensionless parameters has been proposed. This includes the two parameters initially proposed by Gaster [1], namely the adverse pressure gradient and the local Reynolds number

$$\lambda = \frac{\theta_s^2}{\nu} \left.\frac{dU_e}{dx}\right|_s, \qquad Re_{\theta_s} = U_{e,s}\theta_s/\nu \tag{15.1}$$

defined using the free-stream velocity U_e, its streamwise gradient dU_e/dx, the kinematic viscosity ν and the momentum thickness θ, all evaluated at the separation point. The reversed flow relative to the free-stream velocity

$$u_{rev} = -u_{min}/U_e \tag{15.2}$$

is also used to quantify the intensity of the recirculating flow. Finally, a dimensionless frequency

$$St_\theta = f\theta_s/U_{e,s} \tag{15.3}$$

is customarily adopted to compare dynamics among different LSBs.

While the dominance of the scenario based on the convective amplification of external disturbances via inflectional instability is apparent in most of the literature [1–3, 5–7, 9, 10], a careful inspection reveals examples in which the amplifier scenario alone cannot explain the observed phenomena. This includes: numerical simulations recovering transition to turbulence in the absence of incoming disturbances or actuation [6, 11]; 3-D structures with distinct spanwise lengths recovered in the time average of experiments [3, 5, 7, 9]; and low-frequency oscillations of the shear layer (flapping) or the entire separation bubble (breathing) occurring at frequencies substantially smaller than the inflection instability. Finally, the sudden change from a 'short' to a 'long' LSB after a minute change on the inflow conditions, known as *bursting* [1], has also been pointed out as the consequence of the shift from the amplifier behavior to a different kind of scenario.

These observations, together with results from 'global' linear instability analyses (i.e. not resorting to the weakly non-parallel flow assumption and based on partial derivative equations), suggest that additional instability mechanisms must be present. These mechanisms would not rely on their continuous excitation by incoming disturbances, thus being completely self-excited, and under some circumstances, surpass the scenario governed by convective inflectional instability.

This paper summarizes some recent results of linear stability theory (LST) concerning the self-excited instability of subsonic LSBs, showing that a complete transition to turbulence can be triggered even in the absence of incoming disturbances. Then, some recent experimental works (by other groups [5, 7, 9, 10]) are re-visited and discussed in the light of the LST results, to argue that the inherent dynamics described by the self-excited instability might have been present and overlooked. Finally, some caveats are proposed to be considered in future experiments and simulations to ascertain the presence of self-excited dynamics.

15.2 Self-sustained Linear Instability Mechanisms

This section summarizes recent results on the self-excited linear instability of LSB flows by the author and his collaborators. It is acknowledged that additional modal global instabilities of separated flows have been reported in the literature; they are not discussed here due to the limited space.

15.2.1 A Geometrical Criterion for Absolute Inflectional Instability

The presence of the inflectional instability in LSBs is guaranteed by the fact that all velocity profiles representative of a time-averaged separated boundary layer feature one inflection point which is also a maximum in the vorticity magnitude. Rayleigh's and Fjörtoft's theorems ensure that, in the inviscid limit, at least *some* instability waves are amplified. In practical LSBs, the effect of viscosity on the inflectional

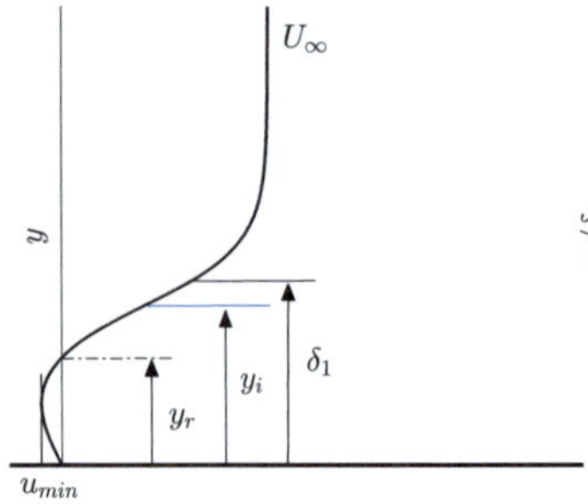

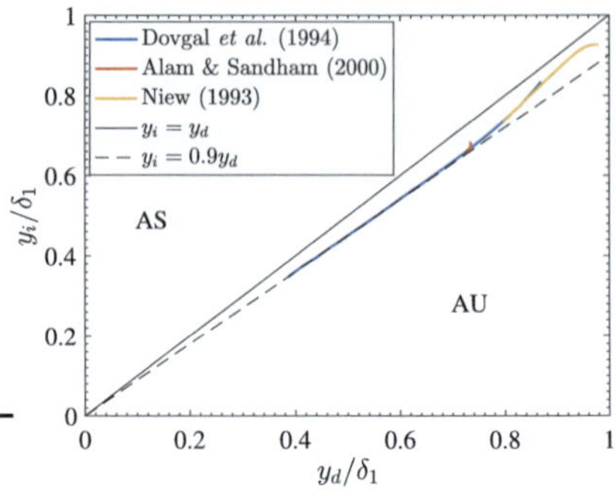

Fig. 15.1 Left: Local 1D velocity profile and characteristic lengths: y_r : end of reverse flow region; y_i coordinate of inflection point; δ_1 displacement thickness; y_d : separation streamline. Right: regions of absolutely unstable (AU) and absolutely stable (AS) inflectional instability, corresponding to different families of analytic velocity profiles

instability is nearly negligible and the associated growth rates exceed those related to other local instabilities.

However, Rayleigh-Fjörtoft's theorems do not provide information regarding the absolute or convective nature of the instability waves. The presence of the reversed flow region suggests that an absolute instability is possible, if the magnitude of the reversed flow u_{rev} is high enough. Different authors analyzed this question employing either spanwise and time-averaged velocity profiles from simulations, or analytical functions that reproduce their main features. Their results agree that a relatively large value of the u_{rev} is required for the absolute instability ($u_{rev} >$ 16–25%, depending on the authors), but the precise threshold value is dependent on the details of the velocity profile.[1]

Avanci et al. [12] revisited the problem. Besides arriving at results consistent with the literature, it was noted that conditions of absolute instability were present for reverse flows as low as $u_{rev} \sim 10\%$. Noting that Rayleigh-Fjörtoft's theorems are based on the geometry of the velocity profile rather than on threshold values, a geometrical criterion was sought for. It was found that absolute instability only happens for mean velocity profiles in which the inflection point lies below the streamline that delimits the recirculating flow region (see Fig. 15.1). This is quantified based on the wall coordinate to the inflection point, y_{in}, such that $U''(y_{in}) = 0$, where $U(y)$ is the base flow, and the coordinate y_d for which the mass flow vanishes, i.e.

$$\int_0^{y_d} U(y)\,dy = 0. \tag{15.4}$$

Accordingly, $y_{in} < y_d$ was proposed as an empirical necessary condition for absolute inflectional instability. This absolute instability can lead to a global oscillator, resulting in self-sustained shedding of K-H vortices.

[1] A complete list of references can be found in [12].

15.2.2 Global Centrifugal Instability

The existence of a 3-D instability acting on nominally 2-D recirculating flows was hypothesized by Dallmann and then demonstrated by Theofilis et al. [13] using a global eigenmode analysis. Since then, it has been consistently found in most 2-D flows featuring a closed recirculation. This is a centrifugal instability, related to the classic Görtler vortices formed on concave walls; however, this instability is global in nature and its spanwise wavelength λ_z scales with the streamwise length of the base recirculation region.

Rodríguez et al. [14] studied, for a series of LSB flows, the possible co-existence of the two self-excited global instabilities mentioned above: a global oscillator arising from absolute inflectional instability and a global centrifugal instability. The baseflows were constructed using an inverse formulation of the 2-D boundary layer equations capable of recovering separated flows with reversed flow $u_{rev} < 13\%$. While absolute inflectional instability was not found for any of the LSB base flows, the centrifugal instability became active for $u_{rev} \sim 6$–7%.

15.2.3 Nonlinear Spanwise Deformation, Self-excited Secondary Instability and Transition

The linear stability results by [14] opened several questions regarding the subsequent development of the LSB flow once the centrifugal instability sets up. Rodríguez et al. [15] addressed some of them.

First, non-linear direct numerical simulations (DNS) were used to study the nature of the bifurcation associated with the centrifugal instability. It corresponds to a pitchfork bifurcation; the linear growth phase followed by a saturation at a finite amplitude, with the amplitude growing monotonously with the departure from the neutral instability conditions. The nonlinearities leading to saturation involve a relatively large number of harmonic spanwise wavenumbers and a strong deformation of the spanwise-homogeneous component. The resulting flows are steady 3-D separation bubbles, with fundamental spanwise wavelengths dictated by the linear instability analysis. At saturation, the peak reverse flow depends on the spanwise coordinate z: at certain $z-$planes, u_{rev} is increased to values surpassing 15% even close to the bifurcation conditions. Conversely, considering the spanwise-averaged flow, the peak reverse flow is reduced with respect to the base flow, remaining below 7% for all cases.

The intense deformation of the LSB and increase of u_{rev} at certain planes suggested that a self-excited secondary instability might appear. An absolute-convective linear instability analysis based on cross-stream (constant x) planes was applied to the 3-D saturated flows, revealing that inflectional instability indeed became of absolute nature as a result of the spanwise deformation of the bubble. These results were also consistent with the geometrical criterion described in Sect. 15.2.1: at each

(x, z)−point, the local inflection point y_{in} and the zero streamwise mass flow point y_d were computed. All the streamwise planes sustaining absolute instability satisfied that $y_{in} < y_d$ at least for a range of spanwise locations.

In summary, the spanwise distortion of the separation bubble exerted by the centrifugal instability, once non-linear amplitudes are reached, leads to the appearance of spanwise "pockets" in which the inflection point is located inside the recirculation region. This results in the inflectional instability becoming of absolute kind. The 3-D nature of the bifurcated flow also alters the spatial structure of the inflectional instability waves, which now present a spanwise-periodic arrangement of peaks and troughs, leading to finite-span structures.

Finally, [15] showed that, resulting from the 3-D absolute inflectional instability, a self-excited global oscillator appears. Using DNS, the bifurcation path defined by the centrifugal (primary) and absolute inflectional (secondary) instabilities was shown to effectively trigger transition to turbulence (Fig. 15.2) in the absence of external disturbances or actuation. After the initial transient, the DNS simulation shows complex dynamics including the formation of Λ vortices (with their feet following the spanwise wavelength of the centrifugal instability) and a fully turbulent reattachment, similar to other numerical simulations [6, 11, 16]. It is remarkable that this scenario was obtained departing from baseflows with peak reverse flow $u_{rev} \approx$

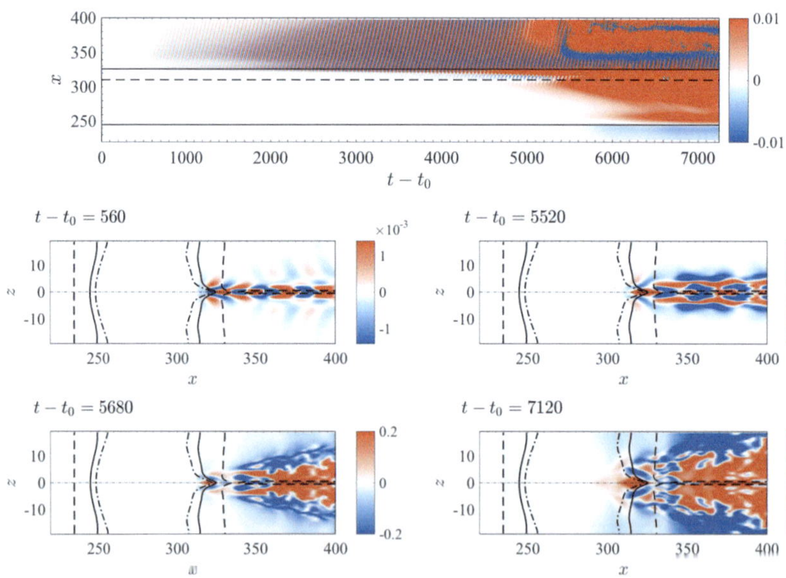

Fig. 15.2 Development of a self-excited instability arising from the 3D distortion of the LSB and triggering laminar-turbulent transition. Top: disturbance streamwise velocity u' at $y = 1.75\delta_1, z = 0$. Bottom: u' at $y = 1.75\delta_1$ and different times. The solid, dashed and dash-dotted lines correspond to the locations of y_d, y_i and y_r, respectively

8%; the spanwise-average of flow after the saturation of the primary instability but before the set-in of the secondary one was reduced to $u_{rev} \approx 6\%$; and this quantity was further reduced in the spanwise and time-average of the fully developed DNS solution.

15.3 Experimental Evidences

It can be argued that the self-excited scenarios discussed in the previous Section would become irrelevant in practical LSBs, which are invariably subject to free-stream disturbances. Even the quietest wind tunnel presents a non negligible turbulence intensity (e.g. Gaster's was reported to be below Tu = 0.05% [1]). These free-stream disturbances would penetrate the pre-separated boundary layer and grow explosively due to the convective inflectional instability, thus bypassing or precluding the self-excited instabilities from manifesting.

The understanding of LSBs as amplifiers of external disturbances was established at an early time in which experimental techniques capable of providing multi-dimensional and time-resolved information simultaneously were not available. Techniques like oil-streak or ink-dot visualizations [3] provided the first evidences of the inherent tendency of LSBs to break up in the spanwise direction originating periodic structures in the time-averaged flow. Complex systems of multiple hot-wires measuring simultaneously, like the one employed by Watmuff [5], capable of providing an impressive degree of spatio-temporal resolution, were not available to most researchers. It has only been with the generalization of particle image velocimetry (PIV) and the appearance of tomographic PIV in the last decade that evidences suggesting self-excited mechanisms became apparent.

Table 15.1 shows a quantitative comparison of some characteristic properties of LSBs measured in low-Tu wind tunnel experiments and those predicted by the self-excited scenario in Sect. 15.2.3. These experimental works studied short LSBs formed

Table 15.1 Comparison of the coherent structures characteristics in LSBs in wind-tunnel experiments and the self-excited scenario in [15]. Reference [5] does not provide accurate numbers for the wavelengths, which were estimated by the present author from figures. This source of inaccuracy might explain the differences with respect the other works for the quantities marked with asterisks

References	u_{rev}	λ_x/θ_s	λ_z/θ_s	λ_z/λ_x	St_θ
[5]	≥7%	50.8*	123*	2.42*	0.008654
[8]	~7.5%	35.1	71.4	2.00	0.017
[7]	~2%	34.8	67.4	1.94	0.01
[9]	~4%	32	72	2.25	0.016
Self-excited scenario					
[15]	7–12%	31.0–37.9	76.3–76.9	2.06–2.41	0.01–0.012

on flat plates [5, 7] or airfoils [8, 9] under Tu $< 0.1\%$, and in the absence of explicit forcing that would favor specific frequencies or structures. The peak reverse flow of the time-averaged flow is shown. Disturbances with dominant wavelengths in the streamwise and spanwise directions (λ_x and λ_z, respectively, shown in the table scaled with the momentum thickness at separation) were observed, exhibiting a dominant dimensionless frequency St_θ. The quantitative agreement with the LSBs resulting from the self-excited scenario is striking.

It has been suggested that the same flow dynamics can also result from the amplifier scenario alone. One explanation suggests that the mainly 2-D vortex rolls resulting from the inflectional instability would experience a secondary instability inducing a spanwise modulation of the vortices and subsequent breakdown. This is a well-known scenario occurring when forced oscillations are imposed upstream of the separation (e.g. [6]), and can lead to streamwise and spanwise wavelengths similar to those in Table 15.1. Another explanation proposes that the dominant spanwise wavelength results from the fact that oblique Tollmien-Schlichting waves are slightly more amplified than the plane ones upstream of the separation [7]. However, these explanations are not without flaw.

Some of these works [5, 7], in agreement with previous experiments, showed that the spanwise waviness of LSBs is also observable in the time-averaged flow. While the convective amplification of disturbances could explain the origin of 3-D disturbances with comparable wavelenghts and frequencies, it is not clear why or how these disturbances would appear consistently at the same spanwise locations, so that their impact would persist in the time-averaged flow field. This, again, points towards the existence of a *global* mechanism, in addition to the convective instability.

A very recent paper by Mohammad Aniffa et al. [10] shows possible evidences of the two different self-excited instabilities discussed in Sect. 15.2. Wind-tunnel experiments were conducted on two different configurations featuring LSBs.[2] The first one, termed geometry-induced separation bubble (GISB), is formed at the blunt leading edge of a flat plate. The second one, termed pressure-induced separation bubble (PISB), is formed downstream on a flat plate by a contoured wall opposite to it, in a set-up analogous to e.g. [7]. Time-resolved planar PIV was used on a vertical (x, y) plane and a spanwise (x, z) plane.

The time-averaged GISB shows an overlap of the line of local inflection points y_i and the separation streamline y_d, suggesting the presence of absolute inflectional instability that would lead to a self-sustained vortex shedding. This is aligned with the sudden appearance of intense vortex rollers soon after separation. These vortices present a 3-D breakdown which is only evident slightly downstream from the location where the vortices are observed in the vertical plane, suggesting that it is due to a secondary instability of the vortex rolls.

The PISB presents a different picture. First, the inflection point is clearly above the separation streamline in the time-averaged vertical plane indicating that absolute instability of 2-D waves should not be expected. The vertical plane shows the formation of vortical structures much farther from the separation point than for the GISB.

[2] Only the cases with low Tu are considered here, denominated as GISB-N2 and PISB-N2 in [10].

Interestingly, the spanwise plane shows a periodic spanwise deformation which is visible upstream of the shedding of vortices in the vertical plane. The spanwise planes corresponding to the peaks and troughs in streamwise velocity remain aligned for different time instants, suggesting that they also persist in the time-averaged flow. Overall, the dynamics displayed by the PISB (including the frequency, $St_\theta = 0.012$) are consistent with the scenario described in Sect. 15.2.3 and other experiments in Table 15.1.

15.4 Conclusions and Outlook

This paper summarizes the main findings of the theoretical/numerical work on the self-excited instability of LSBs performed by the author and collaborators during the last decade. It describes two different mechanisms: (i) a global oscillator arising from absolute inflectional instability, and (ii) a global centrifugal instability of the recirculation flow leading to its spanwise modulation. Whether these instability mechanisms are present in a particular LSB, which one is the primary one, and whether they dominate over the scenario governed by the convective amplification of external disturbances would depend on the characteristics of the LSB: Re_θ, λ (Eq. 15.1) and the characterization of the free-stream turbulence. Some experimental results by other groups are also revisited which, in the opinion of the author, present evidences of the self-excited mechanisms discussed. It is acknowledged that other global instability mechanisms have been proposed, and that many experiments and simulations of LSBs present no evidence of self-excited mechanisms.

The intention of this paper is to emphasize that self-excited instability mechanisms exist, and that they can modify and surpass the amplifier scenario, leading to e.g. well-defined vortex shedding frequencies regardless of the environmental conditions or persistent three-dimensionality of the LSB and coherent structures. Perhaps, phenomena like bursting, not fully understood to date, is rooted on the change on the dominant instability. This notion should be kept in mind in future research, at least to avoid discarding such mechanisms without a proper assessment. To conclude, some caveats and suggestions in this direction are given:

- LSBs, at least near reattachment, are always 3-D. The breakdown of spanwise vortices is not always the cause. If possible, visualize/measure the fluctuations at different spanwise locations or on a spanwise plane. Check for spanwise modulations of the time-averaged flow.
- Using time-averaged and spanwise-averaged flows to assess if self-excited instabilities are present based on predictive criteria can be misleading. The realizable flow is the consequence of the instabilities, and the associated flow structures can modify the mean flow hindering the identification of their origin. For instance, the averaged u_{rev} can be substantially smaller than the one obtained at certain instants and spanwise locations.

- Similarly, the criterion for the absolute inflectional instability based on the relative values of y_{in} and y_d must be applied with caution. The time-average of a LSB exhibiting a self-induced vortex shedding may not strictly show the condition $y_{in} < y_d$ due to non-linear mean flow deformation. The close proximity or overlap of these values (as for GISB in [10]) should be considered as an indication of absolute instability.
- The centrifugal instability is not the only possible origin for the 3-D distortion of the LSB leading to the finite-span absolute inflectional instability in Sect. 15.2.3. Under moderate to high intensities of the free-stream turbulence, streaks and Klebanoff modes appear in the pre-separated boundary layer. These structures can penetrate the separated flow region exerting a spanwise modulation of the shear layer [17], which may trigger an absolute instability localized in time and spanwise coordinate. The resulting finite-span vortical structures could be very similar to those triggered by the centrifugal instability, but they would not leave an imprint in the time-averaged flow, on account of their random appearance.

Acknowledgements The author gratefully acknowledges the fruitful discussions and collaboration with all co-authors of the referenced papers, and especially with Elmer M. Gennaro and Matheus P. Avanci. This work is funded by the Government of the Community of Madrid within the multi-annual agreement with Universidad Politécnica de Madrid through the Program of Excellence in Faculty (V-PRICIT line 3) and by the Spanish Ministry of Science (MCIN /AEI/10.13039/501100011033) and the European Union, grants PID2021-125812OB-C22 and TED2021-129719B-C21.

References

1. Gaster, M.: The structure and behaviour of separation bubbles, Tech. Rep. 3595, NPL Rep- & Memoranda (1966)
2. Dovgal, A.V., Kozlov, V.V., Michalke, A.: Laminar boundary layer separation: instability and associated phenomena. Prog. Aerosp. Sci. **3**, 61–94 (1994)
3. Diwan, S.S., Ramesh, O.N.: On the origin of the inflectional instability of a laminar separation bubble. J. Fluid Mech. **629**, 263–298 (2009)
4. Marxen, O., Lang, M., Rist, U., Levin, O., Henningson, D.S.: Mechanisms for spatial steady three-dimensional disturbance growth in a non-parallel and separating boundary layer. J. Fluid Mech. **634**, 165–189 (2009)
5. Watmuff, J.H.: Evolution of a wave packet into vortex loops in a laminar separation bubble. J. Fluid Mech. **397**, 119–169 (1998)
6. Embacher, M., Fasel, H.F.: Direct numerical simulations of laminar separation bubbles: investigation of absolute instability and active flow control of transition to turbulence. J. Fluid Mech. **747**, 141–185 (2014)
7. Michelis, T., Yarusevych, S., Kotsonis, M.: On the origin of spanwise vortex deformations in laminar separation bubbles. J. Fluid Mech. **841**, 81–108 (2018)
8. Kurelek, J.W., Lambert, A.R., Yarusevych, S.: Coherent structures in the transition process of a laminar separation bubble. AIAA J. **54**(8), 2295–2309 (2016)
9. Kurelek, J.W., Kotsonis, M., Yarusevych, S.: Transition in a separation bubble under tonal and broadband acoustic excitation. J. Fluid Mech. **853** (2018)

10. Mohamed Aniffa, S., Caesar, V.S., Dabaria, V., Mandal, A.C.: Characteristics of geometry-and pressure-induced laminar separation bubbles at an enhanced level of free-stream turbulence. J. Fluid Mech. **957**, A19 (2023)
11. Spalart, P.R., Strelets, M.Kh.: Mechanisms of transition and heat transfer in a separation bubble. J. Fluid Mech. **403**, 329–349 (2000)
12. Avanci, M.P., Rodríguez, D., Alves, L.S.B.: A geometrical criterion for absolute instability in separated boundary layers. Phys. Fluids **31**, 014103 (2019)
13. Theofilis, V., Hein, S., Dallmann, C.: On the origins of unsteadiness and three-dimensionality in a laminar separation bubble. Phil. Trans. R. Soc. Lond. A **358**, 3229–3246 (2000)
14. Rodríguez, D., Gennaro, E.M., Juniper, M.P.: The two classes of primary modal instability in laminar separation bubbles. J. Fluid Mech. **734**, R4 (2013)
15. Rodríguez, D., Gennaro, E.M., Souza, L.F.: Self-excited primary and secondary instability of laminar separation bubbles. J. Fluid Mech. **906**, A13 (2021)
16. Alam, M., Sandham, N.D.: Direct numerical simulation of 'short' laminar separation bubbles with turbulent reattachment. J. Fluid Mech. **410**, 1–28 (2000)
17. McAuliffe, B.R., Yaras, M.I.: Transition mechanisms in separation bubbles under low and elevated-freestream turbulence. J. Turbomach. **132**, 011004 (2010)

Chapter 16
Stability Analysis of Recirculation Bubble Flow in a Viscoelastic Fluid

Beatriz L. Carreira, Elmer M. Gennaro, Daniel Rodríguez, and Leandro F. Souza

Abstract This work investigates the instability of recirculation bubbles formed in a non-Newtonian fluid from the Giesekus model using a global eigenmode stability analysis. The objective is to relate the parameters that control the viscoelastic contribution with characteristics of the recirculation bubble that can be decisive in the emergence of instabilities and to investigate the influence of the non-Newtonian contribution on the hydrodynamic stability of these flows.

Keywords Global stability analysis · Viscoelastic fluid · Recirculation bubble · Giesekus model

16.1 Introduction

Recently, the interest in the transport of fluids containing dissolved polymers has grown, motivated by their potential use in reducing viscous losses and delaying transition to turbulence in internal and external flows. Among the main characteristics of polymer solutions, viscoelasticity stands out, which consists of the presence of viscous and elastic rheological properties at the same time. Viscoelastic materials make up an important class of non-Newtonian fluids that, unlike Newtonian fluids, exhibit a non-linear relationship between the deformation rate and the shear stress, including

B. L. Carreira · L. F. Souza (✉)
Universidade de São Paulo, São Carlos, Brazil
e-mail: lefraso@icmc.usp.br

B. L. Carreira
e-mail: liara.carreira@usp.br

E. M. Gennaro
Universidade Estadual Paulista - UNESP, São João da Boa Vista, Brazil
e-mail: elmer.gennaro@unesp.br

D. Rodríguez
School of Aeronautics, Universidad Politécnica de Madrid (ETSIAE-UPM), Madrid, Spain
e-mail: daniel.rodriguez@upm.es

K. Kato et al. (eds.), *Proceedings of the 10th IUTAM Symposium on Laminar-Turbulent Transition*, IUTAM Bookseries 44, https://doi.org/10.1007/978-981-96-9829-5_16

misalignment of the principal directions and time delays. The process of laminar-turbulent transition is extremely complex and is still not fully understood even for Newtonian fluid flows. For non-Newtonian fluids, even the impact of the complex rheological properties on the comparatively simpler linear instability properties has not been systematically investigated. Interesting studies dealing with stability of viscoelastic fluids are [1–4].

In this work, we consider the Giesekus model for viscoelastic fluids, that provides a wide range of applicability in approximating the behaviour of polymer solutions, and develop a global eigenmode analysis which is valid for this kind of fluids. This approach departs from the classic parallel-flow approximation and thus enables the study of complex flow geometries. As a case of application of a non-parallel flow, we study the modal instabilities that appear on nominally two-dimensional flows featuring closed recirculation zones. In particular, we consider the laminar recirculation region formed on a flat-plate boundary layer subject to consecutive deceleration and acceleration of the free-stream velocity.

The modal instabilities present in comparable recirculating flows of Newtonian fluids are today well understood and documented [5, 6], which enables us to identify the differences introduced by the viscoelastic behaviour and its characteristic parameters. Our results show a strong destabilizing effect of the Weissenberg number, that describes the relative influence of the relaxation time of the fluid. However, this destabilization does not alter the nature of the instability nor the spatial structure of the eigenfunction.

16.2 Mathematical Formulation

The flow is assumed to be unsteady, non-Newtonian, bi-dimensional and incompressible. The conservation of mass (continuity) and conservation of momentum equations governing the flow in the dimensionless form are given by

$$\nabla \cdot \mathbf{u} = 0, \tag{16.1}$$

$$\frac{\partial \mathbf{u}}{\partial t} + \nabla \cdot (\mathbf{uu}) = -\nabla p + \frac{\beta_{nn}}{Re} \nabla^2 \mathbf{u} + \nabla \cdot \mathbf{T}, \tag{16.2}$$

where $\mathbf{u} = [u, v]^T$ denotes the velocity field in the directions x and y, respectively, t is the time, p is the pressure and $\mathbf{T}$ is the non-Newtonian extra-stress tensor (symmetric).

The dimensionless viscosity coefficient $\beta_{nn} = \mu_s/\mu_0$ defines the deviation from the solvent's Newtonian behaviour, where μ_s and μ_p are the Newtonian solvent and polymeric viscosities, respectively, and $\mu_0 = \mu_s + \mu_p$. The Reynolds number is defined as $Re = \rho U_\infty L/\mu_0$, where L and U_∞ denote length and velocity scales, respectively, and ρ represents the density of the fluid. An arbitrary length L was adopted in the present study. As shown later, the Reynolds number based on the length of the recirculation bubbles are $Re_{L_{sep}} = ReL_{sep} \approx 10^4$.

In this paper, we worked with viscoelastic fluid flow governed by the non-linear Giesekus constitutive equation [7], which can be given by

$$\mathbf{T} + Wi\left[\frac{\partial \mathbf{T}}{\partial t} + \nabla\cdot(\mathbf{uT}) - \nabla\mathbf{u}\cdot\mathbf{T} - \mathbf{T}\cdot\nabla\mathbf{u}^{\mathsf{T}}\right] + \frac{\alpha_G Re Wi}{(1-\beta_n)}(\mathbf{T}\cdot\mathbf{T}) = \\ = \frac{1-\beta_{nn}}{Re}(\nabla\mathbf{u} + \nabla\mathbf{u}^{\top}), \quad (16.3)$$

where α_G is the mobility parameter that regulates the shear thinning behaviour of the fluid ($0 \leq \alpha_G \leq 1$), $\mathbf{T}\cdot\mathbf{T}$ is a tensor product. The Weissenberg number is denoted by the dimensionless parameter $Wi = \lambda U_\infty/L$, where λ is the relaxation time of the fluid.

16.3 Global Stability Analysis

Linear Stability Theory assumes that the instantaneous flow can be decomposed as $\mathbf{q} = \hat{\mathbf{q}} + \tilde{\mathbf{q}}$, where $\hat{\mathbf{q}}(x, y)$ describes the steady bidimensional baseflow and $\tilde{\mathbf{q}}(x, y, z, t) = (\tilde{u}, \tilde{v}, \tilde{w}, \tilde{p}, \tilde{\mathbf{T}})$ represents the three-dimensional infinitesimal perturbations. Substituting these approximations into Eqs. (16.1)–(16.3), and neglecting the non-linear terms due to their infinitesimal amplitude, a system of linearized equations for the disturbances is obtained. The disturbances can be written generally in the modal form $\tilde{\mathbf{q}}(x, y, z, t) = \overline{\mathbf{q}}(x, y)e^{i(\beta_z z - \omega t)} + c.c.$, where $\overline{\mathbf{q}}$ represents the spatial structure of the disturbances and $c.c.$ stands for the complex conjugate. Within a temporal analysis, the disturbances have complex frequency ω (to be determined) and a prescribed real wavenumber β_z is associated with the wavelength $L_z = 2\pi/\beta_z$.

Substituting the modal form into the system of linearized equations and considering a temporal analysis, the system can be written as a generalized complex eigenvalue problem $\mathcal{L}(\hat{\mathbf{q}}, \beta_z, Re, Wi, \beta_{nn}, \alpha_G)\overline{\mathbf{q}} = -i\omega\mathcal{R}\overline{\mathbf{q}}$, where the complex eigenvalue ω carries information about the stability and the imaginary part ω_i indicates the growth or damping rate of the disturbances.

The eigenvalue problem is solved here using eighth-order finite difference schemes. From the point of view of linear stability analysis, the most important eigenvalues are those close to the $\omega_r = 0$ axis. Since we are only interested in a window of the eigenspectrum, an in-house implementation of the Arnoldi shift-and-invert algorithm is used. More details on the numerical solution of the eigenvalue problem can be found elsewhere [8].

16.4 Base Flows: Recirculation Bubbles in Giesekus Fluids

Steady two-dimensional boundary-layer flows featuring a recirculation bubble are obtained using direct numerical simulations. A distribution of the free-stream velocity

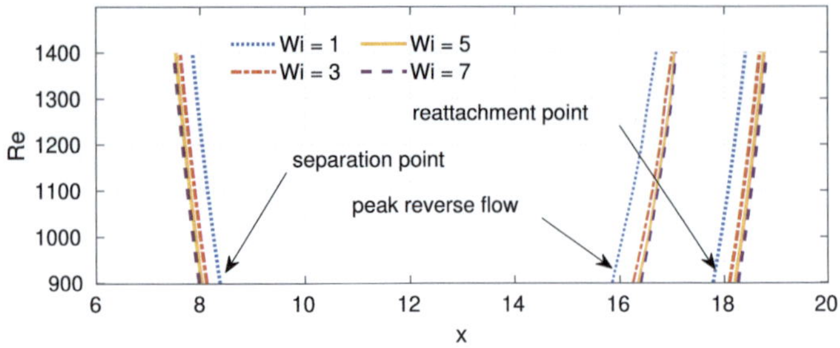

Fig. 16.1 Influence of Weissenberg number in the Bubble characteristics

is prescribed as a boundary condition, that effectively imposes a deceleration and later deceleration of the flow, forcing the boundary layer separation and reattachment [5]. No-slip conditions are imposed at the wall and an analogous to the Falkner-Skan solution consistent the free-stream condition is used as the inlet condition. The high-order solver employed is based on the vorticity-velocity formulation [9], with the viscous terms adopting the Giesekus model. In all the cases considered, the time-marching of the two-dimensional equations recovered asymptotically fully laminar steady solutions. Figure 16.1 shows the influence of Weissenberg Number in the bubble length L_{sep}, showing a moderate reduction with decreasing Wi. This figures also informs a Reynolds number based on the bubble length, $Re_{L_{sep}} = ReL_{sep} = O(10^4)$ for the present flows.

16.5 Results

To generate the baseflows using a high-order numerical simulation code, the domain of numerical integration in the streamwise and normal directions goes from $x_0 = 0.6$ to $x_{max} = 33.2$, and from $y_0 = 0$ to $y_{max} = 3.5$, respectively. The number of points is assumed to be $i_{max} = 1305$ in the streamwise direction and $j_{max} = 193$ in the normal direction. The discretization in the y direction was performed using a mesh with a stretching factor of 1%. However, a less dense mesh is sufficient for the convergence of the stability analysis results, and the baseflow results are interpolated to this new mesh. Here, we consider $Nx = 360$ and $Ny = 190$, where Nx and Ny represent the number of points in the x and y directions, respectively.

In Fig. 16.2, the temporal growth rate ω_i is shown as a function of the spanwise wavenumber β_z for $Re = 1300.5$, and $\beta_{nn} = 0.50$ for different Weissenberg numbers Wi. The eigenmode becomes unstable in a limited range of β_z, reaching its maximum growth rate. This phenomenon is more evident in the cases with the highest values of Wi, demonstrating that elasticity has a destabilizing effect on the separation bubbles. The influence of the Weissenberg number in the separation length L_{sep} and maximum reverse velocity u_{rev} is shown in Table 16.1.

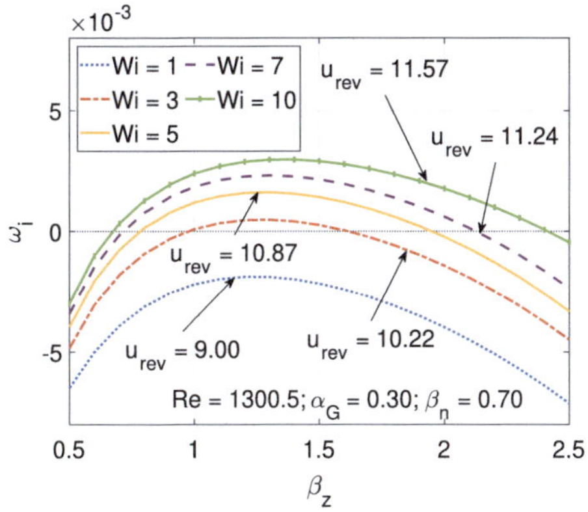

Fig. 16.2 Effect of the Weissenberg number on the growth rate ω_i against spanwise wavenumber β_z for the steady three-dimensional global mode

Table 16.1 Separation length L_{sep} and maximum reverse velocity u_{rev} for each Weissenberg number. $Re = 1300.5$, $\alpha_G = 0.3$, $\beta_{nn} = 0.7$

Wi	L_{sep}	$u_{rev}(\%)$
1	10.36	9.00
3	10.88	10.22
5	11.05	10.87
7	11.14	11.24
10	11.22	11.57

Meanwhile, in Fig. 16.3, the parameters responsible for the viscoelastic behaviour are fixed at $\beta_{nn} = 0.50$ for $Wi = 5$ and 7, and only the Reynolds number varies. It can be seen that for the higher values, the instability range increases, and the highest maximum amplification rates are also reached. The influence of the Reynolds number in the separation length (L_{sep}) and the maximum reverse velocity (u_{rev}) is presented in Table 16.2.

In addition, observing Fig. 16.4, it can be seen that, in a comparison between the different values of Wi, the instability region is larger for $Wi = 7$, that is, when the fluid has greater elasticity. Another fact is that for $Wi = 7$ the flow is already unstable for stable Re values with $Wi = 5$, since the tip of the curve is displaced to the left.

16.6 Conclusions

The results indicate that the non-Newtonian behaviour introduced by the polymer has a destabilizing effect on the recirculating flow, particularly in the centrifugal instability. Increasing values of Wi, related to the elastic fluid behaviour, lead to

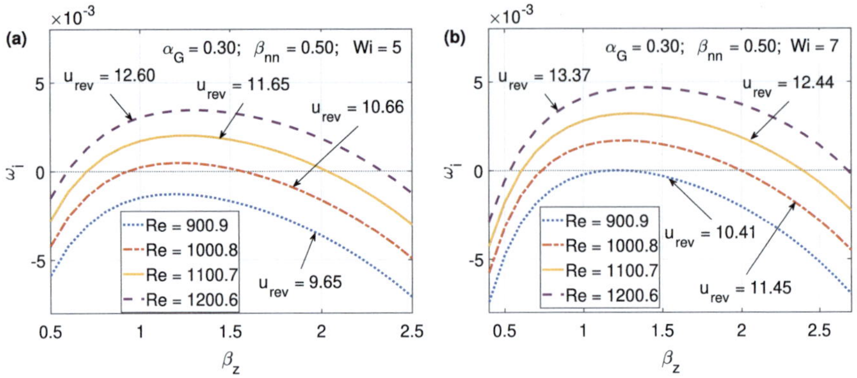

Fig. 16.3 Effect of the Reynolds number on the growth rate ω_i against spanwise wavenumber β_z for the steady three-dimensional global mode for $Wi = 5$ (**a**), and $Wi = 7$ (**b**)

Table 16.2 Separation length L_{sep} and maximum reverse velocity u_{rev} for each Reynolds number. $\alpha_G = 0.3$, $\beta_{nn} = 0.5$

	$Wi = 5$		$Wi = 7$	
Re	L_{sep}	$u_{rev}(\%)$	L_{sep}	$u_{rev}(\%)$
900.9	11.38	9.65	11.59	10.41
1000.8	11.65	10.66	11.84	11.45
1100.7	11.88	11.65	12.07	12.44
1200.6	12.10	12.60	12.28	13.37

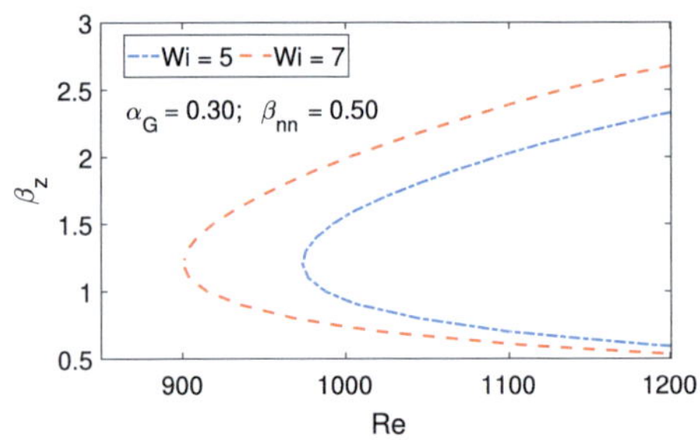

Fig. 16.4 Neutral instability ($\omega_i = 0$) curves for the three-dimensional instability, as a function of the spanwise wavenumber and Re

instability at smaller values of the solvent Re, without altering the preferential spanwise wavenumber β_z or the velocity components of the eigenfunction.

Acknowledgements The authors acknowledge the National Council for Scientific and Technological Development (CNPq) or the research grant received during the development of this study. The research was carried out using the computational resource of the Center for Mathematical Applied to Industry (CeMEAI), funded by FAPESP (Process Number 2013/07375-0). D.R.'s work is funded by the Community of Madrid within the multi-annual agreement with Universidad Politécnica de Madrid through the Program of Excellence in Faculty (V-PRICIT line 3).

References

1. Shokri, H., Kayhani, M.H., Notouzi, M.: Nonlinear simulation and linear stability analysis of viscous fingering instability of viscoelastic liquids. Phys. Fluids **29**(3) (2017). https://doi.org/10.1063/1.4977443
2. Sureshkumar, R., Beris, A.N.: Linear stability analysis of viscoelastic Poiseuille flow using an Arnoldi-based orthogonalization algorithm. J. Non-Newt. Fluid Mech. **56**(2), 151–182 (1995). https://doi.org/10.1016/0377-0257(94)01279-Q
3. Zhang, M., Lashgari, I., Zaki, T.A., Brandt, L.: Linear stability analysis of channel flow of viscoelastic Oldroyd-B and FENE-P fluids. J. Fluid Mech. **737**, 249–279 (2013). https://doi.org/10.1017/jfm.2013.572
4. Wilson, H., Sircar, S., Shukla, P.: Spatiotemporal Linear Stability Analyses. Viscoelastic Subdiffusive Flows: Theory and Computation, pp. 29–65 (2025). https://doi.org/10.1007/978-981-96-0135-6_3
5. Gennaro, E.M., Souza, B.D.P., Rodríguez, D.: The effect of compressibility on the primary global instability of unforced laminar separation bubbles. J. Brazilian Soc. Mech. Sci. Eng. **41**(559) (2019). https://doi.org/10.1007/s40430-019-2063-6
6. Rodríguez, D., Gennaro, E.M., Souza, L.F.: Self-excited primary and secondary instability of laminar separation bubbles. J. Fluid Mech. **906**, A13 (2021). https://doi.org/10.1017/jfm.2020.767
7. Giesekus, H.: A simple constitutive equation for polymer fluids based on the concept of deformation-dependent tensorial mobility. J. Non-Newt. Fluid Mech. **11**, 69–109 (1982). https://doi.org/10.1016/0377-0257(82)85016-7
8. Gennaro, E.M., Rodríguez, D., Medeiros, M.A.F., Theofilis, V.: Sparse techniques in global flow instability with application to compressible leading-edge flow. AIAA J. **51**(9), 2295–2303 (2013). https://doi.org/10.2514/1.J051816
9. Brandi, A.C., Mendonça, M.T., Souza, L.F.: DNS and LST stability analysis of Oldroyd-B fluid in a flow between two parallel plates. J. Non-Newt. Fluid Mech. **267**, 14–27 (2019). https://doi.org/10.1016/j.jnnfm.2019.03.003

Chapter 17
Linear and Nonlinear Response of High-Speed Boundary Layers to Continuous Stochastic Forcing

Neil D. Sandham, Pushpender K. Sharma, and David J. Lusher

Abstract Numerical simulations are used to show the response of a shock-induced separation bubble to stochastic forcing. The forcing is applied in space and time such that it excites all possible response modes for a given computational domain. It is shown how oblique-mode shear layer instability is captured as well as low-frequency breathing of a shock-induced separation bubble. In addition, both Mack and radiating acoustic modes are identified. A study of the effect of amplitude shows that the low-frequency breathing changes from a linear to a quadratically nonlinear response as the amplitude is increased.

Keywords Shock waves · Boundary layer · Separation bubble

17.1 Introduction and Methodology

The interaction of shock waves and boundary layers is a complex problem with practical applications in the context of supersonic engine intakes and control surface deflections. Fully-turbulent interactions are common, but most of the phenomena can be found already in laminar and transitional flow interactions. One phenomenon of interest is a large-amplitude low-frequency motion of the separation point that emerges when the shock is strong enough to create a large separation bubble. Recent work [3] on laminar interactions has provided an interpretation of the low-frequency as the response of a damped linear global mode to forcing. From a related input-output analysis, this can also be identified as a linear resolvent mode with high gain. Previous studies [1, 2] have shown how transition to turbulence develops from the breakdown of oblique Kelvin-Helmholtz (K-H) disturbances that grow in the detached shear layer over the separated flow region and how broadband disturbances

N. D. Sandham (✉) · P. K. Sharma
University of Southampton, Southampton, UK
e-mail: n.sandham@soton.ac.uk

D. J. Lusher
Japan Aerospace Exploration Agency (JAXA), Chofu Aerospace Center, Tokyo, Japan

K. Kato et al. (eds.), *Proceedings of the 10th IUTAM Symposium on Laminar-Turbulent Transition*, IUTAM Bookseries 44, https://doi.org/10.1007/978-981-96-9829-5_17

arising from the breakdown to turbulence drive a weak low-frequency response of the bubble. This response was shown in [1] to be associated with a quadratically nonlinear process.

In the present contribution we consider the reflected shock problem from the point of view of stochastic forcing. Rather than applying forcing as an inflow or wall boundary condition, we apply it continuously throughout space and time as

$$\frac{\partial \mathbf{Q}}{\partial t} = \mathbf{R} + A\mathbf{F}. \tag{17.1}$$

Here $\mathbf{Q}$ is the vector of conservative variables, $\mathbf{R}$ is the residual (containing all the other terms in the compressible Navier-Stokes equations), A is an amplitude and $\mathbf{F}$ is a uniformly distributed random forcing vector in the range $[-1, 1]$. This mimics the method of resolvent mode analysis and, for small amplitude forcing, the linear response of the separation bubble is expected to be recovered by Fourier analysis of the time-dependent response. The methodology is implemented in the open-source OpenSBLI code generation software [4]. The time-domain methodology also permits nonlinear analysis, which we exploit to show the key differences between the shear layer response and the low-frequency breathing mode.

The chosen configuration is a Mach number $M = 2$ interaction with an impinging shock angle of 32.58° that is one of the applications provided in [4]. Inflow freestream quantities are used for normalisation, with lengths scaled with the inflow boundary layer displacement thickness. Here we use an inflow Reynolds number $Re = 950$, with a computational domain $\left[L_x, L_y, L_z\right] = [400, 115, 60]$ and grid $1000 \times 250 \times 100$ (uniform in x and z, stretched in y). In the cases described in this paper the random forcing is only applied to the density field and a WENO-5 scheme is used for the spatial discretisation.

17.2 Results

Figure 17.1 shows a 3D view of a typical response to the random forced flow for amplitude $A = 0.001$. The grey scale contours on the $z = 0$ plane show the incoming shock wave, generated by imposing the shock jump relations on the upper boundary, as well as the reflected compression waves from the separation region, the expansion fan from the apex of the bubble and further compression waves from the reattachment region. The colour contours show the spanwise (ρw) component of momentum and illustrate the development of oblique waves with both positive and negative wave angles relative to the x axis. This is due to the preferential spatial linear amplification of oblique modes in the separated shear layer that encloses the bubble [2]. The effect of forcing amplitude is shown in Fig. 17.2. A laminar response is observed for all A up to $A = 0.01$, where the skin friction rises to a slightly higher value, indicating weak nonlinear effects. At the highest amplitude ($A = 0.1$) a turbulent reattachment is observed and the bubble is significantly shortened. The fact that the stochastically

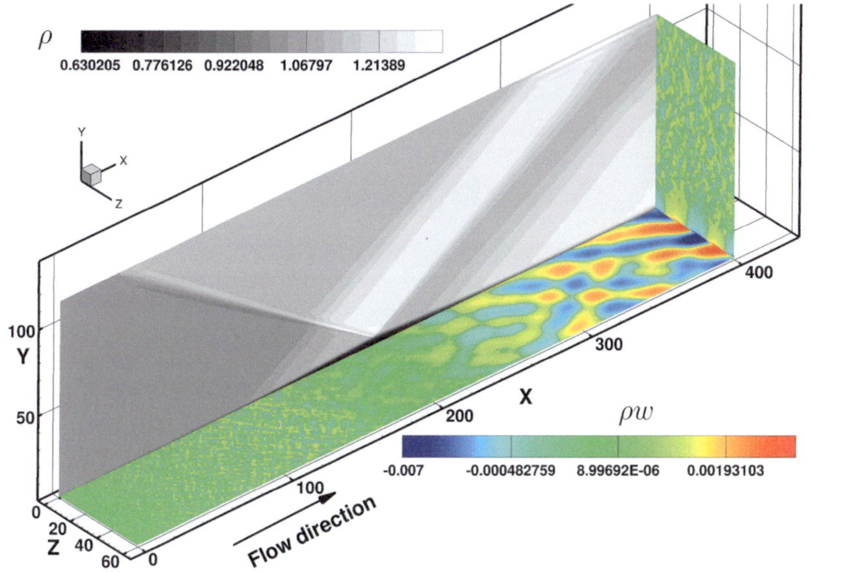

Fig. 17.1 Response to forcing with amplitude $A = 0.001$, showing a side view of ρ contours at $z = 0$ and ρw contours on a wall normal plane at $y = 2$ and on the outflow plane at $x = 400$

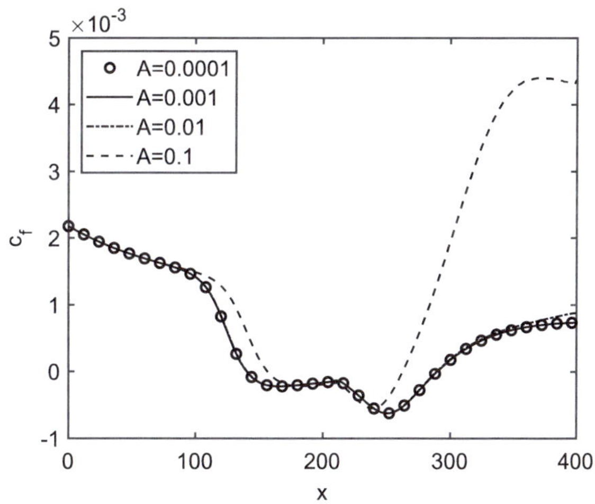

Fig. 17.2 Skin friction coefficient for various forcing amplitudes A

forced cases at low amplitude maintain laminar flow confirms that this configuration is globally stable with respect to both 2D and 3D disturbances.

The form of the response can be analysed by storing time series of the wall pressure and Fourier transforming in time. Figure 17.3 shows contours of the power spectral density (PSD) of wall pressure as a function of dimensionless frequency f and x for the case with $A = 0.01$, which reveals all the features to be discussed.

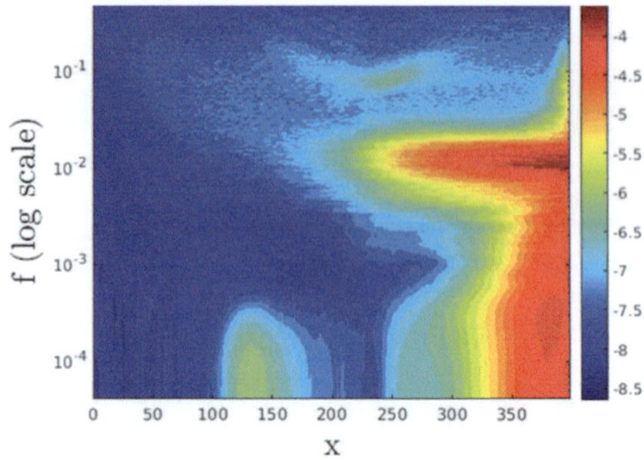

Fig. 17.3 Power spectral density contour plot (on axes of frequency f versus streamwise x location) for $A = 0.01$, based on a long time series of wall pressure fluctuations

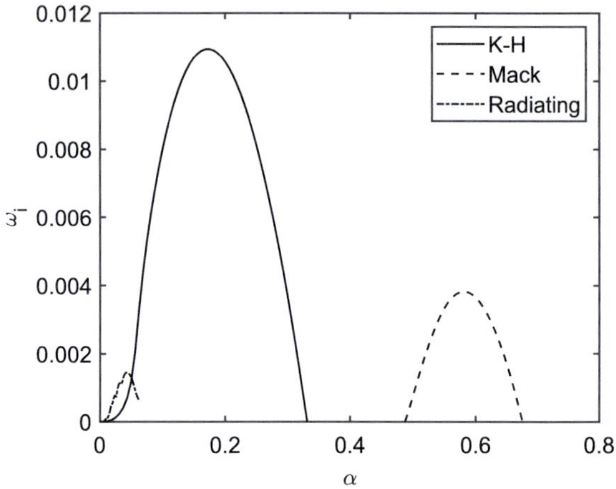

Fig. 17.4 Growth rate as a function of wavenumber, from temporal linear stability theory at $x = 208$

The region with the highest amplitude is for $f \approx 0.015$ which corresponds to the oblique mode instability. The interaction of oblique modes of both signs leads to a rapid broadening of the spectrum after reattachment ($x = 289$), consistent with the start of the process of breakdown to turbulence. Near the separation point ($x = 140$) we see the characteristic low frequency response, which according to [3] would be consistent with the presence of a stable stationary 2D global mode. A similar low-frequency response is seen at reattachment, which is coupled via the global mode to form a 'breathing' behaviour of the bubble. Other more subtle features of the PSD include a distinct region at higher frequencies $f \sim 0.08$, peaking near $x = 240$ and a slight distortion of contours for $f \approx 0.002$.

To understand the origin of some of the features in the PSD, a parallel-flow linear stability analysis was conducted at $x = 208$, which cuts through the growth phase of three of the features noted in the PSD. The results of the temporal analysis are shown on Fig. 17.4 which shows the temporal growth rate ω_i as a function of streamwise wave number α for spanwise wavenumber $\beta = 0$. Two large peaks can be seen, one

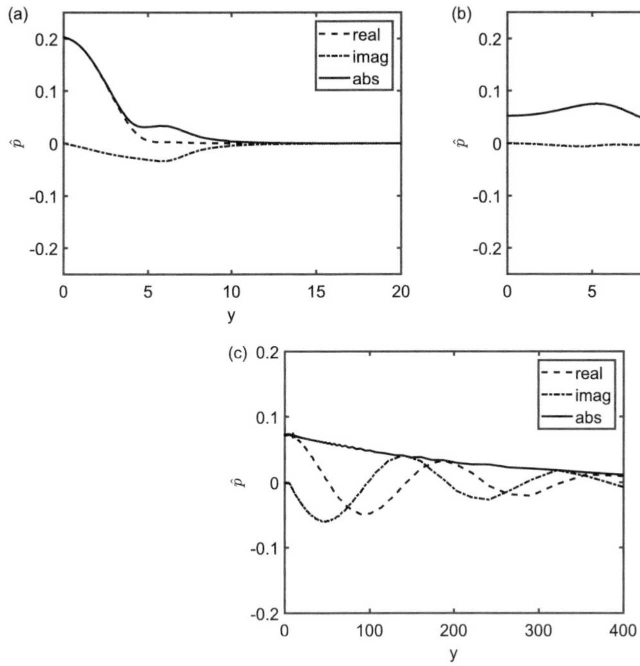

Fig. 17.5 Eigenvectors of wall pressure from LST **a** Mack mode ($\beta = 0$), **b** K-H mode ($\beta = 0.2$), and **c** radiating mode ($\beta = 0$)

at $\alpha = 0.16$, corresponding (via the real part of the frequency at the peak of ω_i) to $f = 0.014$ and one at $\alpha = 0.58$ corresponding to $f = 0.07$. The other feature that can be noted is at very low wavenumbers where another mode, corresponding to $f = 0.0017$, is present. All these features are in close agreement with features in the PSD, confirming that the stochastic forcing is exciting convectively unstable modes. To examine the mode structure in more detail, Fig. 17.5 shows pressure eigenvectors (real, imaginary and absolute parts). Figure 17.5a, b show respectively the high- and medium-frequency modes. The highest frequency mode has peaks at the wall and in the shear layer. Its high phase speed ($c_{\text{ph}} = 0.76$ and peak for $\beta = 0$ suggests that this is a 'Mack' mode, sustained by pressure reflections between the wall and the separated shear layer. The medium frequency mode is the oblique K-H mode, with peak growth rate at $\beta = 0.2$ as shown in Fig. 17.5b. This mode peaks in the shear layer, but also has an influence on the wall pressure fluctuations. The low-frequency pressure eigenfunction is shown in Fig. 17.5c for $\alpha = 0.04375$, here over a much larger range of y. The structure here corresponds to a radiating wave. All three of these linear instability modes resemble those seen in free mixing layers. For high convective Mach numbers the main instability corresponds to oblique K-H waves, with two acoustic unstable waves, one radiating in either direction away from the

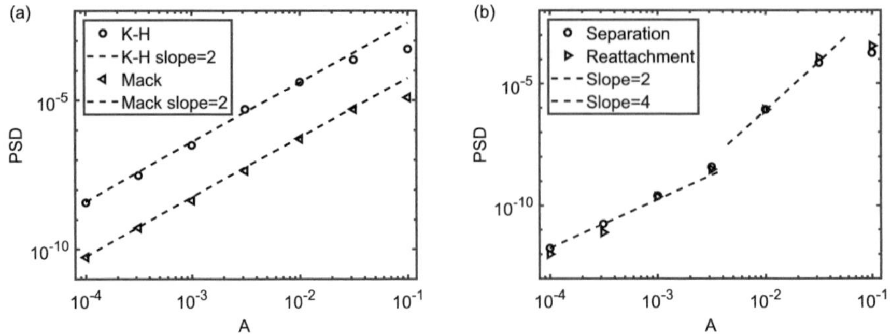

Fig. 17.6 PSD variation with forcing amplitude **a** K-H mode at $x = 300$ and Mack mode at $x = 248$, **b** low-frequency modes at separation and reattachment. Dashed lines show slopes of 2 for linear and 4 for quadratic nonlinear behaviour

mixing layer. In the present context the one radiating downward reflects from the wall and is then known as a Mack mode.

The effect of forcing amplitude on the various features in the PSD is shown in Fig. 17.6. Part (a) shows the oblique K-H and Mack modes, both of which show a linear response behaviour (a slope of 2 since PSD is a squared quantity). In contrast, in Fig. 17.6b the PSD at both $x = 140$ (separation) and $x = 300$ (just after reattachment) shows linear behaviour at low amplitudes, consistent with linear resolvent theory [3], and quadratic nonlinear behaviour (slope 4) at higher amplitudes, consistent with [1]. All the modes show saturation behaviour at the highest amplitudes.

17.3 Conclusions

Stochastic forced simulations of a laminar separation bubble have been used to clarify the various modes that are present. The methodology is easily implemented and clearly reveals the principal oblique-mode convective instability, as well as both Mack and radiating modes that are not usually reported for this configuration. The separation bubble breathing mode is revealed using the same methodology, with a linear response at low amplitude forcing consistent with the known linear resolvent mode. At higher amplitudes a shift to a quadratic nonlinear response is confirmed, motivating further work to identify the nonlinear mechanisms involved.

Acknowledgements NDS and PKS acknowledge support of EPSRC grant EP/W026686/1. DJL was supported by the Japan Society for the Promotion of Science (JSPS) under grant 22F22059.
Disclosure of Interests The authors have no competing interests to declare that are relevant to the content of this article.

References

1. Sansica, A., Sandham, N.D., Hu, Z.: Forced response of a laminar shock-induced separation bubble. Phys. Fluids **26**, 093601 (2014). https://doi.org/10.1063/1.4894427
2. Sansica, A., Sandham, N.D., Hu, Z.: Instability and low-frequency unsteadiness in a laminar shock-induced separation bubble. J. Fluid Mech. **798**, 5–26 (2016). https://doi.org/10.1017/jfm.2016.297
3. Bugeat, A., Robinet, J.-Ch., Chassaing, J.C., Sagaut, P.: Low-frequency resolvent analysis of the laminar oblique shock wave/boundary layer interaction. J. Fluid Mech. **942**, A43 (2022). https://doi.org/10.1017/jfm.2022.390
4. Lusher, D.J., Sansica, A., Sandham, N.D., Meng, J., Siklósi, B., Hashimoto, A.: OpenSBLI v3.0: high-fidelity multi-block transonic aerofoil CFD simulations using domain specific languages on GPUs. Comput. Phys. Commun. **109046** (2025). https://doi.org/10.1016/j.cpc.2024.109406

Chapter 18
Experimental Investigation of Bluntness Effects on Transition on a 7-Degree Blunted Cone

Adam Skora and Stuart Craig

Abstract An experimental investigation into transition on a 7-degree blunted cone was conducted at the University of Arizona. Evidence that the primary instability was suppressed as blunting increased was detected. Additionally, as blunting increased, a subsequent strengthening of energy near the boundary layer edge was detected, and a potential entropy layer type instability was detected using spectral proper orthogonal decomposition.

Nomenclature

M	Mach number
$q_{\inf}$	Dynamic pressure
Q_w	Heat flux
R_{tip}	Tip radius
Re'	Freestream Reynolds number
Re_ϵ	Reynolds number based on streamwise coordinate
Re_{tip}	Reynolds number based on tip radius
S_{pp}	Power spectrum
T_o	Stagnation temperature

18.1 Introduction

Prediction of transition location is of critical importance to the design of hypersonic aircraft due to the effects on heat transfer rates and viscous drag. Realistic flight geometries use blunted leading edges in order to further reduce heat transfer, complicating the transition process since the primary instability may no longer be the dominant instability. Stetson [1] conducted an initial study of the effects of bluntness

A. Skora (✉) · S. Craig
Aerospace and Mechanical Engineering, University of Arizona, Tucson, AZ, USA
e-mail: adamskora@arizona.edu

S. Craig
e-mail: sacraig@arizona.edu

K. Kato et al. (eds.), *Proceedings of the 10th IUTAM Symposium on Laminar-Turbulent Transition*, IUTAM Bookseries 44, https://doi.org/10.1007/978-981-96-9829-5_18

on transition, and concluded that entropy layer swallowing distance was a strong predictor of whether transition reversal would occur. Stetson et al. [2] concluded that for small bluntness, the boundary layer was stabilized by the addition of the favorable pressure gradient, but, for the large bluntness cases, the strengthening of the entropy layer was causing early transition.

Paredes et al. [3] attempted to numerically simulate nonlinear mechanisms to investigate interactions between instabilities in the entropy layer and primary instabilities on a cone, however he was unable to conclusively show how transition could be caused by entropy layer instabilities under realistic conditions. Hartman et al. [4] used direct numerical simulations on a blunt 7 degree cone and found that the entropy layer instability could cause transition through rapid nonlinear growth in the nosetip region. While the exact mechanism by which the entropy layer instability can influence boundary layer stability is starting to be understood, experimental validation of interactions between the boundary layer and the entropy layer is scarce and largely inconclusive.

18.2 Methods

18.2.1 Mach 5 Ludwieg Tube (LT5)

Experiments on the 7-degree blunted cone were conducted in the Mach 5 Ludwieg Tube (LT5) in the Boundary Layer Stability and Transition (BLST) Laboratory at the University of Arizona (UA). LT5 is capable of a range of Re' from 6×10^6 to 22×10^6 m^{-1}). Flow noise levels for LT5 varies between 0.9–1.6% depending on free-stream conditions. A full discussion of the flow quality in LT5 is provided by Bearden et al. [5]. For the experiments listed in this paper, the run conditions are shown in Table 18.1. There are two sets of runs investigated in later sections: a set where unit Reynolds number is kept constant relative to the sharp case and a second

Table 18.1 The experimental conditions run in UArizona LT5 with the 7-degree cone

Run conditions			
Re' [$\times 10^6$ m^{-1}]	T_0 [K]	M	Tip radius [mm]
10.8	415	4.82	.1
11.8	415	4.82	1
10.8	415	4.82	5
10.8	415	4.82	10
13.9	415	4.82	1
13.3	415	4.82	5
16.1	415	4.82	10

set of cases where the flow was kept transitional near the optical fields of view was conducted, hereby referred to as the transitional cases.

Two optical ports on the sides were used for the z-type schlieren measurements, and one of the top ports was used for infrared measurements. This allowed the same region of flow over the top of the cone model to be monitored by all diagnostics.

18.2.2 Diagnostics

The study utilizes a 7-degree modular cone model with four interchangeable tips: “sharp” ($r \approx 90\,\mu m$), 1, 5, and 10 mm. Surface pressure measurements were provided by 7 PCB 132B38 transducers distributed at a regular interval along the surface of the cone. While 11 PCB 132B38’s can be installed, it was found in previous studies that the initial 4 sensors are far upstream of the transition location and were therefore not instrumented to decrease model complexity.

A FLIR X8500sc mid-wave IR camera was employed to measure the surface temperature and calculate heat flux. Additionally, z-type schlieren was employed to provide off-body measurements of the boundary layer. The schlieren system employs a Photron SA-Z camera capturing frames at 300 kHz. A 7.2 W SugarCube Illuminator LED is the source of illumination with an aperture size of 200 microns, meaning the frame rate was limited by the intensity of light. While a smaller aperture could be used to further increase optical resolution, the integration time would have to be sufficiently increased that the framerate would have to be decreased and the image would smear in the flow direction, decreasing resolution more from smearing than the resolution gained from a smaller aperture.

18.3 Results

18.3.1 Pressure Data

The effect of bluntness on stability is first investigated by comparing the power spectra of various bluntness cases with similar freestream Reynolds number. The spectra were normalized by dynamic pressure. As expected, Fig. 18.1 shows how increasing bluntness monotonically stabilizes the boundary layer due to the addition of a favorable pressure gradient.

For the sharp and 1 mm cases, the flow exhibits a broadband excitation over the range expected to be unstable for first mode from linear stability theory (10–100 kHz). Transition for these cases appears to be dominated by broadband excitation caused by the primary instability. As bluntness increases, a series of 20, 33, 55, 70, and 83 kHz sharp peaks begin to appear.

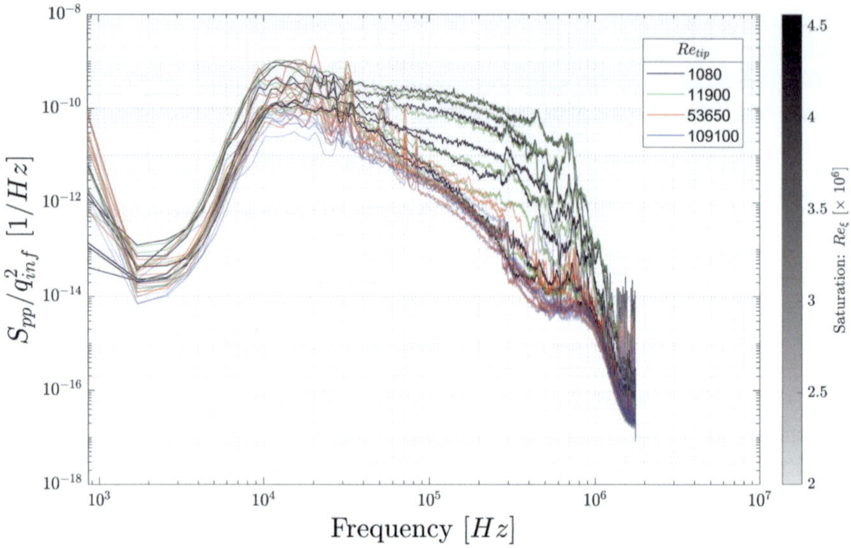

Fig. 18.1 Pressure spectra showing the effect of bluntness on stability. Increased opacity of spectral lines refers to a larger Re_ϵ as indicated in the grayscale color bar

For brevity, only bicoherence estimates from the transitional 5 mm case are shown in Fig. 18.2. This case shows strong evidence of nonlinear interactions far upstream of transition. These interactions disappear at downstream stations, and then the nonlinear interactions reappear at similar frequencies. These results are consistent with predictions by Hartman et al. [4] that entropy layer instabilities could grow rapidly and interact nonlinearly near the nosetip region. A possible explanation for the second set of nonlinear interactions is that if a large enough amplitude to cause transition was not reached in the nosestip region, the instability would decay but could still persist, and may be exciting the first mode instability near transition.

18.3.2 Heat Transfer

Typically, heat transfer can be calculated using infrared thermography to track surface temperatures in conjunction with a numerical scheme to solve the 1d heat conduction equation [6]. Rigid 10k is partially translucent in infrared wavelengths. To correct for this, a numerical correction scheme developed by Kinsey et al. [7] was used in order to integrate the subsurface radiance and then attribute any change in radiance to only the surface in order to estimate the true surface temperature rise. Notably, this assumes that the subsurface is at a constant temperature prior to the run, an assumption which should be valid since the model was allowed to reach thermal equilibrium before each run. Once the surface temperatures are estimated, the numerical formulation of

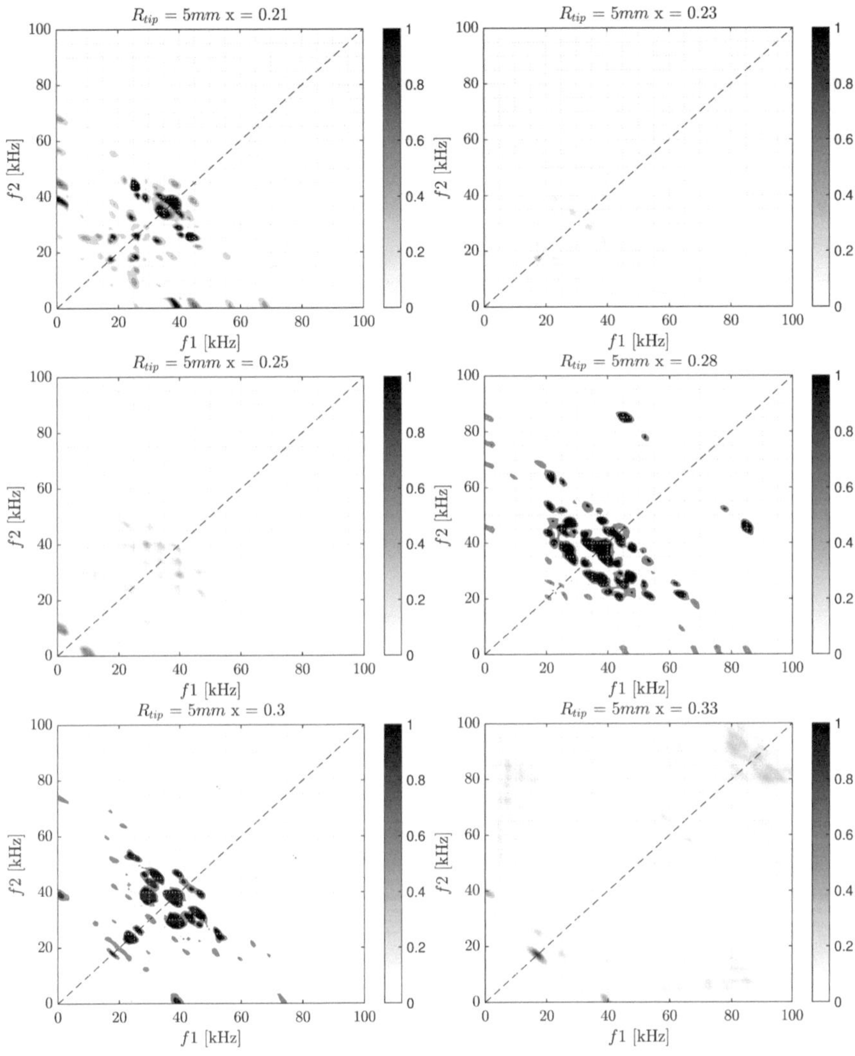

Fig. 18.2 Bicoherence estimates of the 5 mm pressure sensor data

the 1-d heat transfer equation used by Boyd and Howell [6] is then applied with the assumption of a semi-infinite wall. The semi-infinite wall assumption is valid since the thickness of the thermal penetration depth for Rigid 10K is significantly smaller than the wall thickness of the model. Heat transfer traces from expected transitional runs are shown in Fig. 18.3.

While most traces show the expected increase of heating rates, the blunter cases do not reach a steady turbulent value and the 10 mm case only appears to begin transition at the end of the frustum. This is because, at the time of the campaign, the

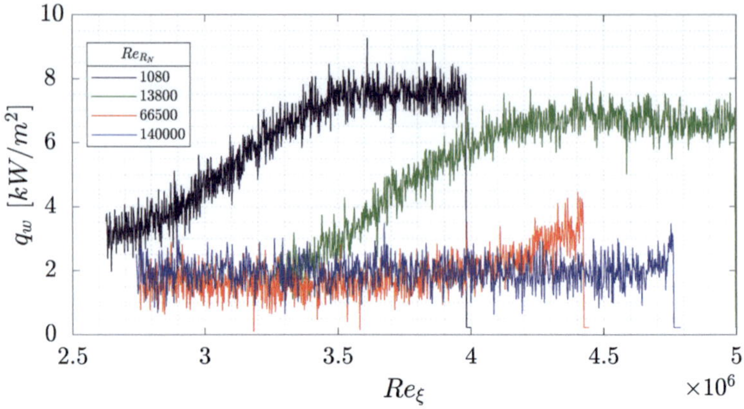

Fig. 18.3 Centerline strip of quantitative heat flux from the transitional cases

heat flux correction code by Kinsey et al. [7] had not been developed. As such, the determination of transitional or turbulent flow during the campaign was determined entirely by the slope of the PCB spectra.

18.3.3 Schlieren

Spectral proper orthogonal decomposition (SPOD) was performed on transitional cases using code from Towne et al. [8] on the schlieren in order to better resolve the shapes and strength of the instability. For brevity, only the transitional sharp and blunter two cases are shown in Fig. 18.4. The frequency of interest for this paper is fixed at 55 kHz because of the peaks in the pressure spectra in blunter cases that were less prevalent in the sharper cases. Figure 18.4 shows a strong weakening of the modal instability as tip Reynolds number increases. Arnal and Délery [9] have shown that increased Re' may result in stabilization of the modal instability due to the increase in tip Reynolds number. A second hypothesis Arnal and Délery [9] proposed was that delay in transition associated with increasing Re' was due to a decrease in tunnel noise and subsequent shift of the tunnel noise to higher frequencies due to the tunnel wall boundary layer. It should be noted that the previous section on the pressure spectra showed that the blunt cases were laminar at similar Re' to the transitional sharp case. Thus the main cause of stabilization shown in Fig. 18.4 is due to the increased bluntness.

Despite the increased stability of the boundary layer for the blunter cases, a new structure that rides the edge of the boundary layer appears in Fig. 18.4b. This mode shape resembles entropy layer type instabilities shown in Berger and Borg [10] and Fedorov and Tumin [11]. Notably, unlike Berger and Borg [10], the modal instability at these conditions is unstable at similar frequencies to the entropy layer instability.

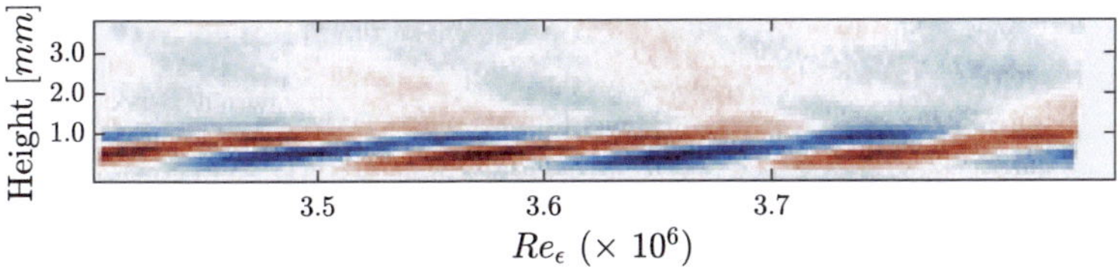

(a) Schlieren frame showing the 55 kHz mode shape for the nominally sharp case.

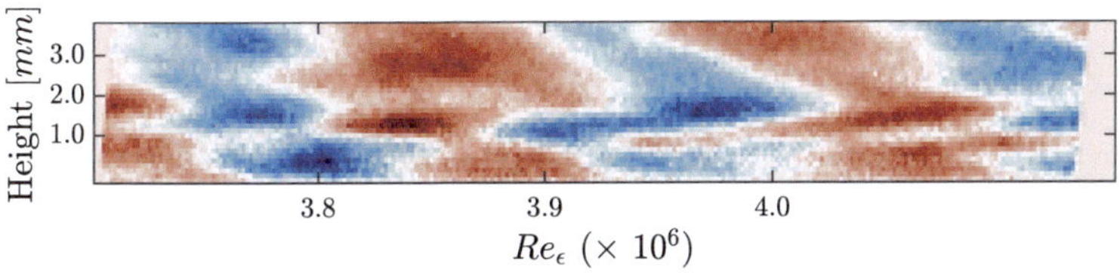

(b) Schlieren frame showing the 55 kHz mode shape for the 5 mm case.

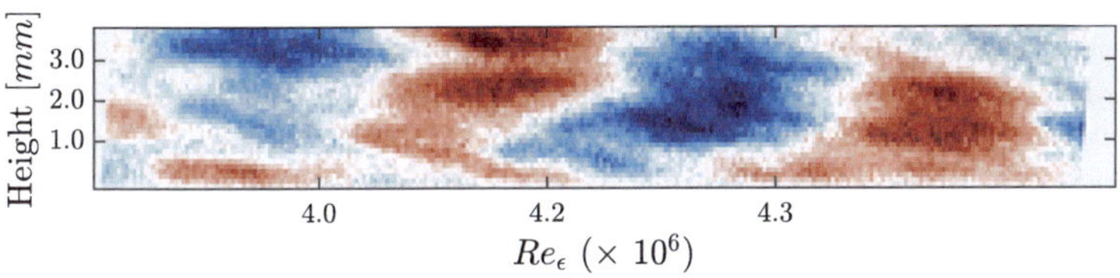

(c) Schlieren frame showing the 55 kHz mode shape for the 10 mm case.

Fig. 18.4 SPOD mode shapes for 55 kHz for a transitional run demonstrating the weakening of the modal instability and an increase in energy above the boundary layer edge as blunting increases

Figure 18.4b seems to show a superposition of a potential modal instability near the wall and an entropy layer instability near the boundary layer edge, indicating that both instabilities have similar energy levels. The presence of the additional instability at the edge of the boundary layer seems to corroborate the trends from pressure bicoherence that an entropy layer instability is persisting until transition. This structure seems to disappear for the 10 mm case, potentially due to further stabilization of the boundary layer as shown in heat flux traces in Fig. 18.3. An alternative explanation is that the entropy layer is much thicker for the 10 mm case, and the resultant entropy layer instability could be dominating the entire field of view.

18.4 Conclusion

An investigation into the effects of bluntness on stability was conducted at UA. Heat transfer was used to confirm transition location, and pressure spectra along with SPOD demonstrated the expected stabilization of the boundary layer as bluntness increased. Bicoherence of pressure data indicated potential evidence of entropy layer type instabilities that decayed near the nosetip region, and then may have grown near transition. SPOD showed evidence of a second instability for the 5 mm case, potentially indicative of a potential entropy layer type instability. As such, this paper shows evidence that an entropy layer type instability was detected, and that it may have contributed to transition to turbulence in some cases.

References

1. Stetson, K.F.: Nosetip bluntness effects on cone frustum boundary layer transition in hypersonic flow. In: 16th AIAA Fluid and Plasmadynamics Conference. AIAA 1983-1763, Air Force Wright Laboratories (1983). https://doi.org/10.2514/6.1983-1763
2. Stetson, K.F., Thompson, E.R., Donaldson, J.C., Siler, L.G.: Laminar boundary layer stability experiments on a cone at mach 8, part 2: Blunt cone. In: 22nd Aerospace Sciences Meeting. AIAA 1984-0006, Air Force Wright Laboratories (1984). https://doi.org/10.2514/6.1984-6
3. Paredes, P., Choudhari, M.M., Li, F.: Mechanism for frustum transition over blunt cones at hypersonic speeds. J. Fluid Mech. **894**, 22 (2020). https://doi.org/10.1017/jfm.2020.261
4. Hartman, A., Hader, C., Fasel, H.F.: Numerical investigation of nonlinear entropy-layer instability waves for hypersonic boundary-layers. In: AIAA Aviation 2020, vol. 1. AIAA 2020-3085, Reston, Virginia (2020). https://doi.org/10.2514/6.2020-3085
5. Bearden, K.P., Padilla, V.E., Taubert, L., Craig, S.A.: Calibration and performance characterizationof a mach 5 ludwieg tube. Rev. Sci. Instrum. **93**, 85104 (2022). https://doi.org/10.1063/5.0093052
6. Boyd, C.F., Howell, A.: Numerical investigation of one-dimensional heat-flux calculations. Technical report, NSWC Dahlgren TR-04/114 (October 1994)
7. Kinsey, B., Skora, A., Mokler, M., Wheeler, M., Dahl, K., Craig, S.A., Portoni, P., Basore, K., Strauss, T., Dean, T., Arndt, A., Wernz, S.: Measurement of heat flux on a 3d printed hemisphere with ir thermography. In: AIAA Aviation 2024. AIAA 2024-4284, Las Vegas, NV (2024). https://doi.org/10.2514/6.2024-4284
8. Towne, A., Schmidt, O.T., Colonius, T.: Spectral proper orthogonal decomposition and its relationship to dynamic mode decomposition and resolvent analysis. J. Fluid Mech. **847**, 821–867 (2018). https://doi.org/10.1017/jfm.2018.283
9. Arnal, D., Délery, J.: Laminar-turbulent transition and shock wave/boundary layer interaction. NATO Research and Technology Organization (2004)
10. Berger, A.R., Borg, M.P.: Study of bluntness-induced elongated structures in hypersonic flow over a 7 degree circular cone. In: AIAA SCITECH 2023 Forum. American Institute of Aeronautics and Astronautics, National Harbor, MD (2023). https://doi.org/10.2514/6.2023-0290
11. Fedorov, A., Tumin, A.: Evolution of disturbances in entropy layer on blunted plate in supersonic flow. AIAA J. **42**, 89–94 (2004). https://doi.org/10.2514/1.9033

Chapter 19
Simulation of K-Type and H-Type Transition Using the Nonlinear One-Way Navier-Stokes Approach

Michael K. Sleeman, Matthew T. Lakebrink, and Tim Colonius

Abstract In principle, transition to turbulence can be studied using direct numerical simulation (DNS) and large eddy simulation (LES), but these approaches are limited by their large computational cost. The Nonlinear One-Way Navier-Stokes (NOWNS) equations have recently been developed and applied to study the early stages of boundary layer transition, and it was demonstrated that they can accurately replicate DNS results with similar accuracy to the nonlinear parabolized stability equations (NPSE) [15]. While having greater computational cost than NPSE, NOWNS is a more robust, convergent parabolization of the governing equations and succeeds for stronger nonlinearity where NPSE fails. In this work, we use NOWNS to reproduce (to the extent possible) the H- and K-type transition scenarios simulated using DNS by Sayadi et al. [14].

Keywords Simulation · Boundary layers · Laminar-turbulent transition

19.1 Methodology

The OWNS procedure was first developed as a method for constructing well-posed one-way approximations of linear hyperbolic systems with slowly-varying coefficients in the direction of integration [16], allowing the equations to be solved as a spatial-initial-value problem, leading to a reduced computational cost. The framework has also been applied to the Navier-Stokes equations, linearized about a slowly-varying baseflow, to perform linear stability analysis of jets and boundary layer flows [7–9, 11, 16, 17]. More recently, the OWNS procedure has been extended to support nonlinear effects [15]. We briefly summarize the NOWNS methodology, and refer to Sleeman et al. for the details [15].

M. K. Sleeman (✉) · T. Colonius
Department of Mechanical and Civil Engineering, California Institute of Technology, Pasadena, USA
e-mail: msleeman@caltech.edu

M. T. Lakebrink
Boeing Research & Technology, Hazelwood, MO, USA

K. Kato et al. (eds.), *Proceedings of the 10th IUTAM Symposium on Laminar-Turbulent Transition*, IUTAM Bookseries 44, https://doi.org/10.1007/978-981-96-9829-5_19

19.1.1 Linear OWNS

We take the compressible Navier-Stokes equations, with constant fluid properties (since we are considering only low-speed flows), as our governing equations. We aim to study disturbances to a time-invariant equilibrium solution (e.g., the Blasius solution), $\bar{\boldsymbol{q}}$. We linearize about $\bar{\boldsymbol{q}}$ and then consider disturbances that are periodic in time with frequency ω and in the spanwise direction with wave number β, such that

$$\boldsymbol{q}'(x, y, z, t) = \sum_{m=-M}^{M} \sum_{n=-N}^{N} \hat{\boldsymbol{q}}_{mn} e^{i(n\beta z - m\omega t)}. \tag{19.1}$$

We assume infinitesimal disturbance amplitudes in the linear case so that each frequency-wavenumber pair $(m\omega, n\beta)$ can be handled separately. Further, we discretize in the wall-normal direction using 4th-order central finite differences which we represent using $D \approx \partial/\partial y$.

Following discretization, we transform to characteristic variables, $\boldsymbol{\phi}$, and obtain the ordinary differential equation (ODE) $\partial_x \boldsymbol{\phi}_\pm = M\hat{\boldsymbol{\phi}}_\pm + \hat{\boldsymbol{g}}$, where $\hat{\boldsymbol{g}}$ is an arbitrary forcing function. The OWNS projection (OWNS-P) approach allows for the construction of a projection operator, P, based on the eigensystem of M, which retains the downstream-going modes, yielding the downstream-going solution $\hat{\boldsymbol{\phi}}' = P\hat{\boldsymbol{\phi}}$. We use P, along with linearity, to obtain an equation that supports only downstream-going modes $\partial_x \hat{\boldsymbol{\phi}}'_\pm = P[M\hat{\boldsymbol{\phi}}'_\pm + \hat{\boldsymbol{g}}]$. In practice, it is not feasible to apply the exact projection operator, so we instead apply it approximately using a recursive filtering approach [16, 17].

19.1.2 Nonlinear OWNS

To obtain the NOWNS equations, we consider a system of linear OWNS equations, for frequency and wavenumber pairs $(m\omega, n\beta)$, coupled together through the nonlinear term. We treat the nonlinear term as a forcing function, $\hat{\boldsymbol{g}}_{mn}(\boldsymbol{\phi}'_\pm)$, to obtain

$$\frac{\partial \hat{\boldsymbol{\phi}}'_{mn,\pm}}{\partial x} = P_{mn}[M_{mn}\hat{\boldsymbol{\phi}}'_{mn,\pm} + \hat{\boldsymbol{g}}_{mn}(\boldsymbol{\phi}'_\pm)], \tag{19.2}$$

for $m = 0, \ldots, M$ and $n = -N, \ldots, N$. We note that $\boldsymbol{\phi}'_\pm$ must be real-valued, which yields the constraint $\hat{\phi}'_{mn,\pm} = \overline{\hat{\phi}'_{-mn,\pm}}$, so that we need only track $(M+1) \times (2N+1)$ Fourier modes. We can additionally introduce a spanwise symmetry condition, as described in Sleeman et al. which further reduces the number of modes to $(M+1) \times (N+1)$ [15]. Lin and Schmidt recently performed a study involing LES of K-type transition where they demonstrated that the flow stops being symmetric prior to the skin friction maximum [10]. With this in mind, we performed the NOWNS calculations for K- and H-type transition both with and without the symmetry condition and found that the calculation was not affected.

19.2 Results

Blowing/suction strips have often been invoked in numerical transition studies (e.g. [4, 12, 13]). NOWNS allows these strips to be included [15], whereas NPSE does not, although the effect of the strip can be modeled using Herbert's secondary stability theory to obtain an inlet boundary condition [1–3, 5, 6]. Here we use this approach to study K-type (fundamental) and H-type (subharmonic) transition, with the goal of reproducing the DNS results of Sayadi et al. [14].

To approach the incompressible limit, we choose $\mathrm{Ma}_\infty = 0.1$, while we use the dimensionless quantities $\mathrm{Re}_x = \frac{\rho^*_\infty U^*_\infty x^*}{\mu^*_\infty}$, $y = \frac{y^*}{\delta^*_0}$, $F = \frac{\omega^* \mu^*_\infty}{\rho^*_\infty U^{*2}_\infty}$, and $b = \frac{\beta^* \mu^*_\infty}{\rho^*_\infty U^*_\infty}$, as well as the u-velocity amplitudes defined in Sleeman et al. [15]. Here the superscript $*$ denotes a dimensional quantity and the subscript ∞ denotes the free-stream value, while ρ is the density, U is the streamwise velocity of the baseflow, μ is the dynamic viscosity, ω is the temporal frequency, and β is the spanwise wave number.

19.2.1 K-Type Transition

For K-type transition, we use the blowing/suction function specified in Rigas et al. [12], which matches that used by Rist and Fasel [13]. We use

$$f(x, z, t) = 5 \times 10^{-3} \sin(\omega t) v_a(x) + 1.3 \times 10^{-4} \cos(\beta z) v_s(x), \tag{19.3}$$

where

$$v_a(x) = \begin{cases} 0, & \mathrm{Re}_x \le \mathrm{Re}_x(x_1) \\ 15.1875\xi^5 - 35.4375\xi^4 + 20.25\xi^3, & \mathrm{Re}_x(x_1) < \mathrm{Re}_x \le \mathrm{Re}_x(x_m) \\ -v_a\big(2\mathrm{Re}_x(x_m) - \mathrm{Re}_x\big), & \mathrm{Re}_x(x_m) < \mathrm{Re}_x \le \mathrm{Re}_x(x_2) \\ 0, & \mathrm{Re}_x(x_2) < \mathrm{Re}_x \end{cases} \tag{19.4a}$$

$$v_s(x) = \begin{cases} 0, & R \le \mathrm{Re}_x(x_1) \\ -3\xi^4 + 4\xi^3, & \mathrm{Re}_x(x_1) < \mathrm{Re}_x \le \mathrm{Re}_x(x_m) \\ v_s\big(2\mathrm{Re}_x(x_m) - \mathrm{Re}_x\big), & \mathrm{Re}_x(x_m) < \mathrm{Re}_x \le \mathrm{Re}_x(x_2) \\ 0, & \mathrm{Re}_x(x_2) < \mathrm{Re}_x \end{cases} \tag{19.4b}$$

where $x_m = (x_1 + x_2)/2$, and $\xi = (\mathrm{Re}_x - \mathrm{Re}_x(x_1))/(\mathrm{Re}_x(x_m) - \mathrm{Re}_x(x_1))$. We choose the parameters $\mathrm{Re}_x(x_1) = 1.36 \times 10^5$, $\mathrm{Re}_x(x_2) = 1.56 \times 10^5$, $F = 110 \times$

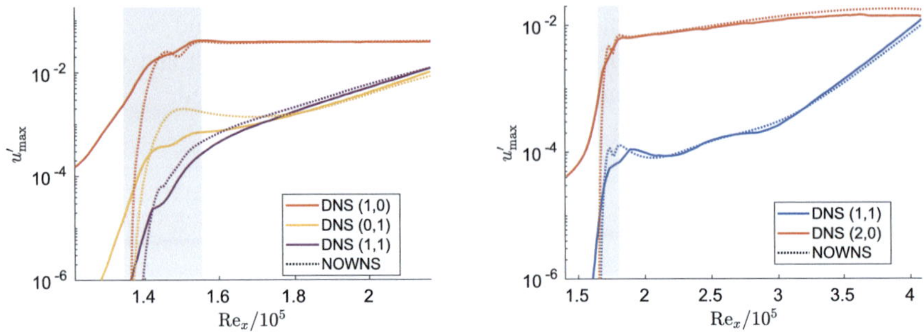

(a) Amplitudes for K-type transition (b) Amplitudes for H-type transition

Fig. 19.1 Amplitudes for K- and H-type transition, compared to the DNS of Sayadi et al. [14]

Fig. 19.2 Skin friction coefficient for K- and H-type transition, compared to the DNS of Sayadi et al. [14]

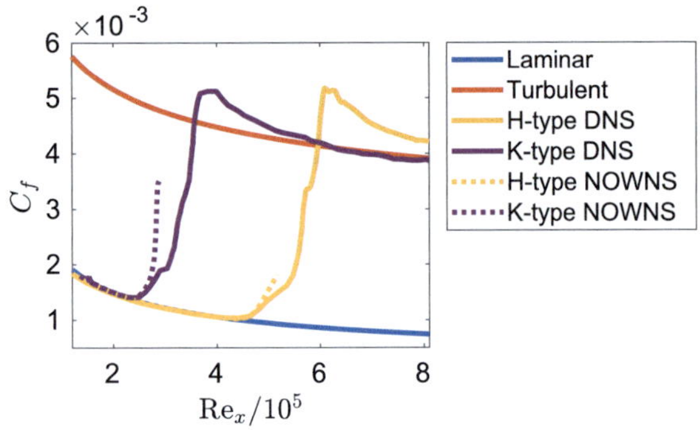

10^{-6} and $b = 0.419 \times 10^{-3}$ to match Sayadi et al. [14]. We further choose $M = 4$ temporal modes and $N = 8$ spanwise modes, with 1300 stations over the streamwise domain $\mathrm{Re}_x \in [1.36 \times 10^5, 2.73 \times 10^5]$, and 100 grid points in the wall-normal direction for $y \in [0, 60]$.

Figure 19.1a shows excellent agreement between the u' amplitudes of NOWNS and the DNS of Sayadi et al. [14]. We note that in the early stages of the march there is disagreement between the DNS and NOWNS calculations because the blowing/suction strip causes upstream effects that NOWNS neglects by construction. However, these disturbances are convective in nature, and the amplitudes predicted by NOWNS rapidly converge to those predicted by DNS as the march progresses downstream. Figure 19.2 shows that NOWNS correctly predicts the initial rise in the skin friction coefficient, but following the initial rise, the NOWNS predicts a much more rapid growth of the skin friction than DNS, before eventually failing to converge. Figure 19.3a shows the u-velocity contours at $y/\delta_{\text{inlet}} = 0.6$, as in Sayadi et al. We observe the aligned Λ-vortex structure that is characteristic of K-type transition, and good qualitative agreement with Sayadi et al. [14].

Fig. 19.3 u-velocity contours at $y/\delta_{\text{inlet}} = 0.6$ for K- and H-type transition

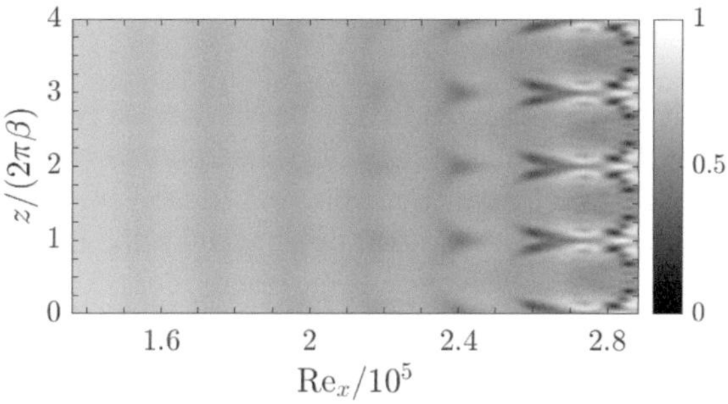

(a) K-type

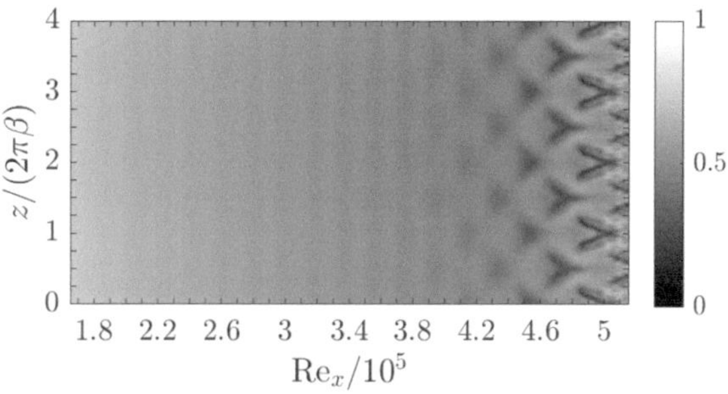

(b) H-type

19.2.2 H-Type Transition

For H-type transition we use the blowing/suction strip function

$$f(x, z, t) = 1.5 \times 10^{-3} \sin(2\omega t) v_a(x) + 1.5 \times 10^{-5} \sin(\omega t) \cos(\beta z) v_a(x), \tag{19.5}$$

where for $v_a(x)$ defined in (19.4a). We choose $\mathrm{Re}_x(x_1) = 1.65 \times 10^5$ and $\mathrm{Re}_x(x_2) = 1.81 \times 10^5$, with $2F = 124 \times 10^{-6}$ and $b = 0.419 \times 10^{-3}$. We choose $M = 8$ temporal modes and $N = 14$ spanwise modes, with 2370 stations over the streamwise domain $\mathrm{Re}_x \in [1.65 \times 10^5, 5.16 \times 10^5]$, and 100 grid points in the wall-normal direction for $y \in [0, 60]$. Figure 19.1a plots the u-velocity amplitudes, while Fig. 19.2 plots the skin friction coefficient. We additionally plot the u-velocity contours in Fig. 19.3b, where we observe the staggered Λ-vortex structure characteristic of H-type transition.

19.3 Conclusions

We have demonstrated that NOWNS accurately simulates the early stages of transition by matching the u-velocity amplitudes predicted by DNS, while correctly predicting the initial rise in the skin friction coefficient, showing that NOWNS can be used to predict transition onset analytically. Doubling the resolution in M and N does not appreciably change the results, nor does it allow NOWNS to march farther downstream. Future work will investigate how the routine can be modified to better agree with DNS in the later stages of transition. In addition, we will apply NOWNS to high-speed boundary-layer flows.

Acknowledgements This work has been supported by The Boeing Company through the Strategic Research and Development Relationship Agreement CT-BA-GTA-1.

References

1. Bertolotti, F.P.: Linear and nonlinear stability of boundary layers with streamwise varying properties. Ph.D. thesis, The Ohio State University, Jan 1991
2. Chang, C.L., Malik, M.R.: Oblique-mode breakdown and secondary instability in supersonic boundary layers. J. Fluid Mech. **273**, 323–360 (1994). https://doi.org/10.1017/S0022112094001965
3. Chang, C.L., Malik, M.R., Erlebacher, G., Hussaini, M.Y.: Linear and nonlinear PSE for compressible boundary layers. NASA Contractor Report 191537. Institute for Computer Applications in Science and Engineering (ICASE), Sept 1993
4. Fasel, H.F., Rist, U., Konzelmann, U.: Numerical investigation of the three-dimensional development in boundary-layer transition. AIAA J. **28**(1), 29–37 (1990). https://doi.org/10.2514/3.10349
5. Herbert, T.: Secondary instability of boundary layers. Ann. Rev. Fluid Mech. **20**(1), 487–526 (1988). https://doi.org/10.1146/annurev.fl.20.010188.002415
6. Herbert, T.: Parabolized stability equations. Ann. Rev. Fluid Mech. **29**(1), 245–283 (1997). https://doi.org/10.1146/annurev.fluid.29.1.245
7. Kamal, O., Rigas, G., Lakebrink, M., Colonius, T.: Input/output analysis of hypersonic boundary layers using the One-Way Navier-Stokes (OWNS) equations. In: AIAA AVIATION, June 2021. https://doi.org/10.2514/6.2021-2827
8. Kamal, O., Rigas, G., Lakebrink, M.T., Colonius, T.: Application of the One-Way Navier-Stokes (OWNS) equations to hypersonic boundary layers. In: AIAA AVIATION 2020 FORUM, June 2020. https://doi.org/10.2514/6.2020-2986
9. Kamal, O., Rigas, G., Lakebrink, M.T., Colonius, T.: Input/output analysis of a Mach-6 cooled-wall hypersonic boundary layer using the One-Way Navier-Stokes (OWNS) equations. In: AIAA AVIATION, June 2022. https://doi.org/10.2514/6.2022-3556
10. Lin, C., Schmidt, O.: Modal decomposition of K-type boundary layer transition. In: AIAA SCITECH, Jan 2024. https://doi.org/10.2514/6.2020-2986
11. Rigas, G., Colonius, T., Beyar, M.: Stability of wall-bounded fows using one-way spatial integration of Navier-Stokes equations. In: 55th AIAA Aerospace Sciences Meeting, Jan 2017. https://doi.org/10.2514/6.2017-1881
12. Rigas, G., Sipp, D., Colonius, T.: Nonlinear input/output analysis: application to boundary layer transition. J. Fluid Mech. **911**, A15 (2021). https://doi.org/10.1017/jfm.2020.982

13. Rist, U., Fasel, H.: Direct numerical simulation of controlled transition in a flat-plate boundary layer. J. Fluid Mech. **298**, 211–248 (1995). https://doi.org/10.1017/S0022112095003284
14. Sayadi, T., Hamman, C.W., Moin, P.: Direct numerical simulation of complete H-type and K-type transitions with implications for the dynamics of turbulent boundary layers. J. Fluid Mech. **724**, 480–509 (2013). https://doi.org/10.1017/jfm.2013.142
15. Sleeman, M.K., Colonius, T., Lakebrink, M.T.: Boundary layer stability analysis using the nonlinear One-Way Navier-Stokes approach. AIAA J. **63**(8), 3145–3159 (2025). https://doi.org/10.2514/1.J064909
16. Towne, A., Colonius, T.: One-way spatial integration of hyperbolic equations. J. Comp. Phys. **300**, 844–861 (2015). https://doi.org/10.1016/j.jcp.2015.08.015
17. Towne, A., Rigas, G., Kamal, O., Pickering, E., Colonius, T.: Efficient global resolvent analysis via the One-Way Navier-Stokes equations. J. Fluid Mech. **948**, A9 (2022). https://doi.org/10.1017/jfm.2022.647

Chapter 20
Generative Machine-Learning Methods to Predict the Wake of a Distributed Roughness Patch in a Hypersonic Boundary-Layer Flow

Friedrich Ulrich and Christian Stemmer

Abstract This study presents a UNet-based conditional generative adversarial network (cGAN) to predict the flow field in the wake of random roughness patches. The model successfully captured the characteristic cross-flow-like vortex, but its fidelity needs to be improved for detailed transition calculations. However, the model proved effective as a screening tool for identifying roughness patches likely to induce or suppress cross-flow-like vortices. By estimating vortex size and location, the model aids in identifying stability-critical vortices. Two roughness patches with a strong and weak vortex were selected based on the model's predictions and subjected to unsteady Direct Numerical Simulations (DNS). The interaction between the cross-flow-like vortex and unsteady perturbations leads to amplification in regions of high shear-flow and vortex breakdown.

Keywords Hypersonics · Generative machine learning · Distributed roughness

20.1 Introduction

Natural laminar-turbulent transition in the hypersonic flow regime for smooth space-capsule geometries cannot be explained by modal instabilities of the high-enthalpy boundary-layer flow. Blunt geometries induce favorable pressure gradients, limiting amplification rates and transient growth [4]. However, transition is observed in the shoulder region [5], suggesting surface roughness as a potential key driver. On blunt bodies, two kinds of roughnesses can be observed—discrete and distributed rough-

F. Ulrich
Research Scientist, Chair of Aerodynamics and Fluid Mechanics, Technical University of Munich, Garching, Germany

C. Stemmer (✉)
Professor, Chair of Aerodynamics and Fluid Mechanics, Technical University of Munich, Garching, Germany
e-mail: christian.stemmer@tum.de

K. Kato et al. (eds.), *Proceedings of the 10th IUTAM Symposium on Laminar-Turbulent Transition*, IUTAM Bookseries 44, https://doi.org/10.1007/978-981-96-9829-5_20

nesses [5]. Di Giovanni and Stemmer [2] identified cross-flow-like vortices in the wake of a random distributed roughness, amplifying unsteady disturbances.

Modern deep learning systems, such as Convolutional Neural Networks (CNNs), are multi-layered neural networks that automatically extract features from data. Generative AI models, including Variational Autoencoders (VAEs), Generative Adversarial Networks (GANs), and latent diffusion models (LDMs), are used for image generation. Autoencoders are used for unsupervised learning and dimensionality reduction.

Generative AI is increasingly applied in Computational Fluid Dynamics (CFD). Lino et al. [7] review physics-driven, data-driven, and hybrid approaches. While physics-driven methods offer potential for analytical solutions, data-driven methods excel at speed but struggle with accuracy and generalization. Future advancements in GNNs, physics integration, and data generation are crucial for reliable end-to-end simulations.

For the re-entry scenario in this, a cross-flow-like vortex forms in the wake of a random-distributed roughness patch. In order to predict the magnitude of the streamwise vorticity of the cross-flow-like vortices in the wake, Ulrich and Stemmer [10, 12] created a Direct Numerical Simulations (DNS) database with over 10,000 simulations. Different machine learning (ML) methods were used to predict the vortex magnitude and other flow parameters characterizing the vortex with a fully connected DNN and CNN.

In this study, we want to use the DNS database and build on our experience on CNNs and (Variational) Autoencoders [12] in order to predict 2D-slices in the spanwise/wall-normal direction of the flow field in the wake of a random distributed roughness patch caused by the ablation of the thermal protection system (TPS). We apply generative ML methods to the database to see the utilization of generative ML in identifying cross-flow-like vortices relevant for transition.

20.2 Methodology

The compressible flow is governed by the Navier-Stokes equations. A five-species air mixture, Park [8], accounts for dissociation below 9000 K. Species viscosities and thermal conductivities, calculated using Blottner's formula [1], are combined via Wilke's mixing rule [14] to determine mixture properties.

The Navier-Stokes-Multiblock Solver (NSMB), a semi-commercial code developed by a consortium of academia and industry [13], is employed for the DNS. This solver has been used in various hypersonic studies [2]. It utilizes structured, finite-volume grids partitioned into blocks for parallel computation via MPI. The solver supports steady and unsteady simulations with various numerical and discretization methods.

An initial axisymmetric 2D simulation with 80,000 grid points provided inflow conditions for a restricted domain behind the bow shock, where the flow remains transonic exceeding 5000 K. The restricted domain, used for ML training, was dis-

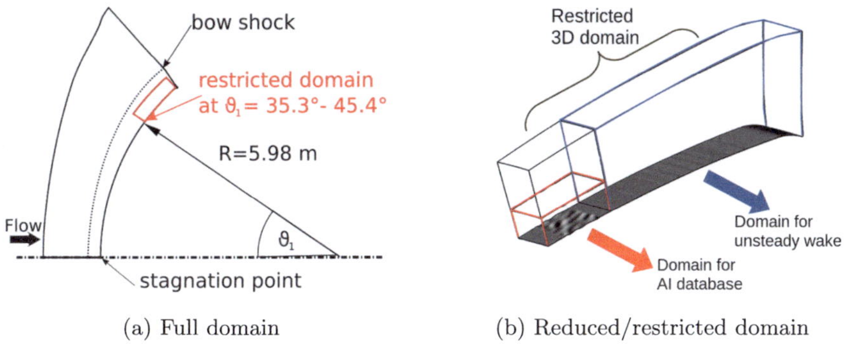

Fig. 20.1 Sketch of the full (**a**), restricted and reduced restricted (**b**) domain

cretized with 1.17 million grid points ($i = 85$, $j = 69$, $k = 200$). A fourth-order central scheme and implicit Euler time-stepping are used for this steady calculation. For unsteady flow analysis downstream of the roughness patch, a 3D computational domain with 46 million grid points ($i = 793$, $j = 145$, $k = 400$) was used, focusing on interactions in the wake region rather than the roughness patch itself [3]. The domain is located in the capsule's shoulder region where the flow becomes transonic ($M_{edge} = 1.1$). Grid convergence is verified by previous studies [2, 3, 11]. A schematic of the domains is provided in Fig. 20.1.

A hybrid five-stage Runge-Kutta method is used for time discretization, while spatial discretization remains a fourth-order central scheme due to the absence of shocks. The inflow conditions are extracted from an initial 2D simulation of the entire capsule, described in detail by Ulrich and Stemmer [11, 12].

For the ML generative method, we used the pix2pix method by Isola et al. [6]. The Pix2Pix network is based on conditional generative adversarial networks (cGAN), a variant of GANs where both the generator and discriminator are conditioned on input data. For the generator, a U-Net architecture [9] takes an input image and outputs a transformed version of it. The U-Net is an encoder-decoder network with skip connections between mirrored layers in the encoder and decoder, allowing the network to preserve fine details from the input image. The cGAN critique is a PatchGAN discriminator that classifies individual image patches as real or fake, rather than evaluating the entire image. This focuses the model on high-frequency details, making it practical for tasks like image-to-image translation.

20.3 Results

In order to establish a transition criterion for blunt body re-entry incorporating roughness geometry, two primary challenges must be addressed: **Geometric complexity** and **transition onset** are intrinsically linked in random roughness flows, where an infinite number of configurations can lead to a multitude of wake flow fields, influencing instability growth and the onset of transition.

To represent an ablating surface, we simplified the roughness patch to a superposition of sinusoidal waves with random amplitudes and phases. Sensitivity analyses confirmed that variations in wavenumbers minimally impact the flow field and vortex formation. This simplified patch can be considered a worst-case scenario. Cross-flow-like vortices emerge within the roughness wake, with the largest one playing a critical role in transition. Unsteady perturbations can grow in regions of significant wall-normal and spanwise shear, leading to vortex breakdown. Larger vortices generally exhibit larger shear and potentially are more destabilizing.

Machine Learning techniques were employed to predict characteristic parameters of the dominant vortex based on geometric roughness features. Neural networks were successfully developed to predict the maximum streamwise vorticity [12], which was chosen as a key parameter, as it effectively quantifies both the intensity of the vortex and destabilizing influence on the wake in a single parameter. The current study demonstrates the potential of generative machine learning models to predict the complete flow field using generative ML models. While generative ML methods cannot produce results accurate enough for transition prediction, they can be used to estimate the stability of the wake of a given roughness geometry. By using cGAN-generated flow fields, we can quickly assess the impact of roughness on the base flow. In Sect. 20.3.1, we describe the results generated by our UNet model implementation and investigate the effects of unsteady perturbations on two selected roughness patches.

20.3.1 ML-Based Flowfield Generation

The U-Net-based model was trained on 3,000 simulations. Increasing the database size yielded little improvements. The training process consumed 1.5 GB of GPU RAM and took 2 h on a NVIDIA Quadro P620. The network was trained for 54,000 epochs. Due to computational limitations, a comprehensive hyperparameter analysis was not feasible. However, we carefully balanced the generator and discriminator strengths to prevent mode collapse.

Figure 20.2 (right) presents a sample flow field generated by the ML model, while Fig. 20.2 (middle) shows a DNS-simulated flow field for a spanwise/wall-normal slice. Both are compared using the same input roughness patch (Fig. 20.2 (left)). Qualitatively, we observe two vortical structures similar to the cross-flow-like vortices found in DNS simulations. However, the ML-generated image lacks the details, sharp gradients and high fidelity of the DNS results.

In a test of 100 simulations, we calculated the average relative error between the ML-generated and DNS-simulated flow fields pixel by pixel. The test sets were predicted with an average error of 23% and a standard deviation of 11%.

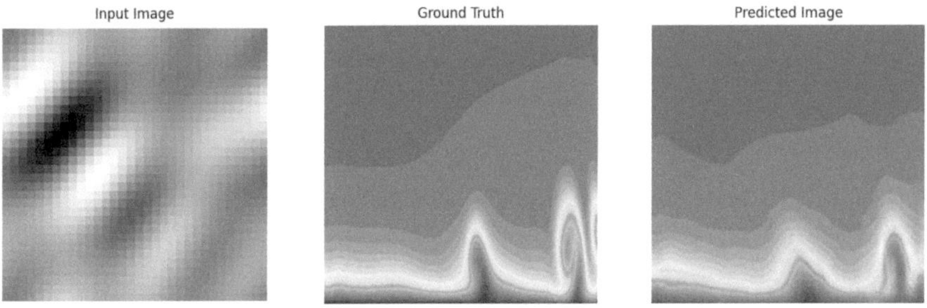

Fig. 20.2 Input roughness image (left), DNS-simulated (middle) and ML-generated (right) sample flow fields

20.3.2 Unsteady Investigation on Two Example Roughness Patches

Two distinct roughness patches were chosen based on predicted flow fields, one featuring one of the strongest vortices recognized and the other a weaker one. Direct Numerical Simulations were conducted with these patches to examine the impact of the cross-flow-like vortex in the wake on transition. The maximum streamwise vorticity downstream as determined by the DNS of roughness patch 1 was 172,797 [1/s] and 213,554 [1/s] for patch 2, respectively.

The Fourier amplitudes of the maximum downstream velocity in wall-normal direction, normalized by the inflow boundary-layer edge velocity ($u_{edge} = 1600$ m/s), are denoted as $\hat{u}$. Figure 20.3 presents the logarithmic values of these normalized amplitudes. We examine the downstream evolution of the perturbation frequency $f_1 = 8.3$ kHz in Fig. 20.3 (left plot). The amplitudes initially decay until $x = 0.14$ m for patch 2 and $x = 0.2$ m for patch 1. Beyond these points, the Fourier amplitudes exhibit exponential growth. The earlier onset of transition in patch 2 can be attributed to its higher streamwise vorticity magnitude linked to larger local wall-normal and spanwise shear.

A detailed discussion of the downstream development can be found in Ulrich and Stemmer [12]. With the help of the generated flow field, we identified two roughness patches with a different effect on transition. In the wake of patch 2, the core of the cross-flow-like vortex is located further away from the wall in normal direction compared to patch 1. The higher core position in combination with a larger vorticity magnitude, the vortex transports more low-momentum fuel upwards and has a greater shear. Hence, this vortex leads to transition onset further upstream, as it can be seen for patch 1 and 2 in Fig. 20.3 (right). In future studies, the onset transition location must be linked with flow parameters of the cross-flow-like vortex to potentially derive transition criteria involving roughness geometry.

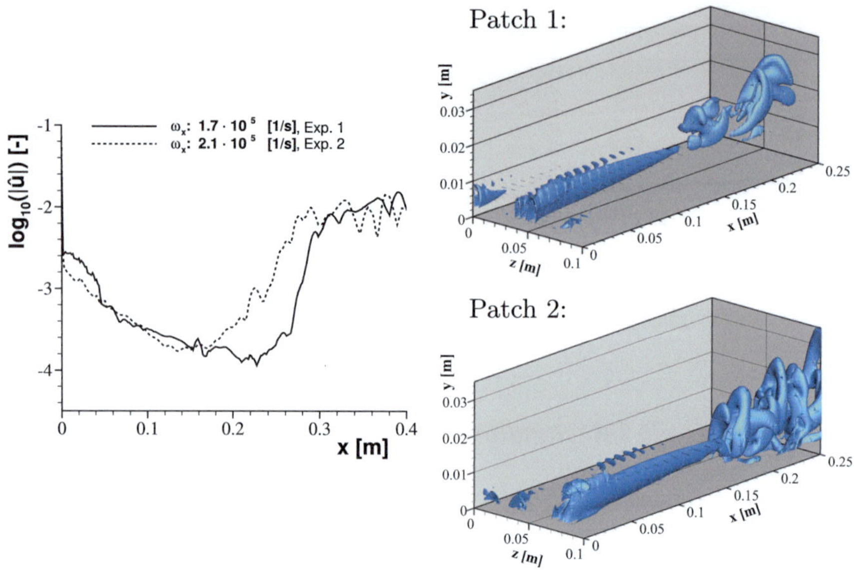

Fig. 20.3 **(left)** normalized maximum amplitudes for the temporal Fouriermodes for $f = 8.3$ kHz for example patch 1 and 2. **(right)** Q criterion $Q = 6.5 \cdot 10^7$ of the two unsteady vortices in the wake

20.4 Conclusions

This study developed a UNet-based cGAN model to predict flowfields induced by random roughness patches. While the model accurately captured cross-flow-like vortices, it fell short of producing high-fidelity solutions for transition predictions. However, it proved useful as a screening tool to identify roughness patches that either induce or suppress these vortices.

A direct estimation of the stability of the flow within the ML algorithm would most likely not improve the accuracy of the approach, but the numerical effort to provide the learning samples for the ML algorithm would increase manyfold.

By estimating vortex size and location, the model aided in assessing their impact on boundary layer stability. Two patches, one with a stronger vortex, were selected based on model predictions and subjected to unsteady DNS. The results demonstrated the interaction between the cross-flow-type vortex and unsteady perturbations, leading to amplification in regions of high shear flow and eventual vortex breakdown.

Acknowledgements This research was supported by the Cusanuswerk e.V. scholarship and founds of the TUM International Graduate School of Science and Engineering (IGSSE). Further, the authors acknowledge the Gauss Centre for Supercomputing e.V. for supporting this project by providing computing time on the GCS Supercomputer SuperMUC-NG. We would also like to thank Mirza Hasovic for his support.

References

1. Blottner, F., Johnson, M., Ellis, M.: Chemically reacting viscous flow program for multi-component gas mixtures. Tech. rep., Sandia Labs, Albuq., NM (1971)
2. Di Giovanni, A., Stemmer, C.: Cross-flow-type breakdown induced by distributed roughness in the boundary layer of a hypersonic capsule configuration. J. Fluid Mech. **856**, 470–503 (2018). https://doi.org/10.1017/jfm.2018.706
3. Di Giovanni, A., Stemmer, C.: Roughness-induced crossflow-type instabilities in a hypersonic capsule boundary layer including nonequilibrium. J. Spacecr. Rocket. **56**(5), 1409–1423 (2019). https://doi.org/10.2514/1.A34404
4. Hein, S., Theiss, A., Di Giovanni, A., Stemmer, C., Schilden, T., Schröder, W., Paredes, P., Choudhari, M.M., Li, F., Reshotko, E.: Numerical investigation of roughness effects on transition on spherical capsules. J. Spacecr. Rockets **56**(2), 388–404 (2019). https://doi.org/10.2514/1.A34247
5. Hollis, B.R.: Distributed roughness effects on blunt-body transition and turbulent heating. In: 52nd Aerospace Sciences Meeting. AIAA paper 2014-0238 (2014). https://doi.org/10.2514/6.2014-0238
6. Isola, P., Zhu, J.Y., Zhou, T., Efros, A.A.: Image-to-image translation with conditional adversarial networks (2018). arxiv.org/abs/1611.07004
7. Lino, M., Fotiadis, S., Bharath, A.A., Cantwell, C.D.: Current and emerging deep-learning methods for the simulation of fluid dynamics. Proc. Royal Soc. A: Math., Phys. Eng. Sci. **479**(2275), 20230058 (2023). https://doi.org/10.1098/rspa.2023.0058
8. Park, C.: A review of reaction rates in high temperature air. In: 24th Thermophysics Conference. AIAA paper 89–1740 (1989). https://doi.org/10.2514/6.1989-1740
9. Ronneberger, O., Fischer, P., Brox, T.: U-net: convolutional networks for biomedical image segmentation (2015). arxiv.org/abs/1505.04597
10. Ulrich, F., Sedlmeyr, T., Stemmer, C.: Methods for neural network based prediction of roughness-induced crossflow-like vorticity in re-entry scenarios. Int. J. Heat Fluid Flow **107**, 109375 (2024). https://doi.org/10.1016/j.ijheatfluidflow.2024.109375
11. Ulrich, F., Stemmer, C.: Investigation of vortical structures in the wake of pseudo-random roughness surfaces in hypersonic reacting boundary-layer flows. Int. J. Heat Fluid Flow **95**, 108945 (2022). https://doi.org/10.1016/j.ijheatfluidflow.2022.108945
12. Ulrich, F., Stemmer, C.: Unsteady evolution of distributed roughness-induced vortices under re-entry conditions. CEAS Space J. **15**(6), 971–988 (2023). https://doi.org/10.1007/s12567-023-00510-2
13. Vos, J., Rizzi, A., Corjon, A., Chaput, E., Soinne, E.: Recent advances in aerodynamics inside the NSMB (Navier Stokes Multi Block) consortium. In: 36th AIAA Aerospace Sciences Meeting and Exhibit (1998). https://doi.org/10.2514/6.1998-225
14. Wilke, C.R.: A viscosity equation for gas mixtures. J. Chem. Phys. **18**(4), 517–519 (1950). https://doi.org/10.1063/1.1747673

Chapter 21
Airfoil Trailing-Edge Tonal Noise Reduction by Roughness Elements

Zhenyang Yuan, Elías Alva, Tiago B. de Araújo, André V. G. Cavalieri, and Ardeshir Hanifi

Abstract In this work, we investigate airfoil tonal noise generation and reduction by the means of streak generators in form of cylindrical roughness elements. Roughness elements attenuate tones in the acoustic field for the case with chord base Reynolds number $Re = 80,000$. Further, the coupling between structures generated by surface roughness and instability modes (Kelvin-Helmholtz) of shear layer has been identified through stability analysis, suggesting stabilisation mechanisms of Kelvin-Helmholtz instabilities by which the sound generation by the airfoil is reduced by the roughness elements.

Keywords Aeroacoustics · Noise control · Kelvin-Helmholtz instabilities

21.1 Introduction

Trailing edge noise is generated by the interaction between an airfoil and the turbulence generated in its own boundary layer and in the vicinity of the wake [1] and is one of the main sources of noise for airfoils moving through non-turbulent flows. Tonal noise, often induced by airfoils at low to moderate Reynolds numbers [2], is well studied in the aeroacoustics community. Tonal noise always appears as sharp peaks in the acoustic spectrum. These peaks are always several dBs higher than the broadband noise and are therefore more unpleasant and damaging to the human ear. A better understanding of the noise generation mechanism can lead to effective control strategies for noise elimination or reduction.

The work of [3] was one of the first to identify ladder-like structures, where the spectrum contains the main tone and secondary tones to the side of the main tone.

Z. Yuan (✉) · A. Hanifi
FLOW, Department of Engineering Mechanics, KTH Royal Institute of Technology, Stockholm, Sweden
e-mail: zhenyang@kth.se

E. Alva · T. B. de Araújo · A. V. G. Cavalieri
Divisão de Engenharia Aeronáutica, Instituto Tecnológico de Aeronáutica, São Jose dos Campus, Brazil

K. Kato et al. (eds.), *Proceedings of the 10th IUTAM Symposium on Laminar-Turbulent Transition*, IUTAM Bookseries 44, https://doi.org/10.1007/978-981-96-9829-5_21

The work of [4] suggested that the ladder-like structure was due to a self-excited feedback loop. The phenomenon can be explained by the following process: the instability waves in the boundary layer, amplified as they pass over the separation bubble, scatter strong tonal acoustic waves at the trailing edge. The scattered acoustic waves then propagate upstream, creating new instability waves in the region near the leading edge, thus closing the loop. The instability waves here are identified as Kelvin-Helmholtz rolls.

To control tonal noise generation, a promising idea is to attenuate spanwise coherent structures (Kelvin-Helmholtz rolls), as tonal noise in airfoils is mostly related to 2D disturbances [5]. In the work of [6], experiments shows that streaks of moderate amplitudes (12% of the free-stream velocity) can stabilize Tollmien-Schlichting waves. The streaks are generated by the means of a spanwise array of cylindrical roughness elements. Inspired by this study, the similar cylindrical roughness elements are placed at the mid of the airfoil chord to generate streaks which can interact with Kelvin-Helmholtz rollers. The cylindrical roughness elements are designed such that their roughness Reynolds number $Re_{kk} = 366$ which is far less than the critical $Re_{kk} = 420$ according to calculation on a flat-plate boundary layer from [6].

Our recent experimental results [7] indicate that tonal noise can be suppressed in the presence of the surface roughness elements. In this study, we focus on the numerical investigation of the tonal noise attenuation or elimination mechanism. A series of wall-resolved implicit large eddy simulations (iLES) are performed to study the problem. The problem setup analysed here is the same as in the experiments [7], namely a NACA 0012 airfoil at zero angle of incidence and chord-based Reynolds number $Re = \rho_\infty U_\infty c/\mu = 80,000$, where ∞ indicates the freestream quantities and c is the chord length. Note that the simulations are performed at a higher Mach number of 0.3 compared to the experiments in order to reduce the convergence time. Both airfoils with/without the surface roughness elements are examined and compared.

21.2 Numerical Simulation

The open-source high-order flux reconstruction numerical framework PyFR [8] is used to solve the compressible Navier-Stokes equations for the above cases. An anti-aliasing technique is applied to remove the oscillations between the elements, which is found to be numerically stable and efficient.

PyFR supports both structured and unstructured mesh topology. To take advantage of this, a wall-resolved o-grid is applied around the airfoil, which provides the DNS level resolution ($\Delta x^+ < 4$, $\Delta y^+ < 0.7$ near the trailing edge). For the roughness element cases, a prism type mesh is applied in the near roughness region to have a better fit of the roughness shapes and to avoid stretching of the mesh. The prism layer has much smaller elements with $\Delta s = 0.002c$, which can reduce the mesh-to-mesh discontinuity due to the discontinuous nature of the flux reconstruction method. In the acoustic domain, unstructured meshes are used to reduce the large aspect ratio and significantly reduce the computational size. Analysis of the simulation data shows that the acoustic waves are still resolved up to Strouhal number $St = 15$. At the end of the computational domain, sponge zones [9] are applied to absorb acoustic waves and reduce possible reflections.

21.3 Results

Figure 21.1 presents the acoustic spectra for both cases studies here. For the clean case, the spectrum is dominated by a main frequency of $St = 5.27$, where $St = fc/U_\infty$. It is evident that secondary tones emerge in the spectrum at frequencies slightly lower and higher than the primary tone. With roughness elements being added to airfoil surface, the main tone in the clean case has been broken down into multiple tones with much lower amplitude and wider frequency bands. Those tones have more or less similar amplitudes. These observations are consistence with the experimental findings.

As an overview of the flow structures, the visualisation of the filed flow is shown in Fig. 21.2 in terms of the Q criterion coloured by the streamwise velocity component u. For the clean case, the dominant structures are K-H type rolls. These spanwise

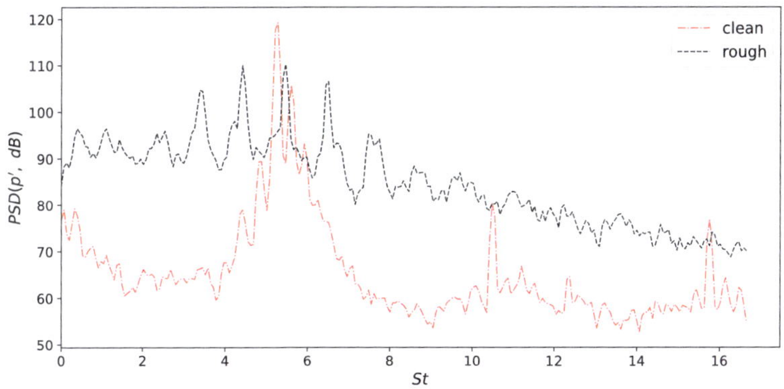

Fig. 21.1 Acoustic spectrum for numerical simulations measured in dB

(a) Baseline clean geometry (b) Airfoil rough geometry

Fig. 21.2 Iso-surface of Q-criterion colored by streamwise velocity u

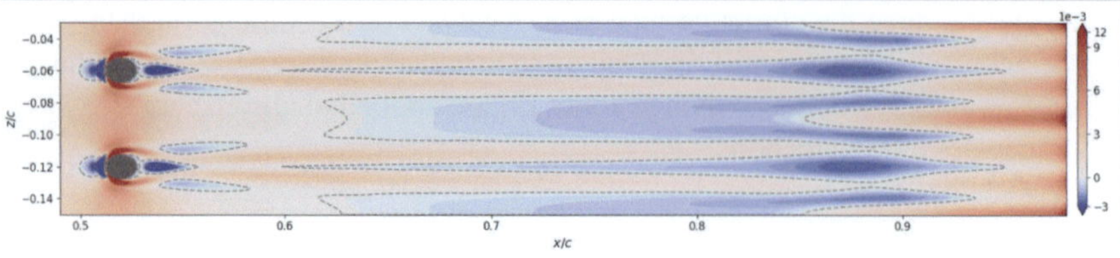

(a) Re = 80k

(b) Experimental visualization of streaks

Fig. 21.3 **a** The C_f map of mean flow field of roughness cases. Red is positive and blue is negative. Dashed grey line is zero-contour line. **b** Oil visualization from the experimental campaign at Re = 150,000

coherent structures are responsible for tonal noise generation at the trailing edge. With roughness elements on the airfoil surface, streaky structures are generated. These streaks develop into lambda structures (also known as hairpins) as they propagate downstream and modulate K-H rolls. Modulation of these rolls by streaks will reduce spanwise coherence and ultimately reduce the acoustic amplitude and shift the tonal frequencies as shown in the Fig. 21.1.

In addition, the time-averaged surface friction coefficient map is shown in the Fig. 21.3, together with the experimental oil visualisation. A long separation bubble (characterised by the negative C_f regions) can be seen downstream of the roughness element locations in both cases. However, the separation bubble has been sliced/modulated into subdomains by the streamwise elongated structures highlighted by the positive C_f values. High speed streaks are induced downstream of the roughness elements. Note that a higher Reynolds number has been chosen to enhance the visualisation effect from the experiment oil visualisation, although the overall mechanism remains the same. This visualisation method captures the fast and slow streaks immediately behind the roughness elements, with the streaks extending up to about $x/c = 0.9$. This result is very consistent with the C_f distribution from the simulation shown in the same figure.

The observations mentioned above confirm the presence and significance of streaks in modulating spanwise coherent structures, ultimately leading to a reduction in tonal noise. Furthermore, we use local spatial stability analysis to quantify the contribution of streaks to reduction of the growth of spanwise coherent structures. In order to accomplish this, we choose the streamwise station $x = 0.75c$ where streaks

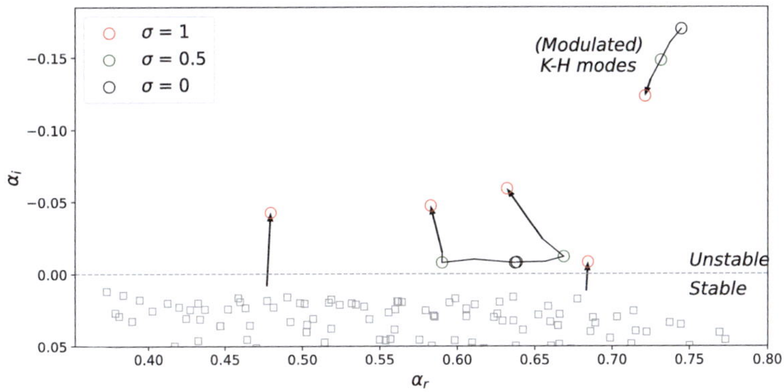

Fig. 21.4 The spectrum traced with gradual increase of mean state modulation parameter σ. Open circles stand for the unstable modes and open squares are stable modes

and K-H instabilities are well developed and amplified. The effect of roughness-generated streaks is investigated through tracking the unstable modes from the mean state without roughness elements to the one with roughness elements [10] by a continuous variation of the mean flow ($\bar{\mathbf{U}}$) as:

$$\bar{\mathbf{U}} = (1 - \sigma)\bar{\mathbf{U}}_{clean} + \sigma\bar{\mathbf{U}}_{rough} \tag{21.1}$$

The tracking mode technique aids in identifying growth changes between two distinct states. Ideally, continuous change of the parameter σ allows us to trace the progression of an unstable mode through these states. The tonal frequencies, i.e. $St = 5.27$ in the clean case and $St = 5.47$ in the rough case, are used in the stability analysis. The spectrum is illustrated in the Fig. 21.4. As can be seen there, there are three unstable modes in the clean case. The most unstable mode stands for two-dimensional K-H instability, with the mode shape depicted in the upper subplots of Fig. 21.5. The other two modes have identical low growth rate and streamwise wavenumber, corresponding to degenerate oblique K-H modes with positive and negative spanwise wavenumbers, as in [10]. While increasing the value of the parameter σ, the growth rate of the most unstable mode is decreasing from $\alpha_i = 0.1699$ to $\alpha_i = 0.1231$. However, the other two unstable modes in the clean state diverge and become more unstable with increasing σ. Two other modes appear from the stable regime, becoming unstable as σ approaches 1. Furthermore, Fig. 21.5 presents mode shapes of the least unstable modes at state $\sigma = 0$ and 1. From the clean state ($\sigma = 0$) to rough state ($\sigma = 1$), a progressive spanwise modulation appears at the middle of domain. The K-H mode is heavily modulated in this region.

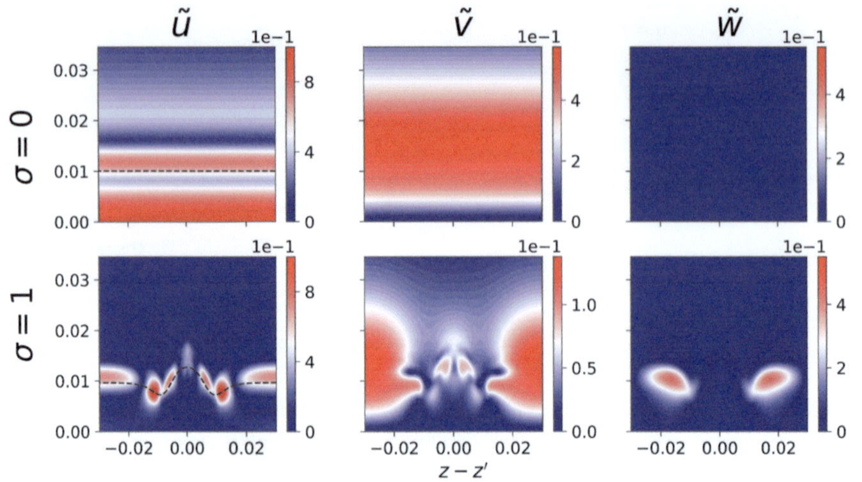

Fig. 21.5 The leading eigen-modes (K-H modes) of mean state modulation parameter $\sigma = 0$ and 1. Black dashed line indicates the critical layer

21.4 Conclusion

The analysis of the simulations presented in this work contributes to a better understanding of the mechanisms behind tone attenuation and suppression. The key aspects are related to the introduction of roughness elements, which induce continuous streaks that render the laminar separation bubble three-dimensional, resulting in a spanwise modulation of Kelvin-Helmholtz modes and a reduction of their growth rates. These effects are expected to weaken the feedback loop leading to tonal noise radiation. Future research can build on these theoretical insights to develop more refined designs of roughness elements aiming at the attenuation of the undesirable tonal noise from airfoils.

Acknowledgements Zhenyang Yuan would like to acknowledge Swedish Research Council for supporting current work under Grant 2020-04084 and the National Academic Infrastructure for Super-computing in Sweden (NAISS) resources at LUMI, CSC Finland.

References

1. Brooks, T.F., Pope, D.S., Marcolini, M.A.: Airfoil self-noise and prediction. NASA Technical Report (1989)
2. Arbey, H., Bataille, J.: Noise generated by airfoil profiles placed in a uniform laminar flow. J. Fluid Mech. **134**(1), 33–74 (1983). https://doi.org/10.1017/S0022112083003201
3. Paterson, R.W., Vogt, P.G., Fink, M.R., Munch, C.L., Aircr, J.: Vortex Noise of Isolated Airfoils **10**, 296–302 (1973). https://doi.org/10.2514/3.60229

4. Tam, C.K.W.: Discrete tones of isolated airfoils. J. Acoust. Soc. Am. **55**, 1173–1177 (1974). https://doi.org/10.1121/1.1914682
5. Ricciardi, T.R., Wolf, W.R., Taira, K.: Transition, intermittency and phase interference effects in airfoil secondary tones and acoustic feedback loop. J. Fluid Mech. **937**, A23 (2022). https://doi.org/10.1017/jfm.2022.129
6. Fransson, J.H.M., Brandt, L., Talamelli, A., Cossu, C.: Experimental study of the stabilization of Tollmien-Schlichting waves by finite amplitude streaks. Phys. Fluids **17**, 1–15 (2005). https://doi.org/10.1063/1.1897377
7. Alva, E., Yuan, Z., Araújo, T.B., Amaral, F.R., Hanifi, A., Cavalieri, A.V.G. : Airfoil tonal noise reduction by roughness elements, Part I—Experimental Investigations (2024). https://doi.org/10.48550/arXiv.2410.05134
8. Witherden, F.D., Farrington, A.M., Vincent, P.E.: PyFR: an open source framework for solving advection–diffusion type problems on streaming architectures using the flux reconstruction approach. Comput. Phys. Commun. **17**, 3028–3040 (2014). https://doi.org/10.1016/j.cpc.2014.07.011
9. Bodony, D.J.: Analysis of sponge zones for computational fluid mechanics. J. Comput. Phys. **212**, 681–702 (2006). https://doi.org/10.1016/j.jcp.2005.07.014
10. Lajús, F.C., Sinha, A., Cavalieri, A.V.G., Deschamps, C.J., Colonius, T.: Spatial stability analysis of subsonic corrugated jets. J. Fluid Mech. **876**, 766–791 (2019). https://doi.org/10.1017/jfm.2019.573

Chapter 22
Numerical and Experimental Study on the Freestream-Turbulence Dependency of Turbulent Transition Processes on a Swept Flat Plate

Kosuke Nakagawa, Takahiro Ishida, and Takahiro Tsukahara

Abstract We conducted direct numerical simulations (DNS) and wind tunnel tests (WTT) to investigate the role of freestream turbulence (FST) in the laminar-turbulent transition of a boundary layer on a swept flat plate with cylindrical roughness. In the DNS, we examined how FST influences the transition process of roughness-induced crossflow vortex. Under short-wavelength FST conditions, hairpin vortices formed behind the roughness even at lower heights compared to long-wavelength FST conditions. Here, the long wavelength is the same length as the most unstable wavelength of a stationary crossflow vortex. Based on the roughness height for hairpin vortex generation, the critical roughness Reynolds number under the short-wavelength FST condition was found to be comparable to that observed in WTT. These findings indicate that high-frequency FST components penetrate the boundary layer in WTT environments. We clarified that the receptivity of high-frequency fluctuations plays a key role in the laminar-turbulent transition process caused by roughness and FST.

Keywords Crossflow instability · Freestream turbulence

22.1 Introduction

On swept wings, a three-dimensional boundary layer is caused by a difference between the directions of the freestream and the pressure gradient. This difference leads to the crossflow instability in the three-dimensional boundary layer [1]. The crossflow instability is sensitive to small disturbances such as wall roughnes, which

K. Nakagawa (✉) · T. Tsukahara (✉)
Department of Mechanical and Aerospace Engineering, Tokyo University of Science, Chiba, Japan
e-mail: kousuke.n091396@gmail.com

T. Tsukahara
e-mail: tsuka@rs.tus.ac.jp

T. Ishida
Aviation Technology Directorate, Japan Aerospace Exploration Agency, Tokyo, Japan

K. Kato et al. (eds.), *Proceedings of the 10th IUTAM Symposium on Laminar-Turbulent Transition*, IUTAM Bookseries 44, https://doi.org/10.1007/978-981-96-9829-5_22

induce crossflow vortices (CFVs). During the laminar-turbulent transition process, high-frequency fluctuations called secondary instabilities occur in the CFV and play a role in triggering the collapse of the CFV [2, 3].

We have studied how the transition process of roughness-induced CFVs is changed by freestream turbulence (FST). We aimed to classify the transition process by focusing on the fluctuation generated around CFVs and clarify the cause of the process modulation between CFV transition and hairpin vortex transition [4–6]. In this study, we investigate changes in laminar-turbulent transitions due to FST by means of both direct numerical simulations (DNS) and wind tunnel tests (WTT) to obtain insights into FST contributions that are independent of specific methodologies.

22.2 Method

In DNS, we used a Falkner–Skan–Cooke (FSC) boundary layer as the base flow, which simulates a flow field on the swept flat plate. We employ the same numerical scheme as in our previous study [4]. We used an external chordwise velocity (U_0) and an displacement thickness (δ_0^*) at inlet for non-dimensionalization. On the swept flat plate, the chordwise direction is x, the spanwise direction is y, and the wall-normal direction is z. We investigate two domain sizes, namely, $(L_x, L_y, L_z) = (400\delta_0^*, 50.28\delta_0^*, 27\delta_0^*)$ and $(800\delta_0^*, 50.28\delta_0^*, 27\delta_0^*)$. Note that, unlike our previous studies, the spanwise domain length is twice the most unstable wavelength of CFV ($25.14\delta_0^*$). The objective of this setup is to capture the effects of longer wavelengths than CFV and eliminate potential biases in the fluctuation distribution within the FST region. The turbulence intensity of FST was set as $Tu = 0.1\%$. The spectral distribution of FST was defined as $E(\mathbf{k}) = \mathbf{k}^4 \exp(-2\mathbf{k}/k_{\max})$ with respect to a peak wavelength of either $\lambda_{\max}/\delta_0^* = (L_z/\delta_0^*)/k_{\max} = 5.0$ or 25.14.

In the experimental investigation, we used a low-turbulence wind tunnel at the Japan Aerospace Exploration Agency, and the nominal turbulence intensity was $Tu = 0.1$–0.16%. We employed a swept flat plate model which has a displacement body to provide a pressure gradient on the plate. The coordinate system is the same as that of the DNS. Further details of experiment were described in our previous paper [7]. We used a hot-wire anemometer to measure the flow field of the CFV induced by the cylindrical roughness array. We arranged the roughness elements such that their spacing matches the most unstable wavelength in the flow field of this swept flat plate model ($\lambda_y = 7.62$ mm). The cylindrical roughness has a diameter of $d = 1.27$ mm and is installed at $x/c = 0.1$ (i.e., 10% of the chordwise length $c = 600$ mm of swept flat plate model). The wind tunnel generates the freestream velocity with magnitude $Q_\infty = 36.9$ m/s, and the boundary layer thickness at the roughness position was $\delta_R = 0.994$ mm.

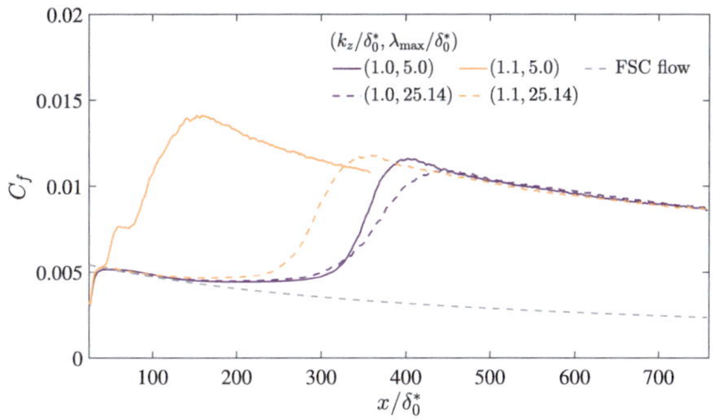

Fig. 22.1 Friction coefficient distribution for different roughness heights

22.3 Result

Figure 22.1 shows time-averaged friction coefficient distributions for all DNS cases. The friction coefficient is defined as the combined shear stress of the chordwise and spanwise directions. In the long-wavelength FST condition ($\lambda_{\max}/\delta_0^* = 25.14$), the transition position shifts gradually as the roughness height increases. This result is consistent with the case in which the most unstable wavelength aligns with the spanwise computational domain [6]. On the other hand, in the short-wavelength FST condition ($\lambda_{\max}/\delta_0^* = 5.0$), a rapid laminar-turbulent transition occurs $1.0 < k_z/\delta_0^* < 1.1$. Figure 22.2 visualizes these two flow fields. The laminar-turbulent transition near the cylindrical roughness is caused by hairpin vortices, which are induced by high-frequency fluctuations that form short-wavelength FST [5]. The roughness Reynolds number ($Re_{kk} = \sqrt{(u_{\mathrm{FSC}}^2 + v_{\mathrm{FSC}}^2)}k_z/\nu$) using this roughness height ($k_z/\delta_0^* = 1.1$), the laminar-flow velocity at the roughness height (u_{FSC} and v_{FSC}), and the kinematic viscosity (ν) is determined as the critical Re_{kk}. Even under low roughness conditions shown in Fig. 22.2a, hairpin vortices occur only near the cylindrical roughness. They disappear downstream, and eventually, the CFV grows. We classify this roughness condition as sub-critical Re_{kk}.

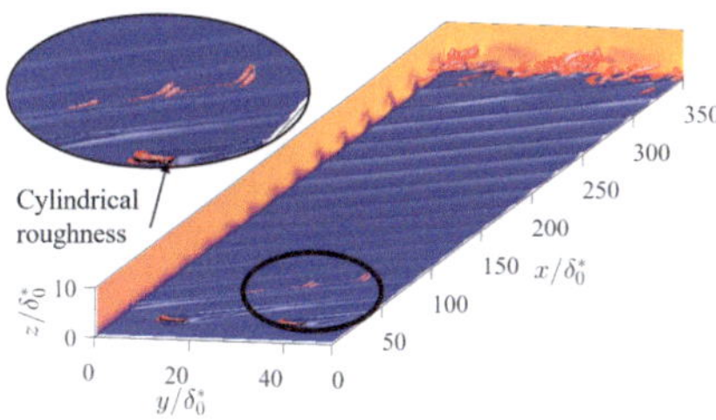

(a) $k_z/\delta_0^* = 1.0$ ($Re_{kk} = 290$)

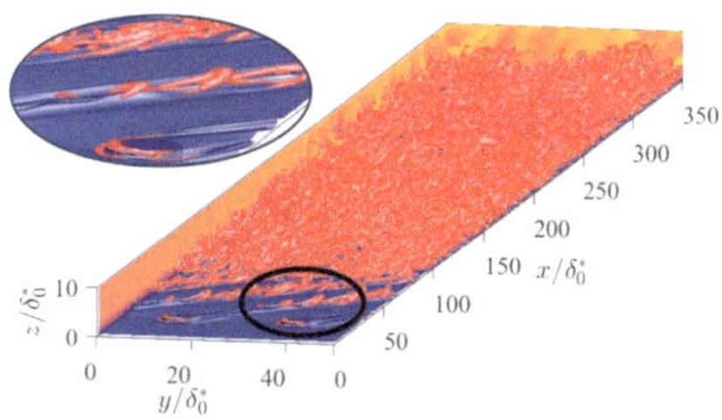

(b) $k_z/\delta_0^* = 1.1$ ($Re_{kk} = 342$)

Fig. 22.2 Instantaneous flow fields under the short-FST condition. Red isosurfaces indicate vortex structures ($Q = 0.2$) and blue isosurfaces indicate low-velocity region ($u/U_0 = 0.3$)

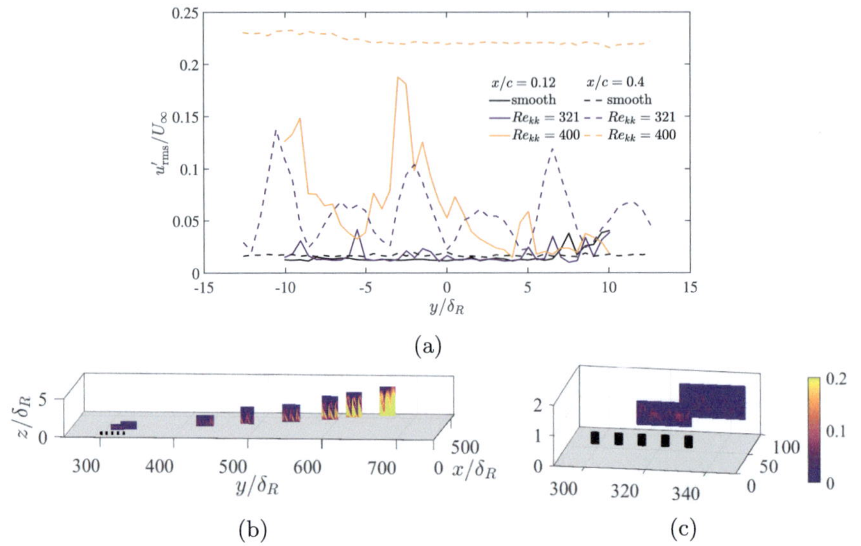

Fig. 22.3 WTT measurements for **a** fluctuation distribution for spanwise direction at $x/c = 0.12$, $z/\delta_R = 0.5$ (solid lines) and $x/c = 0.4$, $z/\delta_R = 1.1$ (dashed line), **b** spatial distribution for $Re_{kk} = 321$, and **c** shown near roughness region

The critical Re_{kk} for the short-wavelength FST is $Re_{kk} = 342$. In a DNS study without FST, the critical value was $Re_{kk} \approx 700$, while in an investigation of WTT that is assumed to be affected by FST, the result was $Re_{kk} \approx 200$–300 [6, 8, 9]. In addition, the present results suggest that the critical Re_{kk} varies with the FST wavelength, and that the result in shorter FST conditions is closer to the result in the WTT. Our WTT investigation was carried out to clarify the cause of the similarity between the results of the DNS with the short-wavelength FST and the results of the WTT. To determine the critical Re_{kk} in the WTT, the spanwise distributions of velocity fluctuations near the roughness ($x/c = 0.16$) at two roughness conditions are shown in Fig. 22.3a. High amplitude fluctuations occur clearly for $Re_{kk} = 400$. These fluctuations, distributed immediately behind the roughness, can be considered a proof of the generation of hairpin vortices, as shown in Fig. 22.2b. These fluctuations spread the whole spanwise direction for $x/c = 0.4$. In contrast, the fluctuations are low for $Re_{kk} = 321$. Figure 22.3b, c indicates the spatial development of the fluctuations for this condition. The fluctuations induced by the roughness increase downstream and are distributed along the shape of CFVs. Eventually, high-amplitude fluctuations spread, and at $x/c = 0.6$–0.7, they cover the whole flow field. The transition position changes rapidly between these Re_{kk} conditions. Together with the comparison at the critical Re_{kk}, i.e., $Re_{kk} = 342$ for DNS and $Re_{kk} = 400$ for WTT, rapid shift in the transition position is the evidence that the flow field in WTT is similar to the environment of the short-wavelength FST in DNS.

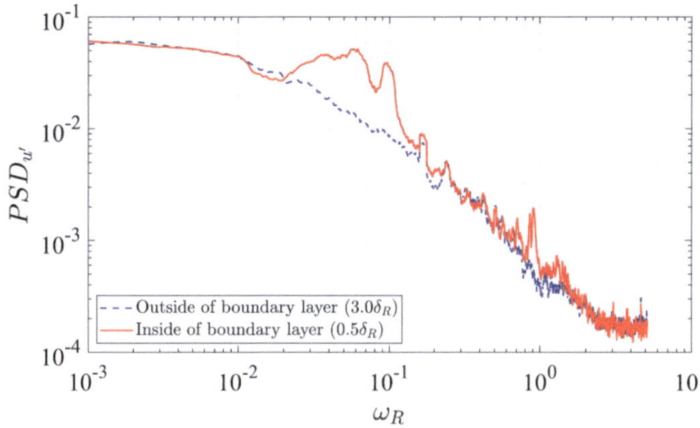

Fig. 22.4 Spectrum distributions of velocity fluctuations at $x/c = 0.1$ under non-roughness condition in WTT

Consequently, we focus on the FST receptivity to the boundary layer developed on the swept flat plate under non-roughness condition. Figure 22.4 shows the frequency distribution of fluctuations both outside ($z/\delta_R = 3.0$) and inside ($z/\delta_R = 0.5$) the boundary layer. The frequency is defined as $\omega_R = 2\pi/(tU_0/\delta_R)$, where $U_0 = 13.4$ m/s in WTT. The fluctuation was measured at $x/c = 0.1$, where the roughness elements were distributed. As for the spectral distribution of velocity fluctuations outside the boundary layer, we observe a conventional amplitude decrease towards the high-frequency side. In contrast, two amplified frequency bands appear within the boundary layer. One of the bands in the low-frequency side of $\omega_R = 0.02$–0.1 coincides to the frequency of traveling CFV [10], and thus we can confirm that traveling CFV is amplified even over a short distance from the leading edge. The other band in the high-frequency side of $\omega_R = 0.8$–1.7 is similar in frequency to the high-frequency secondary instabilities that occur in CFV or hairpin vortices [6, 8, 10]. It suggests that high-frequency fluctuations penetrate easily into the boundary layer.

The characteristics of FST receptivity have also been reported in two-dimensional boundary layers [11]. Under smooth surface condition, this high-frequency fluctuation decreases downstream, and low-frequency traveling CFVs occur [5]. On the other hand, a local high-shear region induced by roughness modifies the frequency dependence. As a result, the amplitude of high-frequency fluctuations increases along the roughness wake structure, and they take a form of hairpin vortices. This phenomenon causes the decrease of critical Re_{kk} in WTT concerning critical value under non-FST conditions.

22.4 Conclusion

To elucidate laminar-turbulent transition processes in a swept flat plate boundary layer, we used DNS to investigate the wavelength dependency of FST, the correspondence between the results of DNS and the WTT, and their causing factors. Under sub-critical Re_{kk} conditions, cylindrical roughness induces CFV, which then collapses and causes laminar-turbulent transition. In contrast, under the critical Re_{kk} conditions, the laminar-turbulent transition is caused by hairpin vortices generated near the cylindrical roughness. The threshold between the sub-critical and critical transitions depends on the FST wavelength, and the critical Re_{kk} for short-wavelength FST is lower than that for long-wavelength FST. Our WTT exhibited the critical Re_{kk} similar to that in the short-wavelength FST condition in the DNS. This similarity is owing to the FST receptivity to the boundary layer, and fluctuations of specific frequencies can penetrate easily into the boundary layer. Among these fluctuations, the high-frequency fluctuation induces hairpin vortices even at low roughness heights. These vortices provide the critical Re_{kk} reduction, and the characteristics of the flow field become similar to that under short-wavelength FST. These results highlight the importance of FST modeling in DNS and FST measurement in WTT for future investigations.

Acknowledgements First author K.N. was supported by Support for Pioneering Research Initiated by the Next Generation (SPRING) of Japan Science and Technology Agency (JST), Grant Number JPMJSP2151. Some of the present numerical simulations utilized supercomputer resources at the Cybermedia Center, Osaka University, and the Cyberscience Center, Tohoku University. The wind tunnel test was carried out using Low-Turbulence Wind Tunnel in cooperation with the Facilities Engineering Research Unit, Aerospace Engineering Directorate, JAXA.

References

1. Saric, W.S., Reed, H.L., White, E.B.: Stability and transition of three-dimensional boundary layers. Annu. Rev. Fluid Mech. **35**, 1, 413–440 (2003)
2. Malik, M.R., Li, F., Choudhari, M.M., Chang, C.-L.: Secondary instability of crossflow vortices and swept-wing boundary-layer transition. J. Fluid Mech. **399**, 85–115 (1999). https://doi.org/10.1017/S0022112099006291
3. White, E.B., Saric, W.S.: Secondary instability of crossflow vortices. J. Fluid Mech. **525**, 275–308 (2005)
4. Nakagawa, K., Tsukahara, T., Ishida, T.: DNS study on turbulent transition induced by an interaction between freestream turbulence and cylindrical roughness in swept flat-plate boundary layer. Aerospace **10**(2), 128 (2023)
5. Nakagawa, K., Ishida, T., Tsukahara, T.: Effects of freestream turbulence on the secondary instability of the roughness-induced crossflow vortex in swept flat plate boundary layers. Int. J. Heat Fluid Flow **102**, 109161 (2023)
6. Nakagawa, K., Ishida, T., Tsukahara, T.: Changes in the transition process of roughness-induced crossflow vortices due to freestream turbulence. Flow Turbul. Combust. (2024)

7. Ishida, T., Ohira, K., Hosoi, R., Tokugawa, N., Tsukahara, T.: Numerical and experimental analysis of three-dimensional boundary-layer transition induced by isolated cylindrical roughness elements. Int. J. Aeronaut. Space Sci. **23**(4), 649–659 (2022)
8. Brynjell-Rahkola, M., Shahriari, N., Schlatter, P., Hanifi, A., Henningson, D.S.: Stability and sensitivity of a cross-flow-dominated Falkner-Skan-Cooke boundary layer with discrete surface roughness. J. Fluid Mech. **826**, 830–850 (2017)
9. Zoppini, G., Michelis, T., Ragni, D., Kotsonis, M.: The near wake of discrete roughness elements on swept wings. J. Fluid Mech. **960**, A11 (2023)
10. Högberg, M., Henningson, D.: Secondary instability of cross-flow vortices in Falkner-Skan-Cooke boundary layers. J. Fluid Mech. **368**, 339–357 (1998)
11. Brandt, L., Schlatter, P., Henningson, D.S.: Transition in boundary layers subject to free-stream turbulence. J. Fluid Mech. **517**, 167–198 (2004)

Chapter 23
Computations of Turbulent Transition in Hypersonic Boundary Layer with Reduced Artificial Dissipation

Minjae Jeong, Suhun Cho, Jiseop Lim, and Solkeun Jee

Abstract Turbulent transition in hypersonic boundary layer is numerically investigated with a focus on effects of artificial dissipation in the computation. The artificial dissipation should be carefully controlled by adequate shock sensor for stable and accurate simulation of high-speed flows. In this study, the original shock sensor of the current flow solver, which is widely used in shock capturing methods, is modified to reduce the numerical dissipation in the transitional boundary layer. The current study indicates that well-controlled artificial dissipation allow major instabilities to grow properly and turbulent flow to be better resolved in hypersonic boundary layer flow.

Keywords Turbulent transition · Boundary layer · Hypersonic flow · Artificial dissipation · Shock capturing method

23.1 Introduction

A low-dissipative scheme is required for a high-fidelity simulation of turbulent transition in hypersonic boundary layer. The artificial dissipation is essential in the vicinity of discontinuities, but it can deteriorate the growth of instabilities and small-scale turbulence in the transitional boundary layer. Therefore, the shock sensor must be carefully selected to ensure the artificial dissipation is applied only where necessary, such as near shock region.

In this study, a low-dissipative flow solver, rhoEnergyFoam [7], is used for a simulation of hypersonic boundary layer. The flow solver adds the numerical dissipation of the AUSM^{+}-up scheme [6] where the shock sensor senses shock. The solver originally uses a classical shock sensor [3], which is widely used in shock cap-

M. Jeong (✉) · S. Cho · J. Lim · S. Jee (✉)
Gwangju Institute of Science and Technology, Gwangju, South Korea
e-mail: mjjeong96@gm.gist.ac.kr

S. Jee
e-mail: sjee@gist.ac.kr

K. Kato et al. (eds.), *Proceedings of the 10th IUTAM Symposium on Laminar-Turbulent Transition*, IUTAM Bookseries 44, https://doi.org/10.1007/978-981-96-9829-5_23

turing method. However, the sensor is activated in the boundary layer unexpectedly in the current simulation. Therefore, the sensor is modified to reduce the artificial dissipation in the transitional boundary layer.

23.2 Methods

Turbulent transition in hypersonic boundary layer on a 7° half-angle cone is simulated in this study. The flow condition is set to the freestream Mach number $M_\infty = 6$ and the unit Reynolds number $Re_\infty = 10.5 \times 10^6$/m [8]. A combined approach of direct-numerical simulation (DNS) and linear stability theory (LST) method is used [1, 5]. The major instability, 2D Mack 2nd mode (denoted as $[1, 0]$), and a pair of oblique modes (denoted as $[1, \pm 1]$) are obtained from LST. The obtained instabilities are superposed on the laminar solution at the inlet of DNS domain, as illustrated in Figure 23.1. The parameters of the instabilities are matched with the previous study [8]. The frequency of the instabilities is $f = 210$ kHz and the amplitudes are $|u'_{[1,0]}|_{max}/u_\infty = 0.7\%$ and $|u'_{[1,\pm 1]}|_{max}/u_\infty = 0.002\%$ at the DNS inlet. The DNS domain starts at $x = 0.3$ m and the wall-parallel length of the domain is $L_{x,DNS} = 0.25$ m. The wall-normal length of the domain is set to $L_y = 0.123$ m, which corresponds to 90 times the boundary layer thickness at the inlet $\delta_{99,in}$. The azimuthal extent of the domain L_ϕ is the azimuthal wavelength of the oblique modes $\lambda_{\phi,[1,\pm 1]}$ with the azimuthal periodic boundary condition. The buffer domain starts at $x = 0.55$ m and extends to $x = 0.6$ m. Two grids in the current study are summarized in Table 23.1 showing the grid size Δ at the inlet $x_{in} = 0.3$ m and the outlet $x_{out} = 0.55$ m. The superscript $^+$ denotes turbulence wall unit and $\lambda_{x,[1,0]}$ is the streamwise wavelength of the Mack 2nd mode.

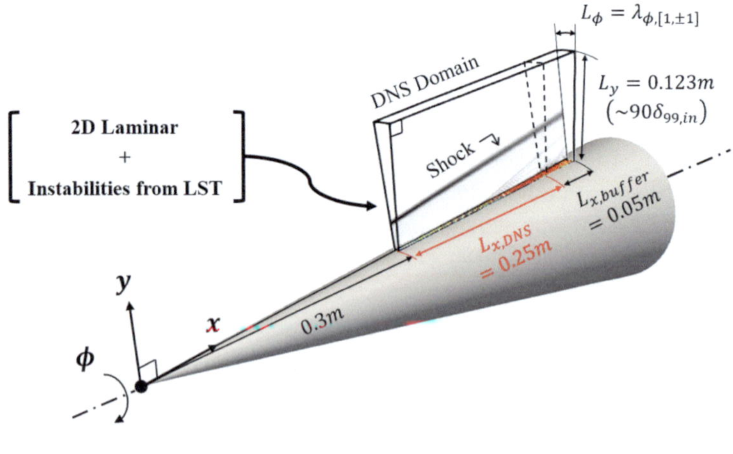

Fig. 23.1 Computational domain of the current simulation

Table 23.1 Two grids in the current study. The wall unit is evaluated with the viscous length scale of turbulent flow at the outlet $x_{out} = 0.55$ m

	$N_x \times N_y \times N_\phi$	$\Delta^+_{x,in}$	$\Delta^+_{x,out}$	$\Delta^+_{y,in}$	$\Delta^+_{y,out}$	$\Delta^+_{\phi,in}$	$\Delta^+_{\phi,out}$	$\frac{\Delta_{x,in}}{\lambda_{x,[1,0]}}$	$\frac{\Delta_{x,out}}{\lambda_{x,[1,0]}}$
Medium	2800 × 132 × 32	5.2	2.6	0.4	0.4	3.2	5.8	1/32	1/64
Fine	3400 × 300 × 64	3.8	2.0	0.2	0.2	1.4	2.5	1/38	1/70

The low-dissipative flow solver, rhoEnergyFoam [7], is used for the current DNS computation. The solver artificially adds numerical dissipation of the AUSM$^+$-up scheme [6], and the amount of the dissipation is controlled by the shock sensor θ as follows [7],

$$f^{\mathrm{D}} = H(\theta - \theta_{th})\ f^{\mathrm{AUSM}} \tag{23.1}$$

$$P^{\mathrm{D}} = \theta\ P^{\mathrm{AUSM}} \tag{23.2}$$

where f^{D} is convective dissipation and P^{D} is pressure dissipation. f^{AUSM} and P^{AUSM} are the numerical dissipation of the AUSM$^+$-up scheme, and θ_{th} is a threshold of the Heaviside function H. The solver originally uses the Ducros shock sensor θ_{Ducros} [3], based on the following formula:

$$\theta_{Ducros} = \max\left(\frac{-\nabla \cdot \mathbf{u}}{\sqrt{(\nabla \cdot \mathbf{u})^2 + |\nabla \times \mathbf{u}|^2 + c_\varepsilon}}, 0\right) \in [0, 1] \tag{23.3}$$

where $\nabla \cdot \mathbf{u}$ is dilatation, $\nabla \times \mathbf{u}$ is vorticity and the coefficient c_ε is set to a small value just to prevent division by zero. However, the current simulation indicates that the Ducros sensor θ_{Ducros} generates excessive artificial dissipation in the boundary layer, not only in the vicinity of shocks. Therefore, the Lele sensor θ_{Lele} [2] is tested to avoid nonphysical damping of instabilities in the boundary layer. The θ_{Lele} is expressed as follows,

$$\theta_{Lele} = \frac{1}{2}\left[1 - \tanh\left\{c_1 + c_2 \frac{\Delta}{a}(\nabla \cdot \mathbf{u})\right\}\right]\theta_{Ducros} \tag{23.4}$$

where Δ is cubic root of cell volume and a is the local speed of sound. The two coefficients are $c_1 = 2.5$ and $c_2 = 10$ following [2].

23.3 Results

Instantaneous vortical structures are visualized by Q-criterion in Fig. 23.2. Near the inlet, two-dimensional structures generated by the 2D Mack 2nd mode are dominant. Lambda(Λ)-shaped vortices emerge around $x = 0.43$ m as the oblique and other higher modes grow. The alignment of the Λ-shaped vortical structures are commonly observed in the fundamental breakdown scenario [1, 5, 8]. Subsequently, hairpin structures appear around $x = 0.48$ m. Finally, the vortical structures break up into smaller eddies and the flow becomes turbulent after $x = 0.52$ m.

Figure 23.3 shows a region where the sensor is activated in the current computation. The Ducros shock sensor θ_{Ducros} activates the artificial dissipation excessively overall in the computational domain, including within the boundary layer. It is due to the amplified instabilities with ultrasonic frequency in the boundary layer. In contrast, the Lele sensor θ_{Lele} significantly suppresses the activation while effectively capturing the oblique shock generated from the leading edge. θ_{Lele} notably reduces the numerical dissipation in the transitional boundary layer.

Streamwise development of selected modes is shown in Fig. 23.4. In the computations with θ_{Lele}, the growth of [1, 0] shows excellent agreement with the LST data up to $x = 0.37$ m, where low-amplitude modes undergo rapid amplification by nonlinear modal interactions. Compared to θ_{Lele}, the growth of the major mode [1, 0] is suppressed by the excessive artificial dissipation with θ_{Ducros}. As a result, the onset of the exponential growth of $[1, \pm1]$ is delayed with θ_{Ducros}. The well-controlled artificial dissipation by θ_{Lele} allows major instabilities to grow properly.

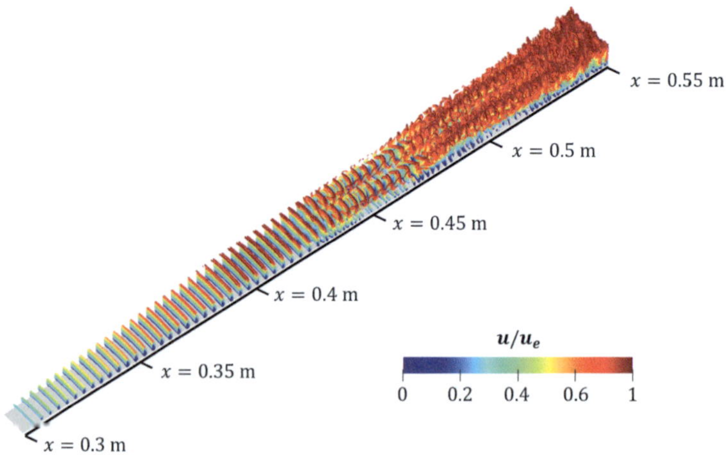

Fig. 23.2 Instantaneous vortical structures visualized by Q-criterion from the fine grid. The Q-criterion is colored by streamwise velocity U, normalized by boundary layer edge velocity U_e. The scale $x : y : \phi = 1 : 4 : 4$ is used just for the visualization purpose

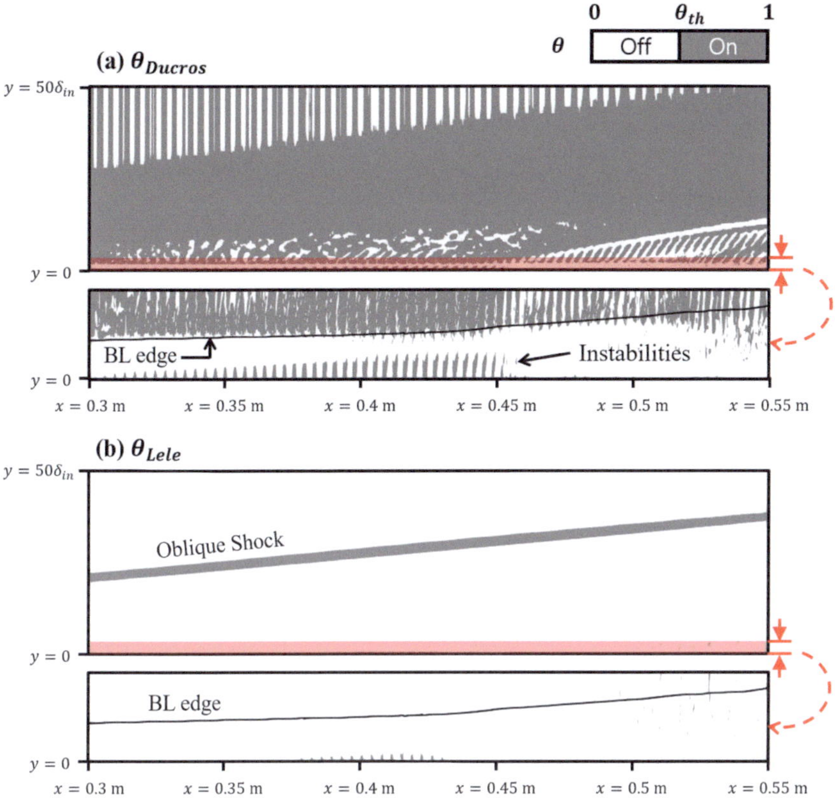

Fig. 23.3 Comparison of the region where the convective numerical dissipation is activated between using **a** the Ducros sensor θ_{Ducros} and **b** the Lele sensor θ_{Lele} [4]

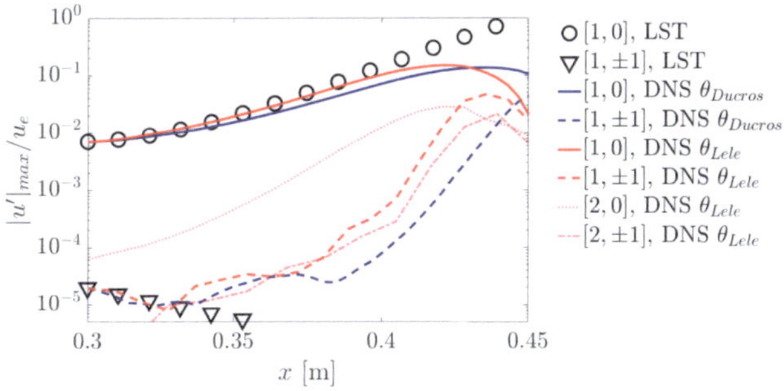

Fig. 23.4 Streamwise developments of selected modes in the current computations [4]

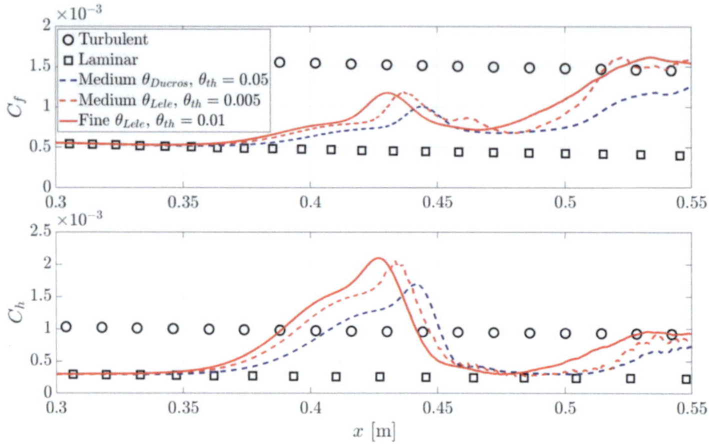

Fig. 23.5 Comparison of (top) the skin friction coefficient and (bottom) stanton number [4]

The skin friction coefficient C_f and the stanton number C_h from the current computations are shown in Fig. 23.5 In the computations with θ_{Lele}, C_f and C_h with θ_{Lele} deviate the laminar C_f and C_h from $x = 0.35$ m and follows the theoretical turbulent C_f and C_h after $x = 0.53$ m. Comparing the results from θ_{Ducros} and θ_{Lele} with the Medium grid, θ_{Ducros} numerically delays the transition and underestimates C_f and C_h in the turbulent region. It is due to the excessive numerical dissipation in the transitional boundary layer by θ_{Ducros} as shown in Fig. 23.3.

23.4 Conclusions

The effect of artificial dissipation on the transitional boundary layer is investigated in this study. The boundary layer transition on a 7° half-angle cone at Mach 6 is simulated using a combined approach of direct-numerical simulation and linear stability analysis. Although the artificial dissipation is required for a stable computation of high-speed flow, excessive dissipation can suppress the appropriate growth of instabilities and small-scale turbulence in the boundary layer. Therefore, the shock sensor must be carefully chosen to control the numerical dissipation adequately. The original sensor of the current flow solver, which is commonly used in shock capturing methods, is modified to reduce the artificial dissipation in the boundary layer. The modification of the sensor allows to properly capture the growth of instabilities and turbulence in the transitional boundary layer.

Acknowledgements This research was supported by the Challengeable Future Defense Technology Research and Development Program through the Agency For Defense Development (ADD) funded by the Defense Acquisition Program Administration (DAPA) in 2024 (No.915067201).

References

1. Bae, H., Lim, J., Kim, M., Jee, S.: Direct-numerical simulation with the stability theory for turbulent transition in hypersonic boundary layer. IJASS **24**(4), 1004–1014 (2023)
2. Bhagatwala, A., Lele, S.: A modified artificial viscosity approach for compressible turbulence simulations. J. Comput. Phys. **228**, 4965–4969 (2009)
3. Ducros, F., Ferrand, V., Nicoud, F., Weber, C., Darracq, D., Gacherieu, C., Poinsot, T.: Large-eddy simulation of the shock/turbulence interaction. J. Comput. Phys. **152**(2), 517–549 (1999)
4. Jeong, M., Cho, S., Kim, Y., Park, J., Jee, S.: High-fidelity simulation of laminar-to-turbulent transition in hypersonic boundary layer on a sharp cone. Phys. Fluids **37**(3), (2025)
5. Lim, J., Kim, M., Bae, H., Lin, R.S., Jee, S.: Turbulent transition control using porous surfaces in hypersonic boundary layer. IJASS **24**(4), 972–984 (2023)
6. Liou, M.S.: A sequel to AUSM: AUSM+. J. Comput. Phys. **129**(2), 364–382 (1996)
7. Modesti, D., Pirozzoli, S.: A low-dissipative solver for turbulent compressible flows on unstructured meshes, with OpenFOAM implementation. Comput. Fluids **152**, 14–23 (2017)
8. Sivasubramanian, J., Fasel, H.F.: Direct numerical simulation of transition in a sharp cone boundary layer at Mach 6: fundamental breakdown. J. Fluid Mech. **768**, 175–218 (2015)

Chapter 24
Experimental Investigation on Boundary-Layer Streaks Induced by Grid-Generated Free-Stream Turbulence

Tristan M. Römer, Markus J. Kloker, Ulrich Rist, and Christoph Wenzel

Abstract This experimental study analyses the influence of various turbulence-grid configurations on Klebanoff modes/streaks developing in a laminar flat-plate boundary layer in a water channel. Grids were designed according to common recommendations to ensure homogeneous free-stream turbulence (FST), which typically induces meandering Klebanoff modes. In a previous study, the authors have shown that the Klebanoff modes do not meander for all grid configurations, a behaviour caused by e.g. small grid imperfections or by unfavourably chosen positions of the wall-normal bars of some grids. The present study focuses on a different Reynolds number and additional measurement positions inside the flat plate's boundary layer. It is also shown that the transition with isolated cylindrical roughness is strongly influenced by oncoming steady streaks through improperly generated 'FST', requiring future FST studies to document the Klebanoff mode's meandering behaviour.

Keywords Free-stream turbulence · Grid · Klebanoff mode · Roughness

24.1 Introduction

Free-stream turbulence (FST) is one of the most significant factors influencing the laminar-turbulent transition process. At higher FST levels the transition process may follow different 'paths to turbulence' [8]. The FST interacts with the boundary layer (BL) through a receptivity process, referred to as *shear-sheltering* [12], which only allows low-frequency disturbances to penetrate the BL. Due to the *lift-up effect*, streamwise velocity streaks develop, often referred to as Klebanoff modes, which algebraically grow in the streamwise direction [4]. These streaks typically meander in the spanwise direction and can undergo secondary instabilities, either sinuous or varicose [2].

T. M. Römer (✉) · M. J. Kloker · U. Rist · C. Wenzel
Institute of Aerodynamics and Gas Dynamics, University of Stuttgart, Stuttgart, Germany
e-mail: roemer@iag.uni-stuttgart.de

K. Kato et al. (eds.), *Proceedings of the 10th IUTAM Symposium on Laminar-Turbulent Transition*, IUTAM Bookseries 44, https://doi.org/10.1007/978-981-96-9829-5_24

In many experimental setups, FST is generated using a rectangular grid placed upstream of the leading edge (LE) of the plate (see Cardesa et al. [3] for near-field analysis behind grids). The grid's spacing M, bar diameter d, its distance to the LE Δx and Reynolds number Re_d (based on d, the free-steam velocity U_∞ and the kinematic viscosity ν) must fall within well-established ranges to induce homogeneous and isotropic turbulence, which have been validated by numerous studies:

$$\text{porosity [6]}: \quad \beta = (1 - d/M)^2 > 0.55 \tag{24.1}$$

$$\text{Reynolds number [7]}: \quad Re_d = U_\infty d/\nu > 80 \tag{24.2}$$

$$\text{distance grid} \leftrightarrow \text{LE e.g. [1,5]}: \quad \Delta x \geq 20M \tag{24.3}$$

24.1.1 Motivation and Objectives

Klebanoff modes are characterized by their meandering and random spatial appearance of low- and high-speed streaks. Over time, this meandering cancels out, leading to a mean flow that is independent of the spanwise location. Based on common criteria from the literature, see, e.g., Eqs. (24.1), (24.2), (24.3), grids were installed in the laminar water channel (LWC) at the Institute of Aerodynamics and Gas Dynamics (IAG). Hereby, the Klebanoff modes did not meander for all grid setups, for some grids they were locked to the mesh width M or $2M$ and were very sensitive to numerous grid differences. Römer et al. [10] have shown that these behaviours can be attributed to inhomogeneities in the FS. It is the main goal of this study to supplement some behaviours reported by Römer et al. [10] by further measurements at different BL positions and different Re. Moreover, the possible impact of FST-induced steady streaks on the roughness-induced transition is addressed.

24.2 Experimental Methods

Experimental facility and measurement equipment All experiments were conducted in the closed-loop LWC at IAG, University of Stuttgart. Single hot-film measurements were taken with a Dantec 55R15 single probe and the Dantec 'Streamline' bridge, based on constant-temperature anemometry. Data was recorded at $f = 100$ Hz using a 16-bit National Instruments USB-6216 A/D converter and converted to velocity using King's law. Visualizations are performed by hydrogen-bubbles, see [9] for details on the experimental methods; measurement uncertainties are addressed in [11].

Experimental Setup
Two grids made of round bars arranged in a square pattern (both $M = 25$ mm) are installed upstream of the leading edge of the flat plate, which differ in their bar connections: (i) grid A ($d = 2$ mm) is a conventionally fabricated grid made of stainless

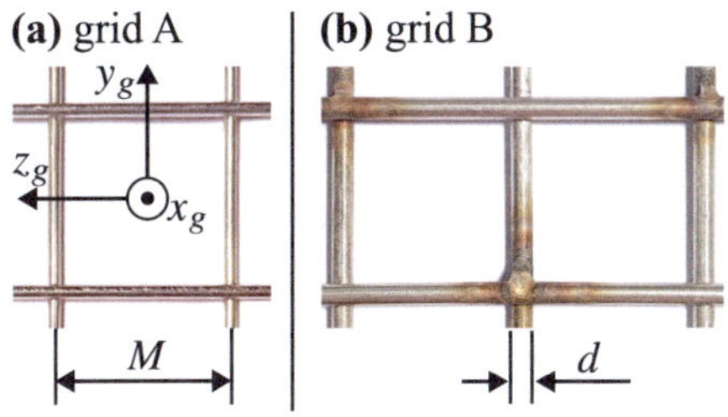

Fig. 24.1 Images of the grids taken from Römer et al. [10]

steel, spot-welded by a machine, see Fig. 24.1**a**; (ii) grid B ($d = 3$ mm) is a custom-made grid made from stainless-steel bars that are welded by hand at every second crossing point, see Fig. 24.1**b**.

Two wall-normal grid positions relative to the flat plate are investigated: (i) LE behind a horizontal bar (Fig. 24.2 (left)), and (ii) LE between two horizontal bars (Fig. 24.2 (right)). All grids are orientated in vertical-horizontal streamwise order; note that the orientation also affects the streaks, see [10]. The flat plate coordinate system (x, y, z) originates at the LE (see Fig. 24.2) and the grid-related coordinate system (x_g, y_g, z_g) is further upstream in the grid cell centre (see Fig. 24.1), with $y_g = 0$ $(z_g = 0)$ between two horizontal bars and $y_g = 0.5M$ $(z_g = 0.5M)$ behind a bar.

24.3 Results

24.3.1 Influence of the Grid Height

Figure 24.2 shows hydrogen-bubble visualizations in the BL observed at $(x, y) = (1400\,\text{mm}, 1.4\delta_1)$, δ_1 being the BL displacement thickness. For both visualizations, the same grid A is used but placed at two different wall-normal positions: in the left visualization, the horizontal bars are placed at the same wall-normal height as the LE, and in the right image the grid is raised by half its spacing $(M/2)$. The upstream position of the vertical bars is indicated by the white dotted lines. In the left figure, the position of the (low-speed) streaks varies across the snapshots in the spanwise direction, clearly showing the expected meandering motion. The spanwise streak spacing varies also, showing streaks of small and large spanwise extent. For the upwards-shifted grid (right figure) the meandering of the streaks is almost completely absent; their spanwise spacing λ_z seems to be locked to $M = 25$ mm. This reveals that the FS is not fully homogeneous in the spanwise and wall-normal direction behind grid A leading to different BL streak behaviours, although the distance between the grid and the LE is $\Delta x = 20M$, a value that is cited often in the literature to be sufficient for homogeneous, isotropic turbulence.

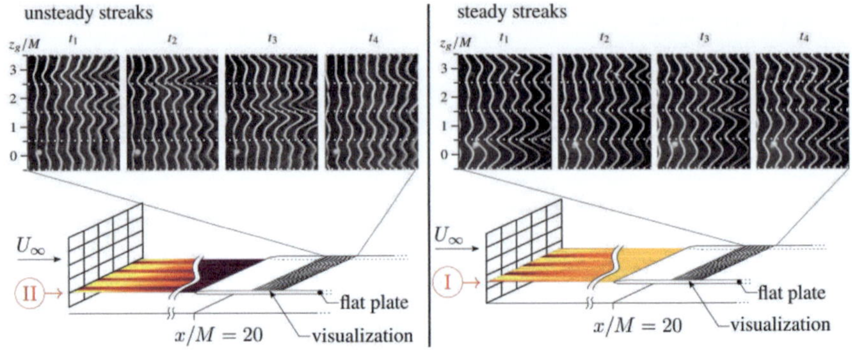

Fig. 24.2 Hydrogen-bubble visualizations with grid A, whereby the flat plate is aligned at the same wall-normal height as the horizontal bars (left) and between two horizontal bars (right). The snapshot sampling period is 20 s. The visualisations are derived from Römer et al. [10]

24.3.2 *Sensitivity of 'FST' to Grid Imperfections*

It is a common practice in wind/water tunnel experiments on FST effects to use pre-fabricated grids, which are typically spot-welded as can be seen in Fig. 24.1. For non-purchasable grid specifications, it is common to manufacture custom grids with welded seams. This section examines grid B which has been welded together at every second grid node, see Fig. 24.1**b**. To evaluate its influence on the BL streaks' behaviour, grid B is installed at $x = -30M$ (LE between two horizontal bars) and the spanwise/wall-normal mean velocity distribution is evaluated from hot-film measurements in the BL at $x = 900$ mm for $Re_d = 290$.

Figure 24.3 reveals that the boundary-layer streaks remain steady at all wall-normal positions without any signs of meandering. Additionally, the high-speed streaks consistently appear at spanwise locations behind the welded grid nodes, while the low-speed streaks form behind the non-welded nodes. This leads to a spanwise wavelength of $\lambda_z = 2M$, indicating a doubling of the streak spacing.

This grid configuration demonstrates that even minor imperfections, such as small welding seams, can significantly affect the homogeneity and stationarity of the boundary-layer streaks. The welding spots act as a hydro-/aerodynamic smoothing mechanism at the grid-bar intersections, reducing the intensity of flow detachment compared to intersections without welding. The inhomogeneity induced by the grid can persist over long spatial distances, even though the grid distance of $\Delta x = 30M$ significantly exceeds the commonly referenced guideline of $\Delta x = 20M$ from the literature [1, 5].

24.3.3 Roughness-Transition with 'FST'-Induced Steady-Streaks

The following section presents a demonstrative setup of how the transition Reynolds number Re_k (based on the roughness height k and U_e) is influenced by a cylindrical roughness element with FST when it is placed in a steady (i) low- or (ii) high-speed streak. Note that Re_k is defined here using the free-stream velocity U_∞, rather than the velocity at cylinder height u_k, as is common in the literature. Consequently, the Reynolds numbers presented in this study may not be directly comparable to those reported in other experimental investigations. However, they are suitable for internal comparison within this specific setup.

Steady streaks with $\lambda_z = 2M = 50$ mm are generated by grid B as demonstrated by Römer et al. [10] and shown in Fig. 24.3. The roughness with height $k = 8$ mm and diameter $d_k = 8$ mm (aspect ratio $\eta = d_k/k = 1$) is placed inside the BL at $x = 900$ mm on the flat plate ($d_k/\lambda_z = 0.16$). The transition Reynolds numbers $Re_{k,tr}$ for both cases (i) and (ii) are identified with hot-film measurements at $x = 2200$ mm by varying the FS velocity U_∞ and evaluating the intermittency function γ, as described by Römer et al. [11] (Note that in [11] the transition location is identified by varying the probe position). The intermittency function varies between 0 and 1, where 0 and 1 denote a laminar and turbulent flow, respectively. The transition Reynolds number is defined where $\gamma = 0.5$.

The evaluated intermittency functions for both spanwise roughness positions are given in Fig. 24.4, where the solid lines give the least-square fit to an exponential function. The results show that the roughness-induced transition varies significantly depending on the spanwise position of the roughness, which is $Re_{k,tr} = 752$ ($(k/\delta_1)_{tr} = 1.66$) in the high-speed streak (HSS) case and $Re_{k,tr} = 801$ ($(k/\delta_1)_{tr} = 1.7$) in the low-speed streak (LSS) case, resulting in a difference of $\Delta Re_{k,tr} = 49$.

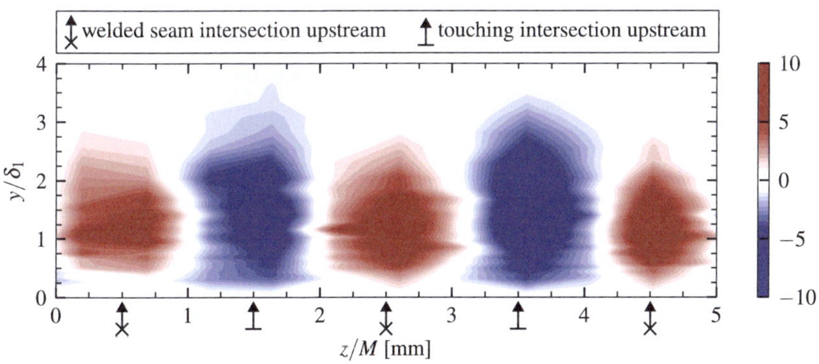

Fig. 24.3 Steady streaks generated by FST at $x = 900$ mm with grid B showing contours of $(\bar{u} - u_{bl})$ being the deviation of the base flow from the Blasius BL flow

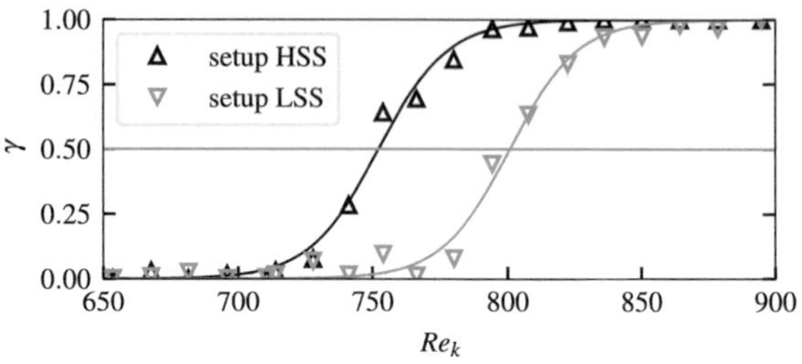

Fig. 24.4 Intermittency functions for a cylinder in a high- and low-speed streak

The finding demonstrates that steady streaks generated by improper FST can greatly influence the roughness-induced laminar-turbulent transition with FST.

24.4 Conclusions

A previous study by Römer et al. [10] investigated the influence of frequently as well as rarely described parameters of turbulence grids on the resulting FST and the Klebanoff modes forming in the BL. Here, some influences were validated by varying the Reynolds number and measuring position. Additionally, the roughness-induced transition under the influence of steady streaks generated by 'FST' has been investigated.

When the LE of the flat plate is mounted at a height between two horizontal bars, the Klebanoff modes that develop in the laminar BL tend to become relatively steady, locking both spatially and temporally to the grid spacing. In contrast, when the LE is aligned with the height of a horizontal bar, the Klebanoff modes exhibit the expected meandering in the spanwise direction, with wavelengths that vary spatially. Small grid imperfections (here welded seams) induce the formation of uniform, steady streaks in the BL. These streaks are fixed at the positions corresponding to the spanwise welding points of the grid. It is demonstrated that the roughness-induced transition Reynolds number with 'FST' changes dramatically depending on whether the roughness is placed in a steady low-speed or high-speed streak, both non-existing in a true FST setup.

The results of this study highlight that FST generated by turbulence grids may bear the memory of the grid, which implies that future studies that rely on turbulence-grid-induced FST would benefit from documenting the actual meandering motion of the generated streaks in the BL.

References

1. Alfredsson, P., Matsubara, M.: Free-stream turbulence, streaky structures and transition in boundary layer flows. AIAA (2000). https://doi.org/10.2514/6.2000-2534
2. Andersson, P., Brandt, L., Bottaro, A., Henningson, D.S.: On the breakdown of boundary layer streaks. J. Fluid Mech. **428**, 29–60 (2001). https://doi.org/10.1017/S0022112000002421
3. Cardesa, J.I., Nickels, T.B., Dawson, J.R.: 2D PIV measurements in the near field of grid turbulence using stitched fields from multiple cameras. Exp. Fluids **52**(6), 1611–1627 (2012). https://doi.org/10.1007/s00348-012-1278-4
4. Ellingsen, T., Palm, E.: Stability of linear flow. Phys. Fluids **18**(4), 487–488 (1975). https://doi.org/10.1063/1.861156
5. Groth, J., Johansson, A.V.: Turbulence reduction by screens. J. Fluid Mech. **197**, 139–155 (1988). https://doi.org/10.1017/s0022112088003209
6. Kurian, T., Fransson, J.H.M.: Grid-generated turbulence revisited. Fluid Dyn. Res. **41**(2), 021–403 (2009). https://doi.org/10.1088/0169-5983/41/2/021403
7. Mehta, R.D.: Turbulent boundary layer perturbed by a screen. AIAA **23**(9), 1335–1342 (1985). https://doi.org/10.2514/3.9089
8. Morkovin, M.V., Reshotko, E., Herbert, T.: Transition in open flow systems-a reassessment. Bull. Am. Phys. Soc. **39**, 1882 (1994)
9. Puckert, D.K.: Experimental investigation on global instability and critical reynolds number in roughness-induced laminar-to-turbulent transition. Ph.D. thesis, Universität Stuttgart, Stuttgart (2019)
10. Römer, T.M., Kloker, M.J., Rist, U., Wenzel, C.: Experimental investigation on boundary-layer streaks induced by grid-generated free-stream turbulence in a water channel. Exp. Fluids **65**(12) (2024). https://doi.org/10.1007/s00348-024-03917-7
11. Römer, T.M., Schulz, K.A., Wu, Y., Wenzel, C., Rist, U.: Delay of laminar–turbulent transition by counter-rotating cylindrical roughness elements in a laminar flat plate boundary layer. Exp. Fluids **64**(2) (2023). https://doi.org/10.1007/s00348-023-03581-3
12. Zaki, T.A.: From streaks to spots and on to turbulence: exploring the dynamics of boundary layer transition. Flow Turbul. Combust. **91**(3), 451–473 (2013). https://doi.org/10.1007/s10494-013-9502-8

Chapter 25
Global Stability Analysis of Installed Jets

Alessandro Franchini, Nicolas Alferez, and Jean-Christophe Robinet

Abstract Installation noise caused by the interaction between jet flow and structures like wings or flaps has become a significant concern for commercial aircraft manufacturers, particularly due to the emergence of pronounced tonal peaks in the acoustic spectrum. To better understand this phenomenon, we developed a global stability analysis tool that combines the compressible nonlinear solver dNami with the source transformation automatic differentiation tool Tapenade. The analysis involves solving the unsteady Reynolds-averaged Navier-Stokes (URANS) equations using the Spalart-Allmaras turbulence model. A 2D global stability analysis at high Reynolds numbers will be presented and the results will be compared with experimental data obtained from wind tunnel by *CNRS-PPrime* laboratory.

Keywords Subsonic jets · Tonal noise · Global instability

25.1 Introduction

In recent times, jet noise during takeoff has become a significant concern for commercial aircraft manufacturers. To mitigate the health impacts of jet noise in densely populated areas, stringent regulations on aircraft noise have been implemented [2]. This, along with the need to reduce fuel consumption and green house gas emissions, has driven to substantial modifications of turbofan engines. Specifically, the jet diameter is increased, introducing installation noise, which arises from the interaction between the jet flow and solid surfaces such as wings and flaps. In order to understand these groups of phenomena, many studies are being pursued within the aeroacoustics community from many different perspectives: experimental studies, mathematical models and CFD simulations. Some studies [1] looked at the installed jet problem by exploring the interaction between large coherent turbulent structure

A. Franchini (✉) · N. Alferez · J.-C. Robinet
DynFluid Laboratory, Arts et Métiers Institute of Technology, Paris, France
e-mail: alessandro.franchini@ensam.eu

N. Alferez
DynFluid Laboratory, Conservatoire national des arts et métiers, Paris, France

K. Kato et al. (eds.), *Proceedings of the 10th IUTAM Symposium on Laminar-Turbulent Transition*, IUTAM Bookseries 44, https://doi.org/10.1007/978-981-96-9829-5_25

and the aircraft wing, by means of geometrically complex CFD simulations with hybrid RANS/LES models. Recent experimental studies held in anechoic room with wind tunnel by *Institut Pprime* [3] explored a vast range of different parameters, generating a large database and insights on the fundamental structure of the interactions.

In this work, experiments from three particular configurations will be compared with global stability analysis performed over the compressible unsteady Reynolds average Navier Stokes equations with a Spalart-Allmaras [7] model for turbulent viscosity μ_t.

25.2 Global Stability Analysis of URANS Equations

The compressible unsteady Reynolds averaged Navier Stokes equations with Spalart Allmaras modelisation for turbulent viscosity formally reads:

$$\begin{cases} \dfrac{\partial \mathbf{q}}{\partial t} + \nabla \cdot (L(\mathbf{q}) + N(\mathbf{q}, \mathbf{q}) - \mu_t G(\mathbf{q})) = \mathbf{0} \\ \dfrac{\partial \mu_t}{\partial t} + H(\mathbf{q}, \mu_t) = 0, \end{cases} \tag{25.1}$$

where $\mathbf{q} = (\rho, \rho u, \rho v, \rho e_t)^T$ is the state vector of conserved variables, $\mu_t = \rho \nu_t$ is the scalar turbulent eddy viscosity field, $L(\mathbf{q})$ is a liner operator representing the linear contribution of the flux vector, $N(\mathbf{q}, \mathbf{q})$ is the quadratic advection term of the flux, $H(\mu_t, \mathbf{q})$ is the operator corresponding to the Spalart-Allmaras model and $G(\mathbf{q})$ is the operator that models the mean of the stress tensor:

$$\overline{N(\mathbf{q}', \mathbf{q}')} = -\overline{\mu}_t G(\overline{\mathbf{q}}).$$

Let's perturb the quantity $(\mathbf{q}, \mu_t)^T = (\overline{\mathbf{q}}, \overline{\mu}_t)^T + \varepsilon(\mathbf{q}', \mu_t')^T$, where $(\overline{\mathbf{q}}, \overline{\mu}_t)^T$ is the fixed point solution of Eq. (25.1). By neglecting high order terms for ε, the dynamics of the perturbation reads:

$$\begin{cases} \dfrac{\partial \mathbf{q}'}{\partial t} + \nabla \cdot (L(\mathbf{q}') + N(\mathbf{q}', \overline{\mathbf{q}}) + N(\overline{\mathbf{q}}, \mathbf{q}') - \overline{\mu}_t G(\mathbf{q}') - \mu_t' G(\overline{\mathbf{q}})) = \mathbf{0} \\ \dfrac{\partial \mu_t'}{\partial t} + \dfrac{\partial H}{\partial \mu_t}(\overline{\mathbf{q}}, \overline{\mu}_t)\mu_t' + \dfrac{\partial H}{\partial \mathbf{q}}(\overline{\mathbf{q}}, \overline{\mu}_t)\mathbf{q}' = 0. \end{cases} \tag{25.2}$$

By discretising Eq. (25.2), the following iterative problem is obtained:

$$\begin{pmatrix} \mathbf{q}' \\ \mu_t' \end{pmatrix}_{k+1} = e^{\mathbf{J}\tau} \begin{pmatrix} \mathbf{q}' \\ \mu_t' \end{pmatrix}_k \tag{25.3}$$

where $\mathbf{J}$ is the discretised Jacobian of the system Eq. (25.2) and τ is a sampling period. In the time-stepping framework [8], the spectrum of the Jacobian is extracted by solving over time the discretised linear equations and constructing a reduced order matrix that will approximate the spectrum of the Jacobian. The non linear generative solver dNami [4] is used to discretise the URANS model. To get the linearised solver capable of solving Eq. (25.3), the source transformation automatic differentiation tool Tapenade [5] is used. We solved the fixed point and eigenvalue problem in a matrix free manner by using the linear algebra solving packages PETSc and SLEPc [9, 10]. In the following, the eigenvalues of **J** are denoted $\lambda = \sigma + i\omega$ where σ is the temporal growth rate and ω the circular frequency.

25.3 Numerical Configuration

The 2D URANS model with Spalart-Allmaras is solved by using explicit centered finite differences of 4th order with a high order flux dissipation scheme of 9th order from [12] without shock capturing sensors ($k_2 = 0$). A multi-block structured grid is used with a total number of nodes $N_x \times N_y = 833 \times 898$, where N_x, N_y represents the number of points in the streamwise and spanwise directions respectively. The computational domain is defined by a rectangle spanning from coordinates $(x_0, y_0) = (-20D, -30D)$, to coordinates $(x_1, y_1) = (150D, 30D)$, with physical domain spanning from coordinates $(x_0, y_0) = (-10D, -10D)$ and $(x_1, y_1) = (50D, 10D)$. The rest of the domain is filled with sponge layers to reduce spurious oscillations. Non reflective boundary conditions are imposed in the far-field, and no-slip adiabatic boundary conditions are imposed on the nozzle walls and on the flat plate. Figure 25.1 shows a sketch of the geometrical configuration of the jet-plate. Table 25.1 shows the parameters of the flow for each case studied. Each quantity has been adimensionalised by reference values taken inside the nozzle, $(U_{jet}, \rho_{jet}, T_{jet})$ and the diameter of the nozzle D.

25.4 Global Stability Analysis Results

At first, Eq. (25.1) is solved for all three configurations. In all cases, the URANS calculations converged to a stationary solution. By using a Newton-Krylov method as

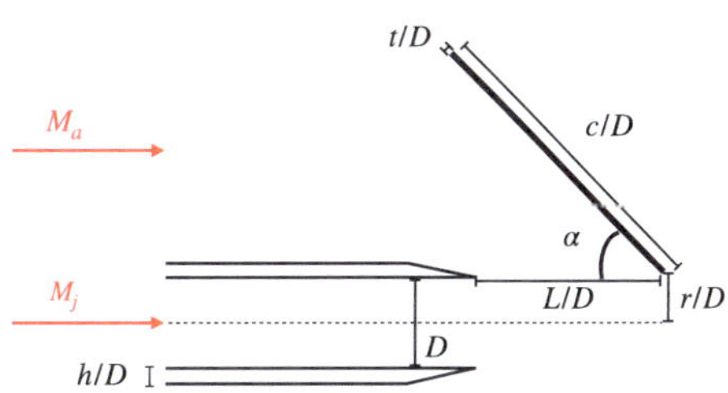

Fig. 25.1 Sketch of the jet plate interaction with important geometrical parameters.

Table 25.1 Cases configuration parameters. D is the nozzle height, L/D is the nozzle plate horizontal distance, c/D is the plate length, t/D is the plate thickness, r/D is the nozzle height from the middle line to the plate, α is the plate incident angle, M_j is the jet Mach number and M_a is the co-flow Mach number

	L/D	c/D	t/D	r/D	α	M_j	M_a	Re
Case 1	3	5	0.025	0.6	45°	0.7	0.1	$7.5 * 10^5$
Case 2	3	5	0.025	0.7	0°	0.8	0.1	$7.5 * 10^5$
Case 3	–	–	–	–	–	0.7	0.1	$7.5 * 10^5$

a convergence accelerator, fixed point solutions are calculated. Each fixed point has been converged until the normalised residual error satisfied the following inequality:

$$\frac{||\overline{\mathbf{q}}(t+\tau) - \overline{\mathbf{q}}(t)||_2}{N_{phy}} < 10^{-6}$$

where t is the adimensionalised time, τ is the sampling time and N_{phy} is the total number of points in the physical domain (without counting the sponge layer). The adimensionalised sampling time $\tau = 0.18$ is chosen so that each period of the instability is resolved by approximately 10 snapshots. We performed global stability analysis by perturbing the fixed point in all state variables $\mathbf{q}' = \left[\rho', (\rho u)', (\rho v)', (\rho e_t)', (\rho \nu_t)'\right]^T$. Figure 25.2 shows the eigenvalues for the three different case studied. All eigenvalues are stable for the cases considered, coherently with the stationary result of the URANS simulations. The plate affects the spectra, since the vertical discrete branches of modes at approximately $St \approx 0.53, 1.1, 1.6$, emerges only in cases where the plate

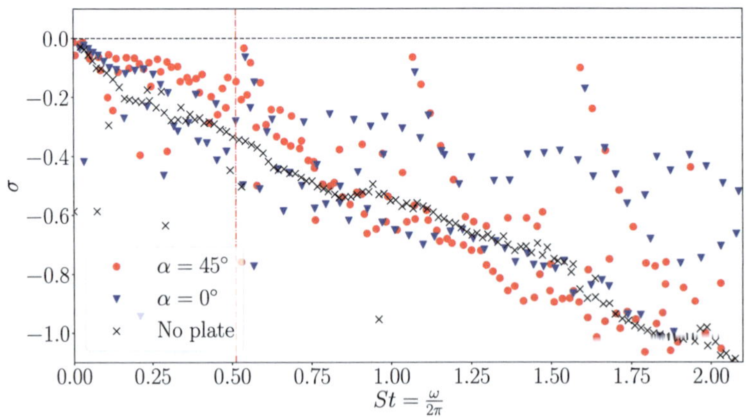

Fig. 25.2 Eigenvalues for different plate incidence angle configurations. Black cross is the case where no plate is present. Blue triangles is the $\alpha = 0°$, red dots is the $\alpha = 45°$ and the vertical dashed line is the peak Strouhal number measured in the experiment for $\alpha = 45°$, $St_{expe} = 0.511$

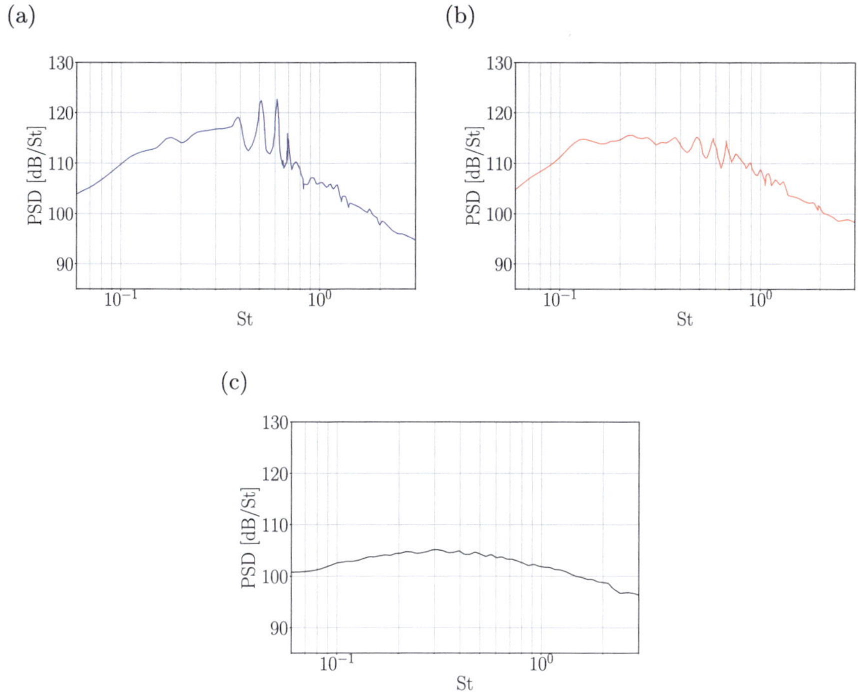

Fig. 25.3 Experimental data collected at R = 14.3D radial distance from the nozzle. The parameter values are in Table 25.1. **a** strong tones. **b** Weak tones. **c** Broadband noise spectrum. In (**c**) no plate is present. Proprietary data from [3]

is present. The modes have a tendency to become more stable when the plate passes from oblique to parallel, indicating that the angle of the plate has an effect on the amplification rate of the modes of the discrete vertical branch. This shows that linear analysis can qualitatively capture this dependency on the angle of the plate that has been observed in the experiments (e.g. Fig. 25.3). When the plate is not present, the spectrum contains a continuous branch of convective modes, with no frequency or discrete vertical branch observed. A comparison between stability results and experiments from [3] has been performed. Figure 25.3 shows the acoustic spectrum measured from experiments for the same configurations as the simulations performed in this work. The highest peak of the acoustic spectrum measured from the experiment is compared with the least stable mode of the branches calculated from global stability analysis. In the experiment, the highest tone appears at $St_{e,obl} = 0.511$ for the oblique plate and $St_{e,hor} = 0.49$ for the horizontal plate. In stability analysis, the least stable modes from the oblique and horizontal plate configurations have almost identical oscillation frequency at $St_{st,obl} = 0.53$, $St_{st,hor} = 0.531$. Even if we do not experience this shift in frequency as in the experiment, the two Strouhal numbers remain quite close to the ones measured in the experiment. However, this comparison has to take into account that the simulations performed here are two dimensional and

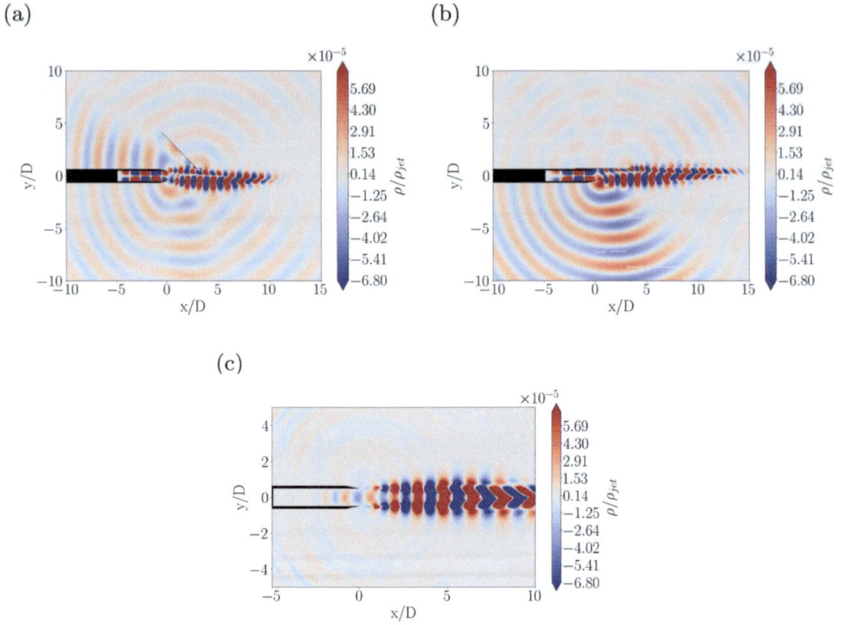

Fig. 25.4 **a** Real part of the density distribution of the eigenvector corresponding to the most unstable mode of the branch for $\alpha = 45°$. **b** Real part of the density distribution of the eigenvector corresponding to the most unstable mode of the branch for $\alpha = 0°$. **c** Real part of the density distribution of the eigenvector corresponding to the mode of the branch at $St \approx 0.5$ for the free jet

consider a planar jet, while the experiments where performed over a rounded jet. Figure 25.4 shows the modes for the cases $\alpha = 45°$, $\alpha = 0°$ and for the free jet. The spatial distribution of the modes for the plate configuration presents clear acoustic waves scattering from the plate. A Kelvin-Helmholtz antisymmetric instability propagating downstream can be observed, impacting the plate and generating acoustic waves. An upstream wave can be observed entering the nozzle when animating the mode by combining the real and imaginary part using the following formula:

$$\rho'(x, y, t) = \Re\left[\rho(x, y)e^{\lambda t}\right] = e^{\sigma t}\left[\rho_r(x, y)\cos(\omega t) - \rho_i(x, y)\sin(\omega t)\right] \quad (25.4)$$

where ρ_r, ρ_i are respectively the real and imaginary part of the mode ρ. This upstream wave can be linked to the k^- mode described in [1, 11]. This mode, called the "trapped" mode, originates in the potential core of the jet and it is responsible for the closure mechanism of the feedback loop. In the free jet case, there is no visible upstream mode, reinforcing the idea that the tonal noise is supported by a feedback mechanism. The modes also highlights the acoustic shielding effect of the plate. There are stronger acoustic waves directed upstream and in the negative y direction for the oblique plate, and in the negative y direction for the horizontal plate case. The shielding effect is stronger in the horizontal plate case, as observed in experiments.

25.5 Conclusion

A numerical framework composed of an high order finite difference solver, capable of automatically generate linearised equations of turbulent models and to perform global stability analysis on geometrically complex turbulent cases is presented. Part of the the acoustic components are captured using low fidelity methods like URANS, being able to determine the impact of the plate in the eigenspectrum. Further work will be needed to generalise this framework in 3D and being able to simulate the combination of a round jet and a flat plate. Exploring if URANS could be destabilised by descending the plate further down into the jet is needed to improve the consistency of the method.

Acknowledgements This research was funded by Airbus and the DGAC through the MAMBO Project. We are indebted to Drs. Jérôme Huber and Grégoire Pont for the many fruitful insights.

References

1. Huber, J., Pont, G., Jordan, P., et al.: Wavepacket Modelling of Jet-Flap Interaction Noise: from Laboratory to Full-Scale Aircraft. Flow Turbulence Combust **113**, 773–802 (2024)
2. Pepper, C.B., Nascarella, M.A., Kendall, R.J.: A review of the effects of aircraft noise on wildlife and humans, current control mechanisms, and the need for further study. Environ Manage. **32**(4), 418–32 (2003)
3. Amaral, F.R., Lebedev, A., Jordan, P. (2023). Experiments on installed jet noise. AIAA AVIATION 2023 Forum
4. Alferez, N., Touber, E., Winn, S., Ali, Y. (2022). dNami: a framework for solving systems of balance laws using explicit numerical schemes on structured meshes. Zenodo. https://doi.org/10.5281/zenodo.6720593
5. Hascoet, L., Pascual, V.: The Tapenade automatic differentiation tool: Principles, model, and specification. ACM Trans. Math. Softw. **39**, 3 (2013)
6. Frantz, R., Loiseau, J.C., Robinet, J.C.: Krylov Methods for Large-Scale Dynamical Systems: Application in Fluid Dynamics. Applied Mechanics Reviews. **75**, 1–30 (2023). https://doi.org/10.1115/1.4056808
7. Spalart, P., Allmaras, S. (1992). A one-equation turbulence model for aerodynamic flows. AIAA 1992-439. 30th Aerospace Sciences Meeting and Exhibit
8. Bagheri, S., et al.: Matrix-Free Methods for the Stability and Control of Boundary Layers. AIAA Journal **47**(5), 1057–1068 (2009)
9. Balay, S. et al.: PETSc Web page (2023). https://petsc.org/
10. Hernandez, V., Roman, J.E., Vidal, V.: SLEPc: A scalable and flexible toolkit for the solution of eigenvalue problems. ACM Trans. Math, Software (2005)
11. Jordan, P., Jaunet, V., Towne, A., Cavalieri, A.V.G., Colonius, T., Schmidt, O., Agarwal, A.: Jet–flap interaction tones. Journal of Fluid Mechanics **853**, 333–358 (2018)
12. Sciacovelli, L., Passiatore, D., Cinnella, P., Pascazio, G. (2021). Assessment of a high-order shock-capturing central-difference scheme for hypersonic turbulent flow simulations, Computers & Fluids,Volume 230

Chapter 26
Infrared Thermographic Evaluation of Wall-Temperature Streak in the Free-Stream Turbulence Transition Scenario

Shumpei Hara, Shinya Nagino, and Kyoji Inaoka

Abstract The present study experimentally investigated streaky structures of wall temperature in the free-stream turbulence transition scenario. An infrared thermographic camera acquired time series data of the wall-temperature distribution on a flat plate employing a heating section made of 5 μm titanium foil installed in a wind tunnel with a turbulence grid. Streamwise-oriented wall-temperature streaks collapsed with increasing free-stream Reynolds number. High- and low-temperature streaks are discussed in comparison with velocity streaks reported in previous studies.

Keywords Wall-temperature streak · Experiment · Boundary-layer transition · Free-stream turbulence

26.1 Introduction

It is necessary to understand the convective heat transfer in a boundary layer with free-stream turbulence (FST) together with the flow state, for example considering turbine blade cooling problems and the heat exchanger inlet section. In particular, the information about the laminar-turbulent boundary layer transition with the FST is useful for the prevention of thermal fatigue fracture or thermomechanical loading of a material. The spatial pattern and magnitude of the wall-temperature fluctuation determine the crack network configuration and growth [1]. The FST-transition is successively caused by the growth and continuous collapse of low-velocity streaks near the wall [2, 3]. At turbulent intensity Tu over 1% the FST transition triggered by the collapse of the streaky structure is expected to occur before the linear growth of the two-dimensional unstable waves. The FST-transition generally known as the bypass transition or K-mode is an appropriate representation, involving the receptivity process of disturbance from FST. Fransson and Shahinfar [4] have proposed a predictive formula of the transition Reynolds number on streamwise integral length and Tu

S. Hara (✉) · S. Nagino · K. Inaoka
Doshisha University, Kyoto, Japan
e-mail: shhara@mail.doshisha.ac.jp

K. Kato et al. (eds.), *Proceedings of the 10th IUTAM Symposium on Laminar-Turbulent Transition*, IUTAM Bookseries 44, https://doi.org/10.1007/978-981-96-9829-5_26

level in a zero-pressure gradient boundary layer in the FST-transition. Convective heat transfer in FST-transition scenario in literature was examined from only Tu level view, and so it is necessary to revisit the issue from a new perspective including the length scale [5]. The streak magnitude and its collapse process due to its secondary instability greatly affect the temperature gradient on the material surface. The wall-temperature distribution of a passive scalar reflects features in the velocity field [6]. However, it is tough to obtain the wall distribution of wall-temperature streaks corresponding to velocity streaks because of low thermal response. The temperature attenuation due to thermal inertia of heating surface contaminates the wall-temperature distribution in the experiment [7]. The present study experimentally investigates the wall-temperature distribution in FST-transition in a zero-pressure-gradient boundary layer and employs a heating section made of 5 μm titanium foil of a small heat capacity.

26.2 Flow Configurations and Experimental Setup

The measurement was performed in a closed-circuit wind tunnel with a 1 m test section and 0.6 m × 0.6 m (height × width) cross-sectional area, located at Doshisha University (Fig. 26.1). The experiments were carried out over a 0.75 m flat plate with a 0.02 m thickness. The flat plate was located at 0.18 m from the bottom of the cross-sectional area. It employed an asymmetric leading edge to get a short pressure gradient region without a suction peak [8]. The origin is on the centerline at the leading edge with x, y, and z axis directed along the streamwise, wall-normal, and spanwise direction, respectively.

A 0.3 m heating section in a 0.11 m spanwise width started at 0.28 m from the leading edge and had a multi-layer made of a 5 μm titanium foil, a 1 mm air layer, a

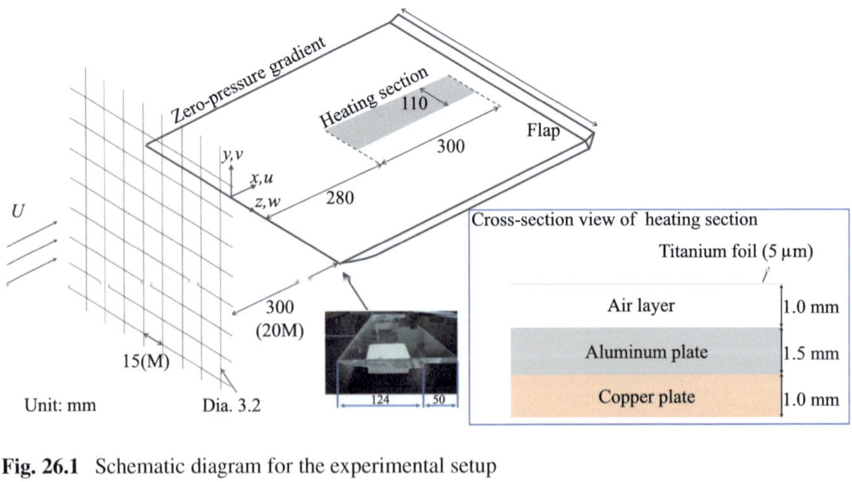

Fig. 26.1 Schematic diagram for the experimental setup

Table 26.1 Free-stream turbulence conditions

U_∞ (ms^{-1})	Tu (%)	Λ_x (mm)	Re_Λ	$Re_{FST} (\equiv TuRe_\Lambda)$
0.80	3.79	10.1	701	26.6
2.0	3.91	10.6	1402	54.8
3.1	3.87	15.8	3150	122

1.5 mm aluminium plate, a 1 mm copper plate, and a balsa wood as an insulator. The titanium foil with electric power (7.3 W) was in iso-heat flux condition. The titanium foil with a small heat capacity provides a high thermal response of around 10 Hz to a wall-temperature fluctuation. Therefore, it is enough to resolve the distribution of wall-temperature streaks. Time series data of the wall-temperature distribution were acquired using an infrared thermographic camera (FLIR SC7500) of 320 × 256 pixels driven at 200 Hz with 60 s sampling time, according to Nakamura and Yamada [7]. The catalogue value of the noise-equivalent temperature difference (NETD) was 0.018 K. Streamwise and spanwise spatial resolutions were 0.70 mm and 0.55 mm. It is noted that the heated titanium foil was stretched by the thermal expansion of the air layer. Nakamura and Yamada [7] have reported that the root-mean-square (rms) value of the vertical displacement of the titanium foil due to the vibration in the flow was less than 1 μm because the thermal expansion coefficient of the titanium was smaller than that of the acrylic resin. Although we did not measure the rms value of the vertical displacement, the same order of magnitude as expected does not affect the free-stream turbulence transition scenario.

Free-stream turbulence was generated using a turbulence grid of 3.2 mm bar diameter and 15 mm mesh width (M), located 300 mm ($= 20M$) upstream from the leading edge. Hot-wire anemometer measurements in the free stream at the leading edge were performed to characterise the inflow condition of Tu and integral length scale (Λ_x). The streamwise turbulence intensity is defined as the ratio between the rms value of the streamwise velocity (u_{rms}) and the free-stream velocity (U_∞). The integral length scale using Taylor's hypothesis was obtained by integrating the autocorrelation function of the velocity signal. Physical properties were estimated at the free-stream temperature. Free-stream Reynolds numbers based on the integral length scale are Re_Λ and Re_{FST}. A hot wire of tungsten had a wire diameter and length of 5 μm and 1.0 mm, respectively. The hot-wire signal in the 10 s sampling time was acquired with a frequency of 10 kHz. Table 26.1 summarises the inflow condition. All FST conditions at the leading edge show Tu around 3.85% and the Λ_x variation of 10.1–15.8 mm.

26.3 Results and Discussion

Figure 26.2 shows time sequences of the distribution of the wall-temperature fluctuation T_w' normalised with the difference between the mean wall temperature $\overline{T_w}$ and free-stream temperature T_∞. Incidentally, the time interval between three contour plots in each condition, when multiplied by the free-stream velocity corresponds to the distance $\Delta x \approx 60$ mm. It is noted that the streamwise Reynolds numbers in Fig. 26.2 were 1.3×10^4–2.2×10^4, 3.3×10^4–5.6×10^4, and 5.1×10^4–8.4×10^4 at $U_\infty = 0.80$, 2.0, and 3.1 m/s, respectively. Streamwise-oriented streaks of high- and low-temperature regions were arranged in the spanwise direction. All wall-temperature streaks move downstream in time and meander in the spanwise direction (See Supplement). High-temperature streaks merge, split, and disappear, while low-temperature streaks only split, and disappear. The wall-temperature streaks appear as accumulated temperature fluctuations that passively follow the flow field. The movement of a low-momentum hot fluid away from the heating wall produces high-temperature and low-velocity streaks. A low-momentum fluid with a wall-normal velocity disturbance is stretched in the streamwise direction and becomes a low-velocity streak, as explained by linear stability theory [9]. The feature of the wall-temperature streaks observed in this study is considered to represent the sinuous mode collapse of the low-velocity streak. It has also been reported that quasi-streamwise vortices appear immediately next to the low-velocity streak during this collapse [10].

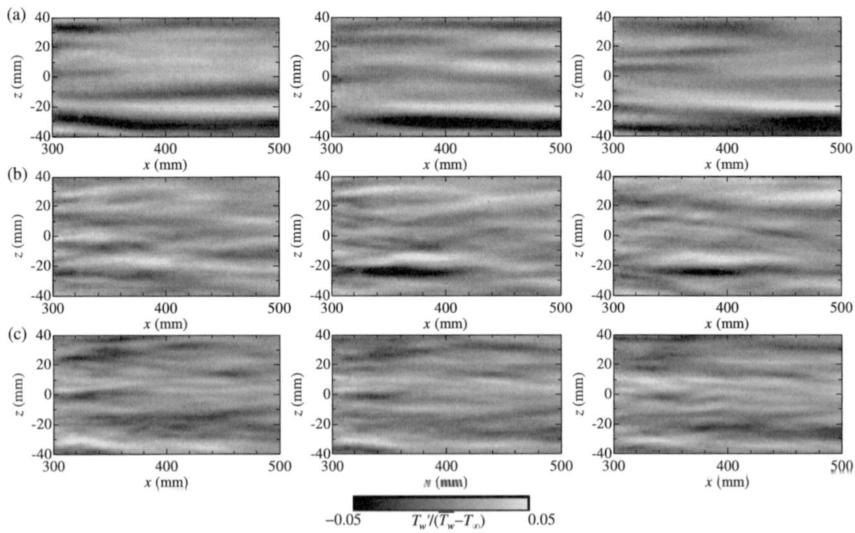

Fig. 26.2 Time sequence showing the evolution of streak of the wall-temperature fluctuation for: **a** $U_\infty = 0.80$ m/s with 25 ms between maps; **b** $U_\infty = 2.0$ m/s with 10 ms; **c** $U_\infty = 3.1$ m/s with 5.5 ms. The flow is from left to right

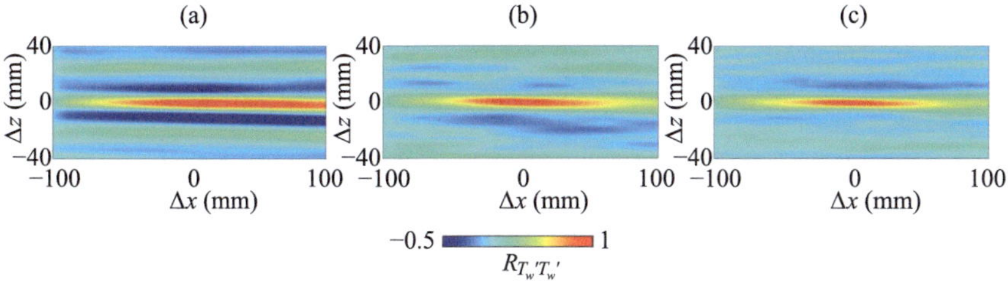

Fig. 26.3 Two-point correlation of the wall-temperature fluctuation at $z = 0$ for **a** $U_\infty = 0.80$ m/s; **b** $U_\infty = 2.0$ m/s; **c** $U_\infty = 3.1$ m/s

This can be interpreted as a pair of high- and low-temperature streaks observed as the free-stream low-temperature fluid that is produced by these vortices (see Fig. 26.2).

To obtain more convincing results, the averaged spatial two-point correlation function of the wall-temperature fluctuation $R_{T_w' T_w'}$,

$$R_{T_w' T_w'}(\Delta x, \Delta z, x_R, z_R) = \frac{\overline{T_w'(\Delta x, \Delta z)\, T_w'(x_R, z_R)}}{\sqrt{\overline{T_w'^2(x_R, z_R)}}}, \tag{26.1}$$

is helpful in discussing streaks in averaged fields. The reference point (x_R, z_R) was set at the centre of the x-z plane. Positive and negative regions indicating streamwise-oriented wall-temperature streaks were also confirmed in the conditional average result, as shown in Fig. 26.3. With increasing free-stream Reynolds number, negative correlation regions with decreasing magnitude became unclear and both regions shrank in actual size. This indicates that wall-temperature streak structures observed in 300 mm $< x <$ 500 mm collapse in the FST-transition scenario instead of the appearance of turbulent spots.

The averaged half-spanwise wavelength $\lambda_z/2$ of wall-temperature streaks was determined from a local minimum value of an averaged spanwise correlation of T_w' at each x. In Fig. 26.4, $\lambda_z/2$ depending on Λ_x increased downstream in the range of about 10-12 mm (Cf. Table 26.1) and matched the spanwise scale of velocity streak [3]. Fransson and Shahinfar [4] proposed the scaling of the spanwise wavelength of the velocity streaks at x with the 0.5 intermittency, namely the laminar-to-turbulent transition location. Figure 26.5 shows the Re_{FST} dependency of λ_z of wall-temperature streaks at $x = 500$ mm, normalized with Tu and Λ_x. The values for $U_\infty = 0.8$ m/s and 2.0 m/s deviated because the FST transition has not yet been completed in the snapshot. The averaged spanwise wavelength of wall-temperature streaks for $U_\infty = 3.1$ m/s was consistent with the empirical equation. In the present experimental condition, it is concluded that the spanwise wavelength of wall-temperature streaks was the same scale as that of the velocity streak. It would be of interest to consider the qualitative difference between both wall-temperature and velocity streaks with widely varying Re_{FST} in future work. Governing equations of momentum and scalar are inherently different, so the scalar field immediately does not reflect the impact of

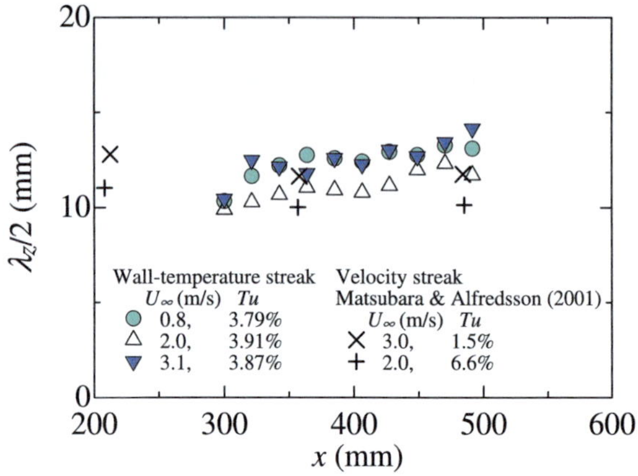

Fig. 26.4 Averaged spanwise wavelength of wall-temperature streaks versus x

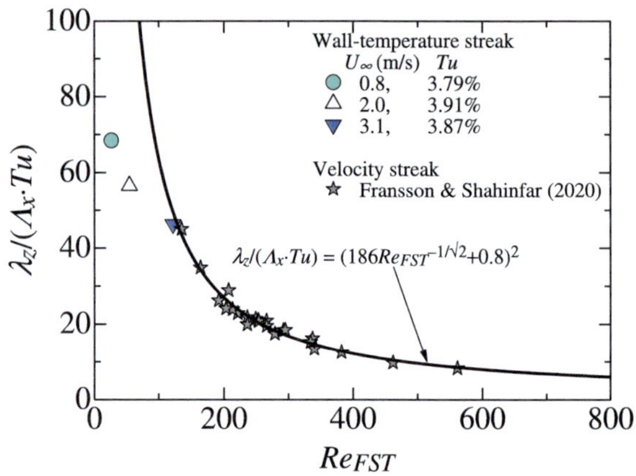

Fig. 26.5 Averaged spanwise wavelength of wall-temperature streaks scaled with Tu and Λ_x versus Re_{FST}

the change in the velocity fields. It is possibility that wall-temperature streaks may be larger than the velocity streaks, depending on the conditions. The collapse of wall-temperature streaks shown in this study will contribute to understanding the FST transition scenario superimposed with curvature (pressure gradient) and cross-flow instability in turbine blade cooling problems.

26.4 Conclusion

In this work, the wall-temperature distribution in FST-transition in a zero-pressure gradient boundary layer is experimentally investigated. An infrared thermographic camera acquired time series data of the wall-temperature distribution on a flat plate employing a heating section made of 5 μm titanium foil.

Streamwise-oriented wall-temperature streaks were confirmed and collapsed with increasing free-stream Reynolds number in the FST-transition scenario. Moreover, high-temperature streaks merged, split, and disappeared, while low-temperature streaks only split, and disappeared. The feature was considered to represent the sinuous mode collapse of the low-velocity streak. In the present experimental condition, the spanwise wavelength of wall-temperature streaks was the same scale as that of the velocity streak [3, 4].

References

1. Maillot, V., Fissolo, A., Degallaix, G., Degallaix, S.: Thermal fatigue crack networks parameters and stability: an experimental study. Int. J. Solids Struct. **42**, 759–769 (2005). https://doi.org/10.1016/j.ijsolstr.2004.06.032
2. Westin, K.J.A., Boiko, A.V., Klingmann, B.G.B., Kozlov, V.V., Alfredsson, P.H.: Experiments in a boundary layer subjected to free stream turbulence. Part 1. Boundary layer structure and receptivity. J. Fluid Mech. **281**, 193–218 (1994). https://doi.org/10.1017/S0022112094003083
3. Matsubara, M., Alfredsson, P.H.: Disturbance growth in boundary layers subjected to free-stream turbulence. J. Fluid Mech. **430**, 149–168 (2001). https://doi.org/10.1017/S0022112000002810
4. Fransson, H.M.J., Shahinfar, S.: On the effect of free-stream turbulence on boundary-layer transition. J. Fluid Mech. **899**, A23 (2020). https://doi.org/10.1017/jfm.2020.444
5. Lienhard, J.H.: Heat transfer in flat-plate boundary layers: A correlation for laminar, transitional, and turbulent flow. J. Heat Trans. **142**, 061805 (2020). https://doi.org/10.1115/1.4046795
6. Stafford, J., Walsh, E., Egan, V.: The effect of global cross flows on the flow field and local heat transfer performance of miniature centrifugal fans. Int. J. Heat Mass Trans. **55**, 1970–1985 (2012). https://doi.org/10.1115/1.4046795
7. Nakamura, H., Yamada, S.: Quantitative evaluation of spatio-temporal heat transfer to a turbulent air flow using a heated thin–foil. Int. J. Heat Mass Trans. **64**, 892–902 (2013). https://doi.org/10.1016/j.ijheatmasstransfer.2013.05.006
8. Fransson, H.M.J.: Leading edge design process using a commercial flow solver. Exp. Fluids **37**, 929–932 (2004). https://doi.org/10.1007/s00348-004-0858-3
9. Landahl, M.T.: On sublayer streaks. J. Fluid Mech. **212**, 593–614 (1990). https://doi.org/10.1017/S0022112090002105
10. Brandt, L., Schlatter, P., Henningson, D.S.: Transition in boundary layers subject to free-stream turbulence. J. Fluid Mech. **517**, 167–198 (2004). https://doi.org/10.1017/S0022112004000941

Chapter 27
Adjoint-Based Approach for Prediction of Non-modal Growth in Spatially Evolving Flows

D. Ohno and M. Karp

Abstract A novel method for approximating transient growth in spatially evolving flows within the framework of local stability theory by using bi-orthogonality is presented. Although the local linear stability analysis is based on the parallel-flow assumption, it is able to accurately predict modal growth of discrete waves in boundary-layer flows using the e^n-method. However, the classical local approach fails for the concept of non-modal disturbances, where the growth is substantially under-predicted. In this study, it is shown that the prediction can be significantly improved by employing local adjoint stability solutions for a sequential downstream projection of the disturbances. While still using local theory with the parallel-flow assumption, this can be physically explained by the redistribution of energy between different scales, which appears to be one of the aspects of non-local disturbance growth. Results for the lift-up mechanism in a Blasius flow are presented and discussed.

Keywords Boundary layers · Transition prediction · Non-modal growth

27.1 Introduction

In flows exposed to high levels of external perturbations, the phenomenon of algebraic or non-modal growth of disturbances can be responsible for the laminar-turbulent transition. In this path to turbulence, which is also known as transient growth, the exponential growth of modal disturbances—for example, Tollmien–Schlichting (TS) waves—is outpaced. This is ascribed to the nature of the eigenmodes from the linearized governing equations, which are decaying individually but may grow when combined due to their non-orthogonality.

D. Ohno (✉) · M. Karp
Faculty of Aerospace Engineering, Technion – Israel Institute of Technology, Haifa, Israel
e-mail: ohno@campus.technion.ac.il

M. Karp
e-mail: mkarp@technion.ac.il

K. Kato et al. (eds.), *Proceedings of the 10th IUTAM Symposium on Laminar-Turbulent Transition*, IUTAM Bookseries 44, https://doi.org/10.1007/978-981-96-9829-5_27

The phenomenon of transient growth in a spatially evolving zero-pressure gradient (ZPG) boundary-layer flow is commonly associated with the lift-up effect. This case, often referred to as optimal growth, is characterized by initial streamwise vortices and was first described by Landahl [8]. The counter-rotating vortices give rise to streamwise elongated streaks via lift-up of low-momentum fluid near the wall (see e.g. Luchini [9]), and may become unstable to secondary instabilities (see e.g. Karp and Cohen [7]). A parameter study employing the parallel-flow assumption by Butler and Farrell [1], showed that the lift-up mechanism exhibits the maximum spatial energy gain for the steady solution.

Transient growth is investigated numerically by solving the linearized Navier–Stokes equations (see e.g. Juniper et al. [5]). In the "classical" local stability theory (LST), a one-dimensional eigenvalue problem is solved with the parallel-flow assumption, in which the base flows' wall-normal velocity and the streamwise derivatives are neglected. In the case of modal growth, e.g. with TS waves, the e^n-method based on the local stability analysis proves to be a robust transition prediction tool even for (weakly) non-parallel flows. However, for non-modal growth in spatial problems, the local theory under-predicts the disturbance growth due to the flow non-parallelism (see e.g. Tumin et al. [11]). One method for calculating transient growth in spatially evolving flows (e.g. the development of the streaks) is to employ direct numerical simulations (DNS).

A less costly alternative is the utilization of parabolized stability equations (PSE), wherein the downstream evolution of disturbances is assumed to be parabolic and solved as an initial value problem (see e.g. Herbert [4] or Juniper et al. [5]). Consequently, this non-local method accounts for non-parallel effects and is therefore well suited for spatial problems. However, this approach requires iterations to satisfy a given auxiliary condition and tends to be numerically unstable for small step sizes in the downstream direction Δx. Furthermore, it requires to solve all involved modes individually.

The present work proposes an approach where the spatial disturbance growth in non-parallel/non-local flows is approximated with local stability analysis using bi-orthogonality. The approach may offer a cost-effective alternative to PSE, since it does not require iterations.

27.2 Mathematical Method

In this work, a flow in a Blasius boundary layer (see Fig. 27.1a) is considered, where the velocity field is denoted $\boldsymbol{u}(t, x, y, z) = (u, v, w)^{\mathrm{T}}$, where u, v and w are the velocity components in the streamwise, wall-normal and spanwise coordinates (x, y, z), respectively, with t as the time. The velocity field is normalized with the free-stream velocity U_∞, while the length scales are scaled by the boundary-layer thickness δ at $0.99U_\infty$ and the time by δ/U_∞. The Reynolds number is defined as $Re = U_\infty\, \delta/\nu$, where ν is the kinematic viscosity.

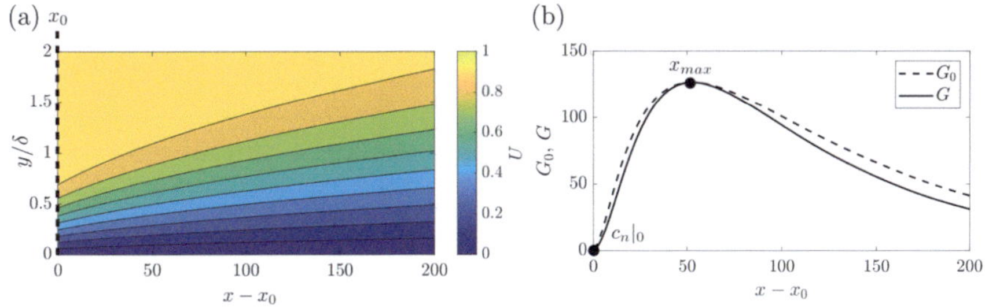

Fig. 27.1 Blasius with $Re = 800$. **a** Streamwise velocity U of base flow **b** Optimal growth envelope, G_0, and disturbance growth, G, based on $c_n|_0$ for $\omega = 0$ and $\beta = 1.85$

A brief introduction to the mathematical framework of the classical local approach to calculate transient growth is provided here (further details can be found in reference [10] by Schmid and Henningson), which is followed by the non-local extension using the adjoint-based projection.

First, local linear stability analysis based on the parallel-flow assumption is used to determine eigenmodes at position x_0. The matrix of the eigenvalue problem is solved with the spectral Chebyshev collocation method (see e.g. Karp and Hack [6]). When solving the eigenvalue problem for the spatial theory with the parameters Re, ω (frequency) and β (spanwise wavenumber), complex eigenvectors $\hat{\boldsymbol{q}}_n(y)$ with respective complex eigenvalues α_n are obtained. A finite number N_m of the most highly amplified (or least damped) modes are selected to construct the perturbation field

$$\boldsymbol{q}' = \tilde{\boldsymbol{q}}(x, y)e^{\mathrm{i}(\beta z - \omega t)}, \qquad \text{with} \qquad \tilde{\boldsymbol{q}}(x, y) \triangleq \sum_{n=1}^{N_m} c_n|_0 \, \hat{\boldsymbol{q}}_n(y) e^{\mathrm{i}\alpha_n x}, \tag{27.1}$$

with $\boldsymbol{q}' = (u', v', w', p')^{\mathrm{T}}$ for the velocities and pressure p, where $c_n|_0$ are the expansion coefficients of the eigenmodes. The streamwise extension of the eigenmode $\tilde{\boldsymbol{q}}(x, y)$ can be considered as the envelope of the disturbance field.

Generally, the transient growth is expressed with the gain $G(x) = E(x)/E_0$ which is calculated with a norm of the kinetic energy, $E(x)$, based on the resulting disturbance field. In order to calculate the $c_n|_0$ for the maximum gain, it is necessary to determine the optimal growth G_0 via Cholesky factorization of the energy matrix; see dashed line in Fig. 27.1b. The specific inflow conditions, $c_n|_0$, for each point on the envelope are determined by employing the singular value decomposition (SVD) (see e.g. Hack and Moin [3]). Finally, the resulting disturbance field with Eq. (27.1) can be constructed from which $G(x)$ is calculated; see solid line in Fig. 27.1b for the maximal gain, which occurs at x_{max}. It should be noted that in this standard approach for calculating transient growth (see Tumin [11]), the stability analysis is only performed at one *fixed* position x_0, which corresponds to the inflow. However, as can

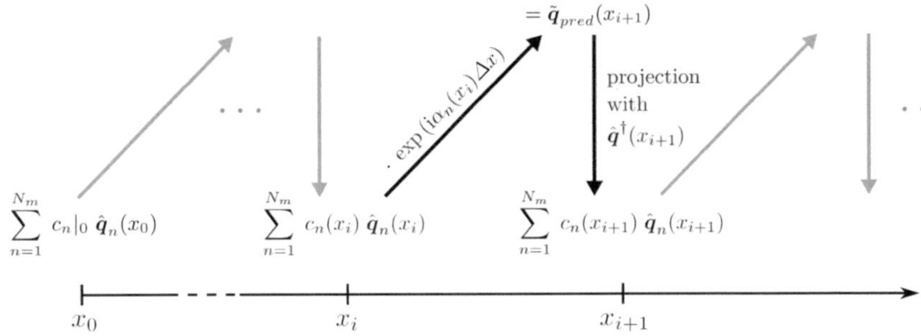

Fig. 27.2 Numerical method of the adjoint-based downstream integration and projection

be seen in Fig. 27.1a, the boundary layer is growing considerably over the course of the amplification process, which is not taken into account here.

The proposed method to predict non-modal growth in spatially evolving flows necessitates a downstream projection using the adjoint eigensolution. Like the regular stability analysis, the adjoint equations (see details in e.g. Hill [2] or Tumin [12]) are numerically solved at every position x. The resulting complex adjoint eigenfunctions are denoted with $\hat{\boldsymbol{q}}^\dagger$. In contrast to the classical local approach in a fixed system, see Eq. (27.1), the disturbance field for the projection approach is formulated with

$$\tilde{\boldsymbol{q}}(x, y) = \sum_{n=1}^{N_m} c_n(x)\, \hat{\boldsymbol{q}}_n(x; y). \tag{27.2}$$

Here, the eigenfunctions $\hat{\boldsymbol{q}}_n$ depend on the local stability solutions at each station in downstream direction x. Furthermore, the x-dependent complex coefficients c_n are introduced, which are acquired via projection with the local adjoint solution at each station x. The coefficients can be regarded as a correction of the modes which physically represents a redistribution of energy between scales. The inflow coefficients from the local formulation, $c_n|_0$, are used. This enables a direct comparison to local solution in the fixed system as well as to the DNS results.

Finally, the performed regular and adjoint local stability analyses can be used for the novel non-local scheme, which is depicted in Fig. 27.2. Starting from the inflow with the initial coefficients $c_n|_0$, the method successively applies two mathematical procedures for each step Δx from x_i to x_{i+1}: an integration in downstream direction by considering the local amplification rate of each mode from LST with $e^{\mathrm{i}\,\alpha_n(x_i)\Delta x}$, resulting in the field $\tilde{\boldsymbol{q}}_{pred}(x_{i+1})$; and the subsequent projection employing the adjoint eigenmodes $\hat{\boldsymbol{q}}^\dagger(x_{i+1})$ onto the predicted disturbance field, yielding new local coefficients at $c_n(x_{i+1})$. In some sense, this can be seen as a predictor-corrector method, where the local solution is first marched based on the LST, and then corrected by the projection method via adjoint solution according to the stability solution at the next station. As elaborated in the next section, enabling the coefficients to evolve with x is essential to account for the redistribution of energy between different scales.

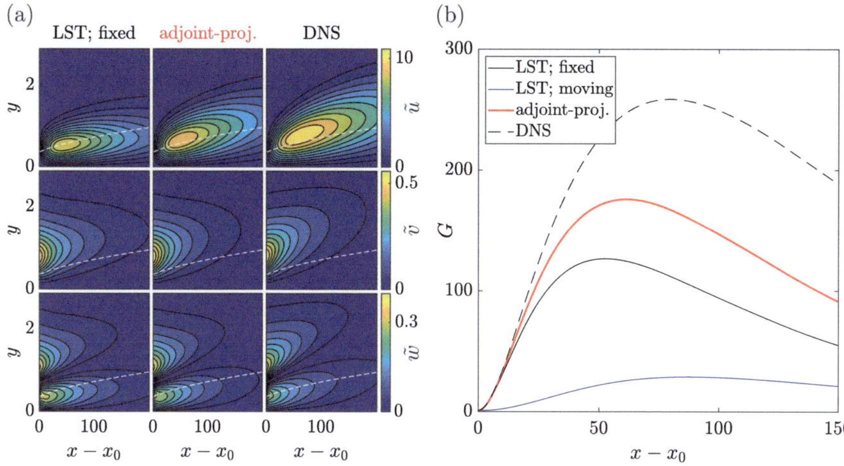

Fig. 27.3 Results of different approaches to calculate transient growth for $Re = 800$ with $\omega = 0$ and $\beta = 1.85$, with given initial conditions $c_n|_0$. **a** Disturbance velocity fields; dashed white line: displacement thickness $\delta^*(x)$. **b** Disturbance energy growth $G(x)$

27.3 Results

The method is demonstrated for the lift-up mechanism with the inflow Reynolds number $Re = 800$ at $\omega = 0$ and $\beta = 1.85$. The selection of parameters corresponds to the maximum gain to be achieved for optimal growth (see e.g. Hack and Moin [3]). The results of different methods shown in Fig. 27.3 are all based on the initial conditions $c_n|_0$ obtained for G_{max} of the envelope G_0 (see Fig. 27.1b), from the purely local approach with $N_m = 100$ modes.

As discussed above, the purely local solution (hence *fixed* at the inflow) underestimates the actual disturbance energy gain represented by the DNS, which is evident in Fig. 27.3b. In contrast to local LST, the DNS considers the boundary-layer growth as well as the non-local modal evolution, which also results in a significantly more pronounced lift-up of the elongated streaks; see streamwise disturbance $\tilde{u}$ in Fig. 27.3a. Another approach is the local analysis in a *moving* system, in which the LST is carried out at all stations over x and, similar to the e^n-method for discrete disturbances, is assembled and integrated in the downstream direction. However, this method significantly reduces the gain in comparison to the fixed approach; see blue line in Fig. 27.3b.

The novel adjoint-based downstream projection method leads to an enhancement in energy gain, especially at the initial growth phase which is most relevant for transition prediction; see red line in Fig. 27.3b. Although this method is based on a local analysis—as in the case of integration in the moving system—the energy exchange between the modes via $c_n(x)$ is permitted here by the downstream projection. This appears to be one of the physical effects in non-local spatial disturbance growth, while the remaining gap between of G_{max} of the adjoint-based projection and the DNS can be explained by the non-parallel base flow and the streamwise gradients, $\partial/\partial x$, of the disturbance evolution, which are not accounted for here. Moreover, as can be observed in the streamwise component, $\tilde{u}$, the lift-up effect is spatially more pronounced in the projection method compared to the result of the fixed local approach and qualitatively almost agrees with the DNS result. Consequently, this result may serve as a better basis for investigations of secondary instabilities.

The perturbation velocity fields shown in Fig. 27.3a allow for a tentative deduction of non-local physical effects involved in the lift-up mechanism. A slight lift-up in the downstream direction can already be seen in the local fixed solution, which results exclusively from the algebraic growth. The adjoint projection leads to a significant increase of the lift-up, which is explained by the redistribution between modes. The consideration of the non-parallel flow and the spatial disturbance evolution, considered in the DNS, finally results to another slight lift-up in wall-normal direction with a significant increase in amplitude.

27.4 Summary and Conclusions

A novel methodology to enhance the prediction of non-modal disturbance growth for spatially evolving flows with local theory, which employs bi-orthogonality, was developed. The method is contrasted with DNS for $Re = 800$, $\omega = 0$ and $\beta = 1.85$. The new approach is found to improve the results of the purely local analysis by taking into account the redistribution of energy between scales, which appears to be a critical aspect of the non-local growth. The scheme may prove to be a cost-effective alternative to PSE.

Acknowledgements This research was supported by the ISRAEL SCIENCE FOUNDATION (grant No. 1522/21).

References

1. Butler, K.M., Farrell, B.F.: Three-dimensional optimal perturbations in viscous shear flow. Phys. Fluids **4**(8), 1637–1650 (1992)
2. Hill, D.C.: Adjoint systems and their role in the receptivity problem for boundary layers. J. Fluid Mech. **292**, 183–204 (1995)
3. Hack, M.J.P., Moin, P.: Algebraic disturbance growth by interaction of Orr and lift-up mechanisms. J. Fluid Mech. **829**, 112–126 (2017)
4. Herbert, T.: Parabolized stability equations. Ann. Rev. Fluid Mech. **29**, 245–283 (1997)

5. Juniper M.P., Hanifi A. and Theofilis, V.: Modal Stability Theory: Lecture notes from the FLOW-NORDITA Summer School on Advanced Instability Methods for Complex Flows, Stockholm, Sweden, 2013, Appl. Mech. Rev. **66**, no. 2 (2014)
6. Karp, M., Hack, M.J.P.: Transition to turbulence over convex surfaces. J. Fluid Mech. **855**, 1208–1237 (2018)
7. Karp, M., Cohen, J.: Tracking stages of transition in Couette flow analytically. J. Fluid Mech. **748**, 896–931 (2014)
8. Landahl, T.: A note on an algebraic instability of inviscid parallel shear flows. J. Fluid Mech. **98**(2), 243–251 (1980)
9. Luchini, P.: Reynolds-number-independent instability of the boundary layer over a flat surface. J. Fluid Mech. **404**, 289–309 (2000)
10. Schmid, P.J., Henningson, D.S.: Stability and Transition in Shear Flows, Springer (2001)
11. Tumin, A., Reshotko, E.: Spatial theory of optimal disturbances in boundary layers. Phys. Fluids **13**(7), 2097–2104 (2001)
12. Tumin A.: The biorthogonal eigenfunction system of linear stability equations: a survey of applications to receptivity problems and to analysis of experimental and computational results. In: 41st AIAA Fluid Dynamics Conference and Exhibit (2001)

Chapter 28
Interscale Energy Transfer in a Transitional Swept-Wing Boundary Layer

Tomotaka Ichikawa, Kosuke Nakagawa, Takahiro Ishida, Takuya Kawata, and Takahiro Tsukahara

Abstract The swept-wing boundary layer flow features a three dimensional base-flow profile that exhibits inviscid instability leading to the formation of crossflow vortices. While the initial growth of these crossflow vortices is well predicted by linear stability analysis, the onset of secondary instabilities and the breakdown of crossflow vortices are inherently nonlinear phenomena, and their mechanisms remain unclear. In this study, we conduct a direct numerical simulation of a swept flat-plate boundary layer to investigate energy transfer during the nonlinear stages of laminar-to-turbulent transition. Specifically, we focus on the interscale energy transfer between the most unstable spanwise (crossflow direction) wavenumber and higher ones, aiming to understand how energy is distributed across various scales in secondary instabilities and breakdown of crossflow vortices.

Keywords Direct numerical simulation · Swept flat-plate boundary layer · Crossflow instability

28.1 Introduction

The boundary layers over swept wings exhibit three-dimensional base flow profiles with an inflection point in the crossflow component. The presence of this inflection point results in an inviscid instability that gives rise to crossflow vortices (CFVs),

T. Ichikawa (✉) · T. Kawata (✉)
Department of Mechanical Engineering, Shibaura Institute of Technology, Tokyo, Japan
e-mail: aa20017@shibaura-it.ac.jp

T. Kawata
e-mail: kawata@shibaura-it.ac.jp

K. Nakagawa · T. Tsukahara
Department of Mechanical and Aerospace Engineering, Tokyo University of Science, Chiba, Japan

T. Ishida
Aviation Technology Directorate, Japan Aerospace Exploration Agency, Tokyo, Japan

K. Kato et al. (eds.), *Proceedings of the 10th IUTAM Symposium on Laminar-Turbulent Transition*, IUTAM Bookseries 44, https://doi.org/10.1007/978-981-96-9829-5_28

whose growth and eventual breakdown lead to a transition to turbulence. While linear instability mechanisms play a certain role in the initial growth of the CFVs, onset of secondary instabilities and breakdown of the CFVs are essentially nonlinear phenomena and their detailed mechanisms are still not fully understood. As the linear instabilities of the CFVs are mainly characterized by the spanwise wavenumber, investigating the energy transfers between different spanwise wavenumbers may give new insights to better understand the mechanisms of the laminar-to-turbulent transition in the swept-wing boundary layer flow. In this study, we perform a direct numerical simulation (DNS) of a swept-flat plate boundary layer and investigate the energy transfers between different spanwise wavenumbers in the laminar-to-turbulence transition process.

28.2 Numerical Simulation

The computational setup is presented in Fig. 28.1. The x-, y- and z-axes denote the chordwise, spanwise and wall-normal directions, respectively, and U, V and W are the velocity components in these directions. The governing equations are incompressible continuity and Navier-Stokes equations with the Reynolds number of $\mathrm{Re}_{\delta_0} = U_0\delta_0/\nu = 338$ (here, U_0 and δ_0 are the freestream velocity and displacement thickness of the boundary layer at the inlet, respectively, and ν is the fluid kinematic viscosity). The domain sizes in each direction are $L_x/\delta_0 = 400$, $L_y/\delta_0 = 25.14$ and $L_z/\delta_0 = 27$, and the inlet flow is a laminar Falkner-Skan-Cooke (FSC) boundary layer with the inflow angle of $\varphi_0 = 55.3°$ without any background unsteady disturbance. Non-slip and periodic boundary conditions are applied on the wall and in y-direction, respectively, and the outlet condition is the convective outflow. The freestream is accelerated in x-direction with the acceleration coefficient $m = 0.34207$.

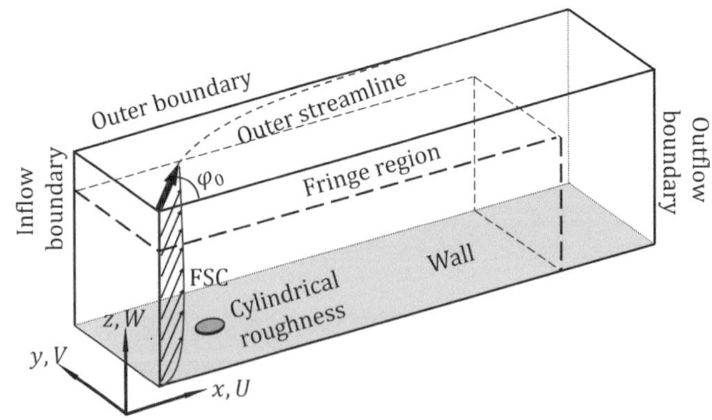

Fig. 28.1 Schematic of a swept-flat plate boundary layer

In order to trigger stationary CFVs, a cylindrical roughness is placed on the wall at the chordwise position of $x/\delta_0 = 20.59$. The height and radius of the roughness element are set to be $h_c/\delta_0 = 1.6$ and $r_c/\delta_0 = 3.0$, respectively. Note here that the spanwise domain size $L_y/\delta_0 = 25.14$ is equal to the spanwise wavelength of the most unstable stationary CFV mode [1]. As will be shown in the next section, the stationary CFVs triggered by the roughness element exhibit secondary instabilities and breakdown at some distance downstream, although no unsteady disturbance is introduced in the inflow. This is because unavoidable numerical errors act as unsteady random disturbances and interact with the stationary CFVs. For further details of the numerical setup, refer to [2, 3] as the present DNS is configured following these previous studies.

28.3 Results and Discussions

A snapshot of instantaneous distribution of chordwise velocity is presented in Fig. 28.2, where the high- and low-speed streaks by the CFVs induced by the roughness element are clearly observed. The CFVs begin to exhibit meandering around $x \approx 180\delta_0$, which is a similar observation to the secondary instabilities associated with velocity fluctuations at relatively high frequencies reported in previous studies [1, 4]. It is also observed further downstream that the CFVs collapse, which eventually leads to a transition to turbulence.

In the following, we introduce the Reynolds decomposition as $U_i = \langle U_i \rangle + u_i$, where $\langle U_i \rangle$ denotes the mean velocities averaged in time and in y-direction and u_i represents the fluctuation from it. In this analysis, we focus on the transport of turbulent kinetic energy $k_t = \langle u_i u_i \rangle / 2$. Figure 28.3a presents the profiles of the spanwise one-dimensional spectrum of turbulent kinetic energy $E_{kt}(k_y)$ at the peak location of k_t for $x/\delta_0 = 180$, 190 and 200 as functions of spanwise wavenumber k_y. The profiles of the energy spectrum indicate a significant energy peak at the wavenumber of the most unstable stationary mode $k_{y,\max} = 2\pi/L_y$ at relatively upstream locations $x/\delta_0 = 180$ and 190, corresponding to the presence of the CFVs. However, at the most downstream position $x/\delta_0 = 200$ the primary energy peak at $k_y/k_{y,\max} = 1$ decreases while the energy at higher wavenumbers increases. The decay of the primary mode has been reported as a sign of turbulent transition due

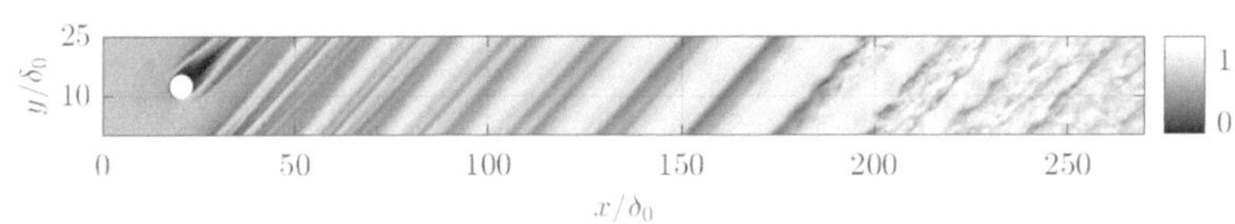

Fig. 28.2 A snapshot of instantaneous distribution of chordwise velocity U/U_0 in xy plane at $z/\delta_0 = 1.41$

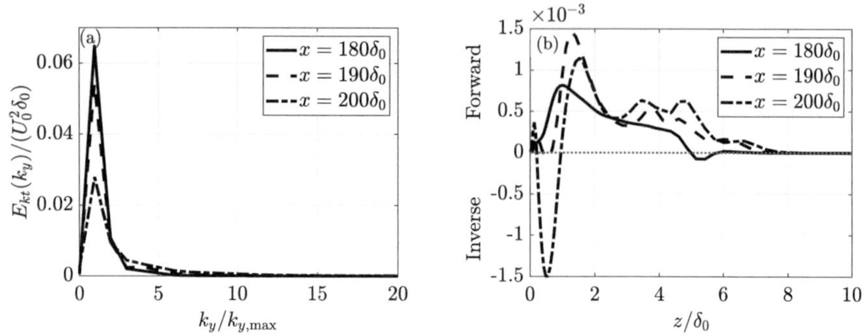

Fig. 28.3 **a** Turbulent kinetic energy spectra $E_{kt}(k_y)$ at z-position of the k_t peak. **b** wall-normal profiles of interscale energy transfer Tr_{kt} across wavenumber $k_y/k_{y,\max} = 1$ at $x/\delta_0 = 180$, 190 and 200

to the onset of secondary instability [2], and in fact, it appears in Fig. 28.2 that the CFVs start to breakdown around $x/\delta_0 \approx 200$.

In order to investigate the interscale energy transfers in the laminar-to-turbulent transition, we introduce a velocity decomposition similar to the analysis by Kawata and Alfredsson [5]: $u_i = u_i^L + u_i^S$, where the large-scale part u_i^L is obtained by applying a spatial low-pass filtering with a sharp cutoff at the wavenumber $k_y/k_{y,\max} = 1$ and the small-scale part u_i^S is defined as $u_i^S = u_i - u_i^L$. With such a decomposition, the cross correlations between the large- and small-scale parts is zero as there is no overlapping wavenumber range between them, and therefore the turbulent kinetic energy k_t is decomposed into its large- and small-scale part as $k_t = k_t^L + k_t^S$, where $k_t^L = \langle u_i^L u_i^L \rangle /2$ and $k_t^S = \langle u_i^S u_i^S \rangle /2$. By deriving the transport equation of each k_t^L and k_t^S, one can see that the interscale energy transport between k_t^L and k_t^S is expressed as

$$Tr_{kt} = -\left\langle u_i^S u_j^S \frac{\partial u_i^L}{\partial x_j} \right\rangle + \left\langle u_i^L u_j^L \frac{\partial u_i^S}{\partial x_j} \right\rangle. \tag{28.1}$$

It should be noted here that the spanwise domain size of the present DNS is equal to the wavelength of the most unstable stationary CFV mode; hence, the decomposed large-scale part of the fluctuating velocity field contains only the spanwise-constant mode ($k_y = 0$) and the most unstable CFV mode ($k_y/k_{y,\max} = 1$), whereas the small-scale part consists of higher harmonics of the most unstable CFV mode. The profiles of Tr_{kt} at $x/\delta_0 = 180$, 190 and 200 are presented in Fig. 28.3b. As shown here, Tr_{kt} is predominantly positive upstream of the breakdown (at $x/\delta_0 = 180$ and 190), indicating energy transfers from the CFVs to smaller-scale structures. Interestingly, as further downstream locations, where the breakdown of the CFVs occurs, the profiles of Tr_{kt} exhibit significant negative values near the wall. This suggests inverse energy transfer from smaller-scale structures to the larger-scale component.

Next, we further decompose the interscale energy transport Tr_{kt} into different terms with respect to the velocity gradients in different directions. In this DNS, the chordwise direction does not coincide with the streamwise direction, and it may, therefore, be appropriate to decompose Tr_{kt} with respect to the coordinates aligned to the flow direction. Hence we perform the coordinate transformation so that x'-axis is in the inflow direction and the y'-axis is transverse to it (note that the z'-axis is identical to z-axis). The interscale energy transport term is then expressed as

$$Tr_{kt} = \underbrace{-\left\langle u'^S_i u'^S \frac{\partial u'^L_i}{\partial x'} \right\rangle}_{Tr_1} \underbrace{-\left\langle u'^S_i v'^S \frac{\partial u'^L_i}{\partial y'} \right\rangle}_{Tr_2} \underbrace{-\left\langle u'^S_i w'^S \frac{\partial u'^L_i}{\partial z'} \right\rangle}_{Tr_3}$$
$$\underbrace{+\left\langle u'^L_i u'^L \frac{\partial u'^S_i}{\partial x'} \right\rangle}_{Tr_4} \underbrace{+\left\langle u'^L_i v'^L \frac{\partial u'^S_i}{\partial y'} \right\rangle}_{Tr_5} \underbrace{+\left\langle u'^L_i w'^L \frac{\partial u'^S_i}{\partial z'} \right\rangle}_{Tr_6} \quad (28.2)$$

where u'_i represents the velocity components in x'_i directions in the transformed coordinates. The balance of these six terms is presented in Fig. 28.4 for two chordwise positions; $x/\delta_0 = 180$ and 200, the upstream and downstream location of the CFV breakdown, respectively. As shown in Fig. 28.4a, the forward energy transfer at $x/\delta_0 = 180$ can be attributed to the crossflow velocity gradient $\partial/\partial y'$. This means that the energy transfer by the spanwise velocity gradients plays a dominant role in the forward interscale transfer. Nakagawa et al. [3] showed that the Type–I secondary instability is associated with the turbulent energy production by the spanwise velocity gradient, and the above-described observation of the forward interscale transfer by Tr_{kt} is consistent with this description of the energy transfer in the Type–I instability. It is also known that the Type I instability occurs on the upwind side of the CFV [6, 7], but the location of the significant forward interscale energy transfer

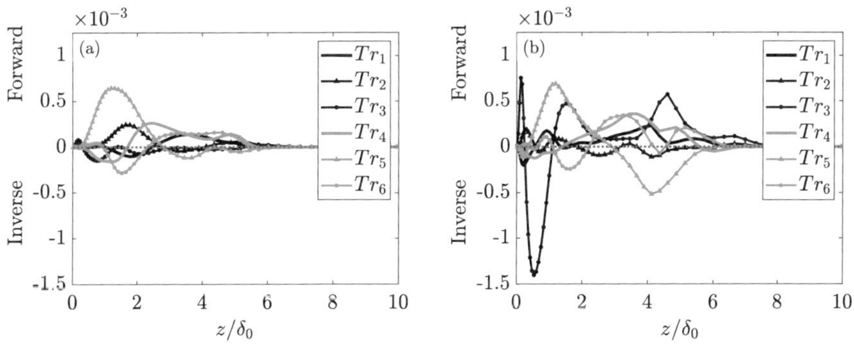

Fig. 28.4 Comparison of various terms on the right-hand side of Eq. (28.2) at **a** $x/\delta_0 = 180$ and **b** $x/\delta_0 = 200$

does not coincide with these description (not shown). This is probably because we investigate the energy transfer focusing on the spanwise wavenumber in the present study, while in these previous studies, the energy transfer between the stationary and non-stationary modes were mainly investigated.

As for the inverse interscale energy transfer observed in the downstream position $x/\delta_0 = 200$, it is shown by Fig. 28.4b that the inverse energy transport in the near-wall region is mainly due to Tr_3, the term associated with the smaller-scale velocity fluctuations multiplied by the wall-normal gradients of the larger-scale velocities as shown by Eq. (28.2). It is still unclear what this inverse interscale energy transport physically indicates, but similar inverse energy transport from shorter to longer spanwise wavelengths has also been reported in fully developed wall turbulence (for example, [8]). It is a certainly interesting observation that such inverse interscale energy transport has been commonly observed in two totally different situations: in the laminar-to-turbulent transition of a three-dimensional boundary layer and in fully-developed, canonical wall turbulence. This may imply that the inverse interscale energy transport observed above is not a phenomenon specific to the crossflow-instability-dominated transition in the swept flat-plate boundary layer, but rather a common feature of turbulent wall-bounded shear flows. Nevertheless, further close investigation is required on the mechanism of the inverse interscale energy transport.

28.4 Conclusion

In this study, we performed a DNS of swept flat-plate boundary layer flow to investigate interscale energy transfer in the process of transition to turbulence. We decomposed the fluctuating velocities to the large- and small-scale parts by a cutoff at the spanwise wavenumber of the most unstable CFV mode; hence, the large-scale part mainly contains only the most unstable CFV mode and the small-scale part consists of the higher harmonics. Then, we evaluated the energy transfer between the large- and small-scale part of the fluctuating velocity fields. As a result, we have shown that forward energy transport from the larger to smaller scale part is observed before the breakdown of the CFVs, while the inverse energy transport takes place in the further downstream region. The forward energy transfer was found to be mainly associated with the velocity gradients in the crossflow direction, suggesting link to the Type-I secondary instability of the CFVs. The inverse energy transfer was, on the other hand, shown to be mainly associated with the wall-normal velocity gradients.

References

1. Brynjell-Rahkola, M., Shahriari, N., Schlatter, P., Hanifi, A., Henningson, D.S.: Stability and sensitivity of a cross-flow-dominated Falkner-Skan-Cooke boundary layer with discrete surface roughness. J. Fluid Mech. **826**, 830–850 (2017). https://doi.org/10.1017/jfm.2017.466
2. Ishida, T., Tsukahara, T., Tokugawa, N.: Parameter effects of spanwise-arrayed cylindrical roughness elements on transition in the Falkner-Skan-Cooke boundary layer. Trans. Jpn. Soc. Aeronaut. Space Sci. **65**(2), 84–94 (2022). https://doi.org/10.1007/s42405-022-00485-0
3. Nakagawa, K., Ishida, T., Tsukahara, T.: Effects of freestream turbulence on the secondary instability of the roughness-induced crossflow vortex in swept flat plate boundary layers. Int. J. Heat Fluid Flow **102**, 109161 (2023). https://doi.org/10.1016/j.ijheatfluidflow.2023.109161
4. Nakagawa, K., Tsukahara, T., Ishida, T.: DNS study on turbulent transition induced by an interaction between freestream turbulence and cylindrical roughness in swept flat-plate boundary layer. Aerospace **10**, 128 (2023). https://doi.org/10.3390/aerospace10020128
5. Kawata, T., Alfredsson, P.H.: Inverse interscale transport of the Reynolds shear stress in plane Couette turbulence. Phys. Rev. Lett. **120**, 244501 (2018). https://doi.org/10.1103/PhysRevLett.120.244501
6. Malik, M.R., Li, F., Choudhari, M.M., Chang, C.-L.: Secondary instability of crossflow vortices and swept-wing boundary-layer transition. J. Fluid Mech. **399**, 85–115 (1999). https://doi.org/10.1017/S0022112099006291
7. Serpieri, J., Kotsonis, M.: Three-dimensional organisation of primary and secondary crossflow instability. J. Fluid Mech. **799**, 200–245 (2016). https://doi.org/10.1017/jfm.2016.379
8. Lee, M., Moser, R.D.: Spectral analysis of the budget equation in turbulent channel flows at high Reynolds number. J. Fluid Mech. **860**, 886–938 (2019). https://doi.org/10.1017/jfm.2018.903

Chapter 29
Roughness-Induced Laminar-Turbulent Transition in Hypersonic Ramp Flows

Giuseppe Chiapparino and Christian Stemmer

Abstract This work investigates the effect of unsteady freestream disturbances on a Mach 6, ramp-induced Shock-Wave/Boundary-Layer Interaction (SWBLI) with discrete roughness elements upstream of the corner. Building on the previous numerical studies investigating the baseflow for the same geometry and flow conditions, the present work analyses the breakdown to turbulence along the ramp of the high-vorticity, "mushroom-like" structure forming in the wake of the roughness as the flow reattaches downstream of the separation bubble. It is found that transition is mainly driven by the amplification of the y-modes promoted by the strong wall-normal gradients present in such high-vorticity structure, especially in the side lobes of the mushroom, where the onset of breakdown is first observed. At the surface, consequence of the transition onset is the localized rise of wall temperature, with a hot streak extending much further along the ramp compared to the steady case.

Keywords Hypersonics · SWBLI · Transition · Roughness

29.1 Introduction

The ramp-induced SWBLI is a typical flow configuration investigated under hypersonic conditions, as it characterizes the flow regions in proximity of control surfaces and engine intakes of high-speed vehicles. If strong enough, the pressure rise imposed by the oblique shock makes the incoming boundary layer separate well upstream of the compression corner. Hence, a complex flow system (comprising separation and reattachment shocks, as well as a recirculation bubble at the corner) forms, affecting the pressure and temperature distributions at the wall. Surface roughness can influence the SWBLI by altering the baseflow properties of the boundary layer and the

G. Chiapparino (✉) · C. Stemmer
Chair of Aerodynamics and Fluid Mechanics, Engineering Physics and Computation Department, Technical University Munich, Garching, München, Germany
e-mail: giuseppe.chiapparino@tum.de

C. Stemmer
e-mail: christian.stemmer@tum.de

K. Kato et al. (eds.), *Proceedings of the 10th IUTAM Symposium on Laminar-Turbulent Transition*, IUTAM Bookseries 44, https://doi.org/10.1007/978-981-96-9829-5_29

flow reattaching on the ramp [1, 4]. In turn, this affects the response of the flow to unsteady disturbances and the laminar-turbulent transition process on the ramp. In [3], the effect of several variants of a discrete roughness patch placed at different locations upstream of a $\theta = 10°$ corner was investigated by performing Direct Numerical Simulations (DNS) of a $M_\infty = 6$ hypersonic flow. Two of the roughness patches were placed upstream of the large ramp-induced separation bubble, and their geometrical parameters were chosen to have "critical" roughness [8], hence only inducing persistent modifications of the baseflow, rather than an immediate boundary-layer transition. Results showed that the roughness elements generate streamwise vortices which are convected downstream in the boundary layer, increasing its resistance to separation compared to smooth-wall conditions. The vortices further convect in the detached shear layer above the shorter separation bubble and eventually on the ramp once the flow reattaches. Downstream of reattachment, a high-vorticity structure forms and dominates the flow. This structure is characterized by large streamwise-velocity gradients in wall-normal and spanwise direction, known to be responsible for the amplification of flow instabilities such as y- and z-modes [6, 11], which can potentially lead to transition further donwstream along the ramp. The present work focuses on one roughness position, whose baseflow is perturbed by injecting freestream pressure disturbances, as typical of wind-tunnel environments, in order to characterize the flow response to unsteady conditions.

29.2 Numerical Setup

The study focuses on a $M_\infty = 6$ flow at $T_\infty = 220$ K over a $\theta = 10°$ ramp geometry and investigates the unsteady effect of isolated roughness elements on the SWBLI. The freestream conditions and roughness parameters are the same considered in [3]. The geometry consists of a flat plate with sharp leading edge ($L_c = 0.300$ m) followed by an inclined ramp ($L_r = 0.390$ m). The domain extends for $L_y = 0.065$ m and $L_z = 0.021$ m in wall-normal and spanwise directions, respectively. Figure 29.1 shows the computational grid ($\simeq 280 \times 10^6$ points distributed over 3600 blocks) with position and shape of the roughness. The latter is the 3D patch "RP1" investigated in [3], consisting of three sinusoidal elements in streamwise direction and placed along the centerline of the domain at $x_k = 0.0461$ m from the inflow. The peak height is $k = 1$ mm, which yields a roughness Reynolds number $Re_{kk} = \rho_k u_k k/\mu_k \simeq 370$, too low to trigger immediate transition [7, 9]. DNS are performed to solve the compressible 3D Navier–Stokes equations under the assumption of a perfect-gas behaviour on the finite-volume solver NSMB [10]. The details of the spatial solution scheme are given in [3]. The steady state baseflow is achieved through a pseudo-time integration with an implicit Euler scheme based on a LU-SGS method. An explicit third-order Runge–Kutta scheme, instead, is used for the unsteady case, with a time step of $\Delta t = 10^{-9}$ s ($CFL \simeq 0.150$). The steady boundary conditions are the same described in [3] for the adiabatic-wall case. 2D pressure disturbances are prescribed at the inflow plane of the domain. The profile shown in Fig. 29.1 is applied uniformly in spanwise direction

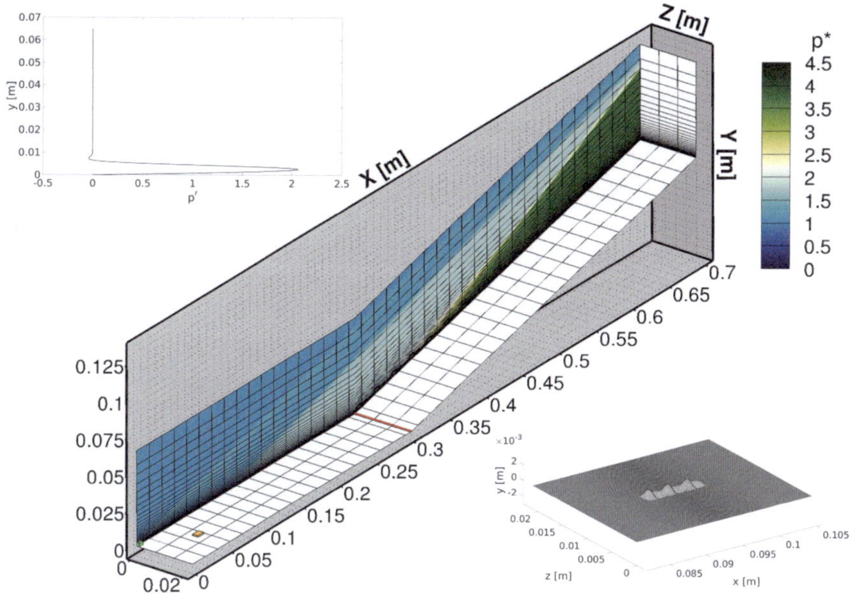

Fig. 29.1 Schematic of the computational domain showing coordinates-system origin (green), patch position (orange) and corner location (red). Background slice shows a cross-section of the unperturbed pressure field. Details picture the roughness-patch shape (the image refers to patch "RP2" in [3], which has the same shape as the patch RP1 considered in this paper) and pressure and pressure disturbances under unsteady conditions

and is added on top of the steady freestream, resulting in following time-dependent Dirichlet inflow boundary condition:

$$p(x=0,y,t) = p_\infty + p'(y,t) = p_\infty + p_0 \cdot sin\left(\frac{2\pi y}{\lambda_y}\right) \cdot e^{-\left(\frac{y}{\delta}\right)^3} \cdot sin(2\pi f t) \tag{29.1}$$

The amplitude p_0 is $p_\infty \cdot 10^{-3} = 2.5$ Pa. In wall-normal direction, the disturbance profile consists of 5 sinusoidal waves ($\lambda_y = L_y/5$) and it is damped away from the wall by the exponential function with $\delta = 0.005$ m, confining the disturbance to the wall region where it interacts with the boundary layer. In [6], a frequency of about 100 kHz showed to be the most amplified in a flat-plate flow at $M = 4.8$. Hence, for the present work, the frequency effect was assessed on a 2D setup over the range [50, 100, 175, 250] kHz. As the frequency f = 100 kHz showed indeed the highest amplification on the ramp, it was selected for the 3D computations.

29.3 Results

Normalized quantities are indicated with "*". Lengths are normalized with L_c, velocities and pressure with freestream values, and temperatures with total value $T_0 = 1804$ K. As already shown in [3], the baseflow is characterized by the formation of a mushroom-shaped structure in the wake of the roughness once the flow reattaches on the ramp. In absence of external disturbances, such a high-vorticity structure is stable enough not to break down on the ramp (Fig. 29.2, top left). Under unsteady conditions, instead, it quickly breaks down and the flow undergoes transition to turbulence (Fig. 29.2, top right). The flow breakdown affects the wall-temperature distribution downstream of reattachment. The bottom row of Fig. 29.2 compares the surface temperature under steady conditions (left) with the time-averaged temperature of the unsteady-inflow case (right) computed over a time window of two disturbance periods. In both cases, the hot streak at the wall has an almost symmetric structure. However, for the unsteady case, the streak shows localized high- and low-temperature spots alternating as the flow starts to break down, with temperature ranging between 0.805 and 1.040, exceeding the steady-case range $0.875 \leq T_w^* \leq 0.945$.

Figures 29.3 and 29.4 show the Fourier-amplitude iso-contour of the streamwise-velocity oscillations $u^{*\prime}$ (first harmonic, 100 kHz) between reattachment and flow breakdown on the ramp, from $x^* = 1.300$ to $x^* = 1.500$. The iso-lines, instead, represent the shear-magnitude (y- and z-components), which measures the effect of the streamwise-velocity gradients in wall-normal and spanwise directions and was computed as in [3]: $\sigma = \sqrt{\sigma_y^2 + \sigma_z^2} = \sqrt{(\partial\hat{u}/\partial\hat{y})^2 + (\partial\hat{u}/\partial\hat{z})^2}$. The Fourier amplitude of $u^{*\prime}$ sharply rises at reattachment, with highest amplification being in the vortical mushroom-shaped structure (both in the "hat" and in the side lobes) forming on the ramp. The amplitude of the first harmonic peaks at around $x^* \simeq 1.45$,

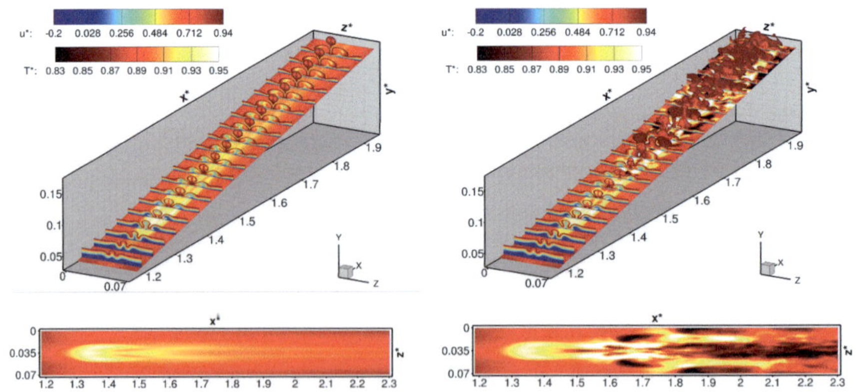

Fig. 29.2 Evolution of the vortical structure forming at reattachment in the roughness wake (top) and time-averaged wall-temperature distribution on the ramp (bottom) for both steady (left) and unsteady (right) cases

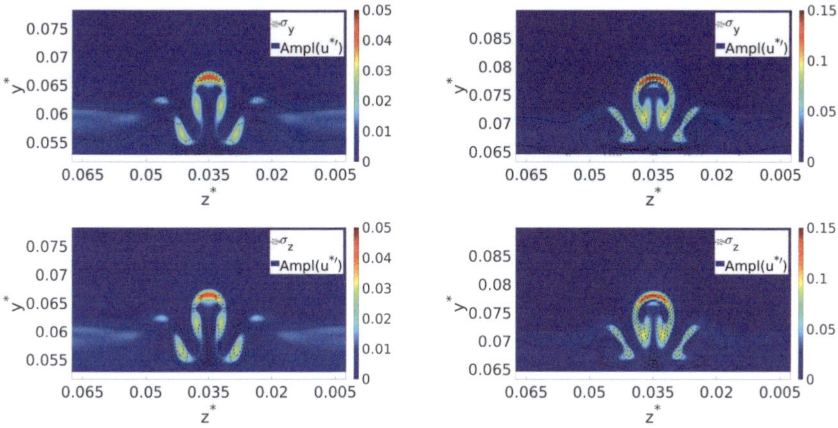

Fig. 29.3 FA vs σ_y and σ_z at $x^* = 1.300$ (left) and $x^* = 1.366$ (right)

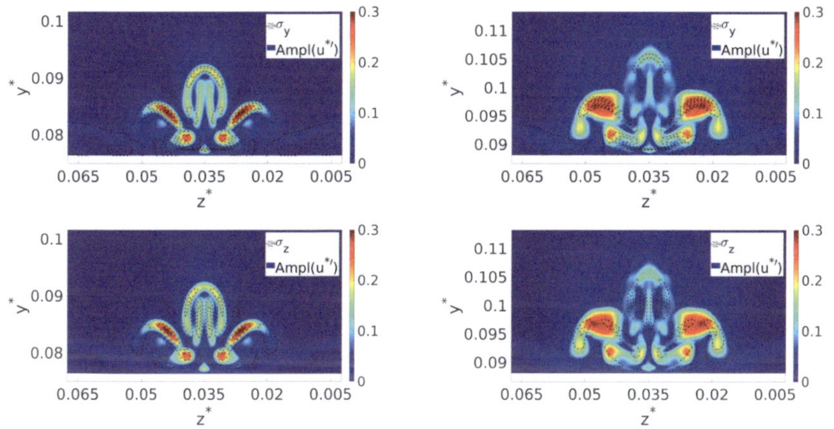

Fig. 29.4 FA vs σ_y and σ_z at $x^* = 1.433$ (left) and $x^* = 1.500$ (right)

just before breakdown, in the side lobes of the structure. Further downstream (not shown here), the $u^{*\prime}$ amplitude away from the wall decreases, while local spots of high-amplitude turbulent flow appear close to the surface. Up until flow saturation, the amplitude of non-linearly generated harmonics is negligible compared to the first one. Figures 29.3 and 29.4 show the high-amplitude regions being confined within the mushroom-shaped structure, where large gradients occur. In particular, the growth rates are larger where the shear-magnitude is stronger. To study the driving mechanism of the disturbances amplification, the distribution of the components σ_y and σ_z (associated to wall-normal and spanwise gradients, respectively) was investigated to identify the areas in the baseflow where y- and z-modes would prevail [5, 6, 11]. Before the shear layer completely reattaches on the ramp ($x^* = 1.300$), the shear-magnitude peaks in the hat of the forming mushroom structure, and is mainly due to the wall-normal-gradient contribution. In this area, the peak of the Fourier amplitude overlaps perfectly with σ_y (Fig. 29.3, top left). As the flow fully reattaches and moves downstream, the $u^{*\prime}$ amplitude in the side structures picks up. At $x^* = 1.366$ (Fig. 29.3, right), the shear-magnitude still peaks in the hat area and is dominated by σ_y. However, progressively larger amplitudes start to characterize the 45°-inclined lateral structures, as well. In these side lobes, both y- and z-gradients contribute equally to the total shear magnitude at $x^* = 1.366$. Further downstream, σ_y decreases in the mushroom hat. Hence, $u^{*\prime}$ in the hat increases only slightly ($\simeq 30\%$) for $1.366 < x^* < 1.433$. On the other hand, over the same streamwise distance, the $u^{*\prime}$ amplitude in the side structures increases by a factor of at least four. Hence, at $x^* = 1.433$, the amplitude in the side lobes is much larger than in the hat and dominates the flow until transition, reaching a peak of $u^{*\prime} \simeq 0.305$. As shown in Fig. 29.4 (left), the contribution from the wall-normal gradients overlaps almost perfectly with the Fourier-amplitude peaks and is much larger (between 50% and 75%) than σ_z, although both gradients are still relevant in those side-lobe areas. Finally, at $x^* = 1.500$ (Fig. 29.4, right), the flow starts to break down. Once again, the highest amplitude characterizes the side areas dominated by strong σ_y gradients under steady conditions. Further downstream, the direct comparison between steady shear-magnitude profiles and Fourier amplitudes becomes more difficult, as the unsteady flow deviates noticeably from the baseflow.

29.4 Conclusions

This work investigated the effect of unsteady disturbances on a ramp-induced SWBLI in presence of discrete roughness elements. The steady baseflow is characterized by a high-vorticity flow structure that forms once the boundary layer in the wake of the roughness approaches and reattaches on the ramp. This "mushroom-like" structure breaks down to turbulence shortly downstream of reattachment. Comparing Fourier-analysis results with the shear magnitude characterizing the high-vorticity baseflow, transition is found to be mainly driven by the amplification of the y-modes promoted by the strong wall-normal gradients in the mushroom structure on the ramp. This happens especially in the side-lobes of the structure, which is where the first onset of breakdown takes place. Noticeable effect of transition is the localized increase of the wall temperatures once the flow starts to break down. Compared to the steady baseflow, high wall temperatures are not only observed reattachment location, but are also detected further downstream on the ramp until a full turbulent flow is achieved.

References

1. Berry, S.A., Auslander, A.H., Dilley, D.A., Calleja, J.F.: Hypersonic Boundary-layer trip development for hyper-X. J. Spacecr. Rockets **38**(6), 853–864 (2001)
2. Chiapparino, G., Stemmer, C.: Numerical investigation of a Mach 6 hypersonic laminar flow on two-dimensional cold-wall compression corners with controlled surface roughness. Int. J. Heat Fluid Flow **94**, 108937 (2022)
3. Chiapparino, G., Stemmer, C.: Numerical investigation of a Mach 6 laminar shock-wave/boundary-layer interaction on a two-dimensional ramp with 3D controlled surface roughness. Int. J. Heat Fluid Flow **103**, 109193 (2023)
4. Choudhari M., Li F., Edwards J.: Stability Analysis of Roughness Array Wake in a High-Speed Boundary Layer. AIAA Paper 2009-0170 (2009)
5. Di Giovanni, A., Stemmer, C.: Cross-flow-type breakdown induced by distributed roughness in the boundary layer of a hypersonic capsule configuration. J. Fluid Mech. **856**, 470–503 (2018)
6. Groskopf, G., Kloker, M.J., Marxen, O.: Bi-global crossplane stability analysis of high-speed boundary-layer flows with discrete roughness. In: 7th IUTAM Symposium on Laminar-Turbulent Transition, Stockholm, 171–176 (2010)
7. Redford, J., Sandham, N., Roberts, G.: Compressibility effects on boundary-layer transition induced by an isolated roughness element. AIAA J. **48**(12), 2818–2830 (2010)
8. Schneider, S.P.: Effects of roughness on hypersonic boundary-layer transition. J. Spacecr. Rockets **45**(2), 192–209 (2008)
9. Van den Eynde, J.P., Sandham, N.D: Numerical simulations of transition due to isolated roughness elements at Mach 6. AIAA J. **54**(1), 53–65 (2016)
10. Vos J., Duquesne N., Lee H.: Shock wave boundary layer interaction studies using the NSMB flow solver. In: 3rd European Symposium on the Aerothermodynamics for Space Vehicles, ESA SP-426 (1999)
11. Wassermann, P., Kloker, M.: Mechanisms and passive control of crossflow-vortex-induced transition in a three-dimensional boundary layer. J. Fluid Mech. **456**, 49–84 (2002)

Chapter 30
Linear and Nonlinear Analysis of Wavy-Surface-Induced Laminar Separation Bubbles

Mohammad Moniripiri, Daniel Rodríguez, and Ardeshir Hanifi

Abstract Linear global stability analysis is performed on a laminar separation bubble formed due to surface waviness. The eigenspectrum shows a globally unstable mode, responsible for the three-dimensionalisation of the bubble, and a family of low-frequency globally stable modes. An adjoint sensitivity analysis shows high sensitivity of the stable modes upstream of the bubble's reattachment point. Direct numerical simulation (DNS) alongside with a linear impulse response analysis are performed. DNS shows that, when transition occurs due to self-excited mechanisms, low-frequency upstream propagating waves form inside the bubble; this is not the case in linear impulse analysis. It is conjectured that these upstream propagating waves correspond to the low-frequency stable modes in the spectrum which become active through nonlinearity when transition occurs.

Keywords Laminar separation bubble · Boundary layer stability · Transition to turbulence

30.1 Introduction

Laminar Separation Bubbles (LSBs) formed on aerodynamic surfaces with waviness and other smooth curvature variations, referred to as geometry-induced LSBs, may affect aerodynamic performance by causing laminar-turbulent transition. LSBs can also form on smooth surfaces if the flow is subjected to strong enough adverse pressure gradient. In this case, the LSB is referred to as adverse-pressure-gradient (APG) separation bubble.

M. Moniripiri (✉) · A. Hanifi
FLOW, Department of Engineering Mechanics, KTH Royal Institute of Technology, Stockholm, Sweden
e-mail: momp@kth.se

D. Rodríguez
ETSIAE-UPM, School of Aeronautics, Universidad Politécnica de Madrid, Madrid, Spain

K. Kato et al. (eds.), *Proceedings of the 10th IUTAM Symposium on Laminar-Turbulent Transition*, IUTAM Bookseries 44, https://doi.org/10.1007/978-981-96-9829-5_30

Active instabilities in an LSB can be categorized into forced and self-excited mechanisms. Forced mechanisms arise due to convective instabilities that can be triggered by free stream turbulence, or propagating noise inside the boundary layer [1]. However, no external disturbance is needed to activate self-excited mechanisms. If insufficiently understood, the dynamics due to self-excited instabilities can be misinterpreted or overshadowed by forced mechanisms. For incompressible LSBs there are two main known linear self-excited mechanisms: a two dimensional global oscillator, and a steady three-dimensional centrifugal instability. The existence of a steady three-dimensional global mode has been shown by Theofilis et al. [2] using global linear stability analysis approach. This mechanism can become active for bubbles with reverse flow intensities as low as $\approx 7\%$ of the freestream velocity [3]. The growth of this mode originates a spanwise velocity component inside the bubble, which leads to spanwise modulation and three-dimensionalisation of the whole recirculation region [4, 5]. The main goal of the present article is to investigate the instability characteristics of a laminar separation bubble caused by surface waviness, and study the role of different global modes that exists in the bubble using direct numerical simulation and impulse response analysis.

30.2 Flow Configuration and Governing Equations

The geometry considered in this work is shown in Fig. 30.1. The geometry consists of a flat plat, where a sinusoidal waviness with height $h = 1.8756$ and wavelength $\lambda = 172.07$ is placed centred at $x = x_c = 300$, as shown in Fig. 30.1. All coordinates are normalized by Blasius displacement thickness at the inlet of the domain ($x = x_0$). The corresponding Reynolds number at inlet of the domain, defined as $Re_{\delta^*_{x0}} = U_\infty \delta^*_{x0}/\nu$

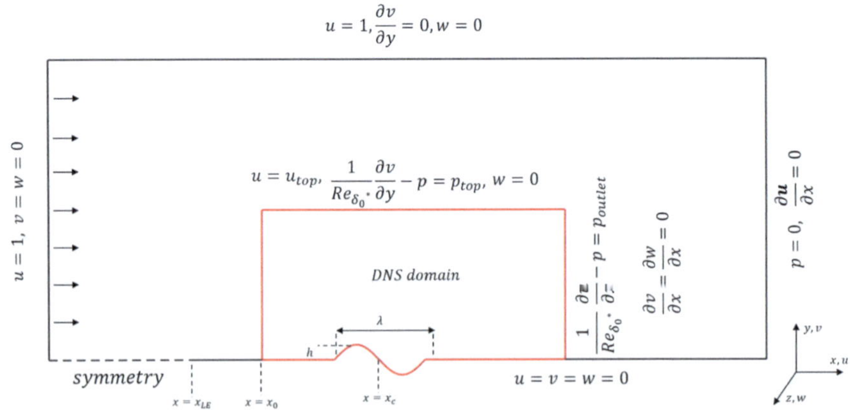

Fig. 30.1 Geometry and boundary conditions used in the simulations

is equal to 1332.93, and the local Reynolds number at the centre of the waviness $(x = x_c)$ is equal to $Re_{\delta^*_{x_c}} = 1720.8$.

30.2.1 Governing Equations

Nonlinear equations The non-dimensional Navier-Stokes and continuity equations for an incompressible, constant viscosity flow are

$$\frac{\partial \boldsymbol{u}}{\partial t} + (\boldsymbol{u} \cdot \nabla)\boldsymbol{u} + \nabla p = \frac{1}{Re_{\delta_0^*}} \nabla^2 \boldsymbol{u} + \boldsymbol{f}, \tag{30.1}$$

$$\nabla \cdot \boldsymbol{u} = 0, \tag{30.2}$$

where $\boldsymbol{u} = [u, v, w]^T$ are velocity components in Cartesian x, y, and z directions, respectively, p is the pressure, $\boldsymbol{f}$ is the forcing term, and $Re_{\delta_0^*} = (U_\infty \delta_0^*)/\nu$ is the Reynolds number based on the reference length, δ_0^*, which is taken as displacement thickness at $x = x_0$; U_∞ is the free-stream velocity and ν is the kinematic viscosity. The corresponding boundary conditions on each boundary are given in Fig. 30.1. The profiles and values for $\boldsymbol{u}_{inlet} = [u_{inlet}, v_{inlet}]$, u_{top}, p_{top} and p_{outlet} are obtained from a previous simulation using a larger domain (enclosed by black lines in Fig. 30.1). This simulation was performed using the commercial software COMSOL, where the steady-state solution of laminar Navier-Stokes equations was obtained. This choice of boundary conditions is made to make the corresponding baseflow independent of domain size. Note that the velocity profile used for $\boldsymbol{u}_{inlet}$ is slightly different than Balsius profile due to the upstream effect of the waviness. The governing equations are solved with the Spectral Element Method (SEM) solver Nek5000 [6]. Note that, grid study has been performed for both simulations performed by COMSOL and Nek5000 to ensure grid-independent results.

Linear equations Considering a steady solution of the governing equations $\bar{\boldsymbol{q}} = [\bar{\boldsymbol{u}}, \bar{p}]^T$ as baseflow, the nonlinear governing equations can be linearized around $\bar{\boldsymbol{q}}$. This reads as

$$\frac{\partial \boldsymbol{u}'}{\partial t} + (\boldsymbol{u}' \cdot \nabla)\bar{\boldsymbol{u}} + (\bar{\boldsymbol{u}} \cdot \nabla)\boldsymbol{u}' + \nabla p' = \frac{1}{Re_{\delta_0^*}} \nabla^2 \boldsymbol{u}', \tag{30.3}$$

$$\nabla \cdot \boldsymbol{u}' = 0, \tag{30.4}$$

where $(')$ denotes perturbation quantities. Homogeneous Dirichlet boundary conditions are used for the perturbation velocity at inlet, upper boundary, and the wall. Homogeneous Neumann condition is imposed at the outlet.

30.2.2 Modal Stability Analysis

Assuming exponential growth for the asymptotic behavior of modal perturbations and periodicity in spanwise direction, the perturbations can be written as

$$\boldsymbol{q}'(x, y, z, t) = \hat{\boldsymbol{q}}(x, y)e^{i(\beta z-\Lambda t)} + \text{c.c.}, \tag{30.5}$$

where β is the spanwise wave number, $\Lambda = \omega + i\sigma$, and $c.c.$ denotes the complex conjugate. By substituting (30.5) into (30.3) and (30.4), a generalised eigenvalue problem (GEVP) of the form

$$\boldsymbol{L}\hat{\boldsymbol{q}} = \Lambda \boldsymbol{R}\hat{\boldsymbol{q}}, \tag{30.6}$$

is found, where Λ and $\hat{\boldsymbol{q}}(x, y) = [\hat{u}, \hat{v}, \hat{w}]^T$ are eigenvalues and eigenvectors of the corresponding eigenvalue problem, respectively. The corresponding terms for linear operators $\boldsymbol{L}$ and $\boldsymbol{R}$ of the resulting eigenvalue problem can be found in [4]. If $\sigma > 0$ for any eigenvalue, the system is globally unstable. Three-dimensional instabilities can be found as the solution of (30.6) by setting $\beta \neq 0$. While the solution of GEVP (30.6), which is referred to as the *direct* problem, gives information about the instability eigenmodes, the information about the spatial regions most sensitive to their external forcing can be obtained by solving the adjoint of (30.6). The adjoint of linearised Navier-Stokes equations and the corresponding linear terms can be found in [4]. The corresponding eigenvalues and eigenvectors of adjoint problem are denoted by Λ^+ and $\hat{\boldsymbol{q}}^+(x, y) = [\hat{u}^+, \hat{v}^+, \hat{w}^+]^T$, respectively.

30.3 Results

The global spectrum of direct and adjoint problems for the resulting LSB at $\beta = 0.4$ is depicted in Fig. 30.2a. Note that this bubble is globally stable with respect to two-dimensional ($\beta = 0$) perturbations. The (unstable) mode with maximum growth rate has a zero frequency ($\omega = 0$). Figure 30.2b shows the change in growth rate of this mode for different values of β. The real parts of the direct eigenfunctions for u (denoted as $\hat{u}_r$) and w (denoted as $\hat{w}_r$), along with the corresponding real parts of the adjoint eigenfunctions $\hat{u}_r^+$ and $\hat{w}_r^+$ are depicted in Fig. 30.3 The figure is plotted for two eigenvalues, labelled as a and b in Fig. 30.2. Eigenvalue a (first row in Fig. 30.3) corresponds to the steady unstable global mode found in pressure-induced laminar separation bubbles on flat plates [4]. As the centrifugal mode a grows over time, the wall-normal extension of the recirculation region and the intensity of the reversed flow changes across different spanwise planes. This happens because at planes where the eigenfunction $\hat{u}_r > 0$, the streamwise perturbation amplifies, leading to an increase in the streamwise velocity component, and vice versa (cf. equation (30.5)). As a result, the separation streamline gets closer to the wall where $\hat{u}_r > 0$, and gets away from the wall where $\hat{u}_r < 0$. At the same time, the eigenmode growth also

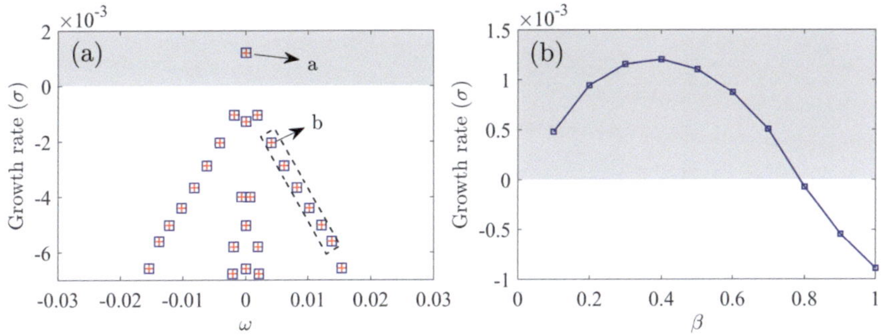

Fig. 30.2 Panel **a**: Eigenspectrum for $\beta = 0.4$. Direct eigenvalues are shown with blue squares, while adjoint eigenvalues are shown by red plus symbols. Panel **b**: Maximum growth-rate for different spanwise wave numbers. The maximum growth rate of the zero-frequency mode is obtained at $\beta \approx 0.4$

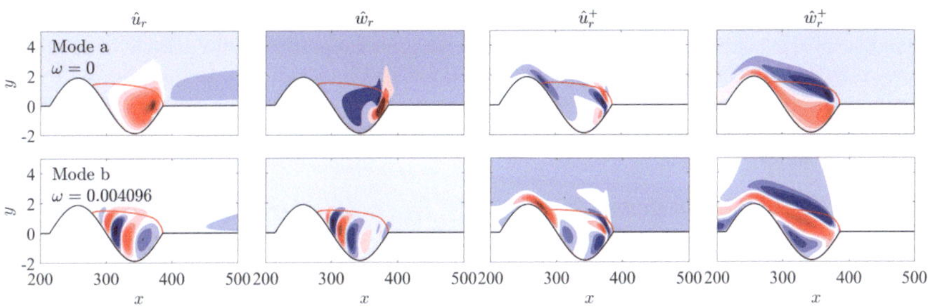

Fig. 30.3 Real part of eigenfunctions for direct and adjoint u, and w components. Panels in the first row correspond to the eigenvalue marked with label a, and second row correspond to the eigenvalue marked with letter b in Fig. 30.2. Red solid streamlines show separation streamline of the two-dimensional bubble

gives rise to a spanwise component inside the recirculation region. Thus, the bubble starts to become three-dimensional. Note that when perturbations inside the bubble become strong enough, non-linear interactions start to occur, and deformation of the bubble might occur due to higher harmonics of perturbations, not predictable with linear global modes. This process might trigger self-excited secondary instabilities and lead to transition [4, 7, 8].

Much less attention has been paid to the family of stable low-frequency modes in the eigenspectrum of LSBs, i.e., mode b and other eigenvalues of the same family enclosed by the dashed rectangular region in Fig. 30.2a. Being globally stable, these eigenmodes cannot appear in the bubble in the absence of an external forcing. However, the adjoint eigenfunctions of these modes show high receptivity near the reattachment point, as depicted in Fig. 30.3. Thus, strong non-linearities occurring just upstream of the reattachment point may force these modes, that subsequently behave as a convective instability propagating upstream inside the reversed flow. This has been tested by running two numerical tests. The size of the three-

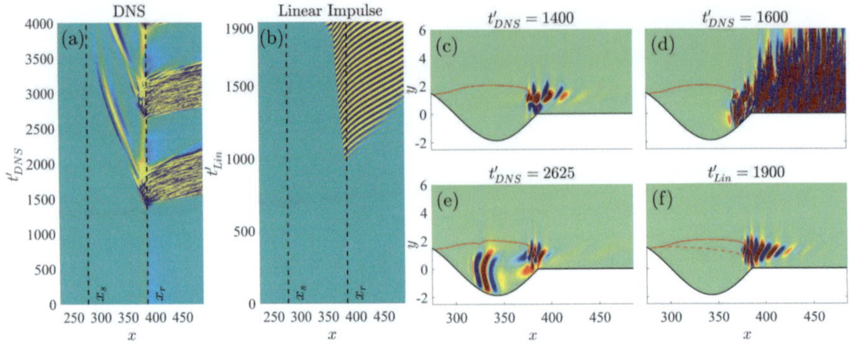

Fig. 30.4 x-t diagram of spanwise perturbation on a line parallel to the wall for DNS **a** and impulse response analysis (**b**). Spanwise perturbations in a x-y plane, at middle of the domain for three time instances in DNS (**c**, **d**, **e**). Panel (f) shows the spanwise perturbations in the linear simulation due to the impulse input. Separation (x_s) and reattachment (x_r) points are marked by vertical black dashed lines in panels (**a**) and (**b**). Red solid streamline in panels (**c**–**e**) shows instantaneous separation streamline of the bubble in DNS. In panel (**f**), separation streamline of the bubble at saturation is shown by red solid streamline, while the dashed red line indicates the same streamline for the initially two-dimensional flow

dimensional domain for these tests is $(H, L, L_z) = (50, 560, 2\pi/0.4)$, where H, L, and L_z are height, streamwise size, and spanwise size of the domain. A periodic boundary condition is applied in the spanwise direction. Note that spanwise size of the domain is chosen based on the spanwise wave length of the most unstable mode (see Fig. 30.2b). In the first test, non-linear DNS is performed, starting from an initially two-dimensional baseflow and marching up to transition. The corresponding x-t diagram is depicted in Fig. 30.4a, where the spanwise perturbation velocity on a line parallel to the wall which passes through the bubble is shown over time. A saturated baseflow (by preventing transition) is also obtained in this case by applying selective frequency damping (SFD) [9]. In the second test, a linear simulation is performed where the saturated baseflow obtained by SFD is used, and a localized momentum injection is added close to reattachment of the bubble as initial condition. The corresponding x-t diagram is shown in Fig. 30.4b. By comparing these two figures, it is clear that in DNS, after excitement of secondary instabilities through a self-excited mechanism, transition to turbulence happens which is followed by relaminarisation of the flow. A quasi-periodic burst cycle reappears for the rest of the simulation. The origin of this needs further investigations. On the other hand, in the linear simulation, a continuous formation of Kelvin-Helmholtz disturbances from the same spatial location where the shedding appears in DNS, occurs. This suggests that the quasi-periodic burst cycle observed in DNS happens due to a non-linear mechanism. Furthermore, when transition starts in DNS, an upstream propagation of low frequency waves is observed. Spanwise velocity perturbations in DNS are shown at three time instances in Figs. 30.4c–e. By comparing the upstream propagating waves in these figures with the shape of the low-frequency globally stable modes in Fig. 30.3, and the fact that these structures do not appear in a linear framework (Fig. 30.4b, f), it can be conjectured that these modes are triggered in the vicinity of the reattachment point through non-linear mechanisms and travel upstream inside the bubble. The role of this on dynamics of the bubble needs to be investigated.

30.4 Conclusions

This study identifies a globally unstable mode as the driver of three-dimensionalisation in a laminar separation bubble, along with low-frequency stable modes that gain significance during transition. Direct numerical simulation reveals upstream-propagating waves forming within the bubble due to self-excited nonlinear mechanisms, which are not captured by linear impulse response analysis. These findings emphasize the critical role of globally stable modes in the transition process, not being studied before.

Acknowledgements This project has received funding from the European Union's Horizon 2020 research and innovation programme under the Marie Skłodowska-Curie grant agreement No 955923 (SSECOID project). D.R. acknowledges funding from the Spanish Ministry of Science, Innovation and Universities and European Union's FEDER (PID2021-125812OB-C22), and from the Government of the Community of Madrid within the Program of Excellence in Faculty (V-PRICIT line 3). The computer resources provided by The National Academic Infrastructure for Supercomputing in Sweden (NAISS) are gratefully acknowledged.

References

1. Lardeau, S., Leschziner, M., Zaki, T.: Large eddy simulation of transitional separated flow over a flat plate and a compressor blade. Flow Turbul. Combust. **88**, 19-44 (2012). https://doi.org/10.1007/s10494-011-9353-0
2. Theofilis, V., Hein, S., Dallmann U.: On the origins of unsteadiness and three-dimensionality in a laminar separation bubble. Philos. trans. R. Soc. Lond. Series A: Math. Phys. Eng. Sci. **358**, 3229–3246 (2000). https://doi.org/10.1098/rsta.2000.0706
3. Rodríguez, D., Gennaro, E.M., Juniper, M.P.: The two classes of primary modal instability in laminar separation bubbles. J. Fluid Mech. **734**, R4 (2013). https://doi.org/10.1017/jfm.2013.504
4. Rodríguez, D., Theofilis, V.: Structural changes of laminar separation bubbles induced by global linear instability. J. Fluid Mech. **655**, 280–305 (2010). https://doi.org/10.1017/S0022112010000856
5. Gallaire, F., Marquillie, M., Ehrenstein, U.: Three-dimensional transverse instabilities in detached boundary layers. J. Fluid Mech. **571**, 221–33 (2007). https://doi.org/10.1017/S0022112006002898
6. Deville, M.O., Fischer, P.F., Mund, E.H.: High-order methods for incompressible fluid flow. Cambridge University Press. **9** (2002)

7. Rodríguez, D., Gennaro, E.M., Souza, L.F.: Self-excited primary and secondary instability of laminar separation bubbles. J. Fluid Mech. **906**, A13 (2021). https://doi.org/10.1017/jfm.2020.767
8. Moniripiri, M., Rodríguez, D. Hanifi, A.: Self-excited instability and transition to turbulence in laminar separation bubbles induced by surface waviness. J. Fluid Mech. **1011**, A27 (2025) (Cambridge University Press)
9. Åkervik, E., Brandt, L., Henningson, D.S., Hæpffner, J., Marxen, O., Schlatter, P.: Steady solutions of the Navier-Stokes equations by selective frequency damping. Phys. Fluids. **18** (2006). https://doi.org/10.1063/1.2211705

Chapter 31
Emergence of Edge State in Suspension of Self-Propelled Particles

Yoshiki Hiruta and Kenta Ishimoto

Abstract We numerically study a model convection system of a suspension of self-propelled particles, motivated by recent experimental findings of localized and bistable bioconvection pattern, being distinct from classical Rayleigh–Bénard convection. Linear stability analysis of the model system reveals that the trivial non-covection state is stabilized by an increase of self-propelled speed in the vertical direction. Through numerical simulations, we found a nonlinear convection state even when the nonconvection state is stable. Applying ideas and tools developed in wall-bounded flows, we numerically identified an edge state, which is an unstable solution on a basin boundary in the model dynamical systems.

Keywords Dynamical systems · Bioconvection · Self-propelled particle

31.1 Introduction

A suspension of swimming microorganisms often exhibits convective flow patterns, known as bioconvection [1–3]. In typical settings, bioconvection patterns sustain for a longtime if the average density of a suspension exceeds a critical value as seen in the Rayleigh–Bénard convection. Recently, Yamashita et al. [3] experimentally manipulated a suspension of microorganisms that swim away from the light source at the bottom of a chamber, and reported an emergence of localized pattern and bistability at the onset of the transition, which cannot happen in Rayleigh–Bénard convection [4].

Y. Hiruta (✉)
Department of Physics and Astronomy, Faculty of Science and Technology, Tokyo University of Science, Chiba, Japan
e-mail: hiruta@rs.tus.ac.jp

K. Ishimoto
Department of Mathematics, Kyoto University, Kyoto, Japan
e-mail: ishimoto@math.kyoto-u.ac.jp

K. Kato et al. (eds.), *Proceedings of the 10th IUTAM Symposium on Laminar-Turbulent Transition*, IUTAM Bookseries 44, https://doi.org/10.1007/978-981-96-9829-5_31

Localization and bistability are well studied in wall-bounded shear flows [5–7] and other flows in a periodic box [8, 9] and analytical methods based on dynamical systems theory have been developed. In particular, the existence of bistable turbulent localized pattern is well characterized by a so-called *edge state* [7, 10], which is an invariant set on a basin boundary. The phase space is then separated into two regions by the basin boundary, with the asymptotic longtime behaviors approaching localized turbulent and laminar states.

Motivated by the experimental findings in bioconvection and similar dynamical structure to the wall-bounded shear flows, in this paper, we investigate the onset of the transition in a model system of a dilute suspension of self-propelled particles (SPP), proposed by Childress et al. [1, 2] for bioconvection phenomena. Although the model equations are close to the Rayleigh–Bénard convection system, a term from the self-motility is added.

In the present study, we will examine the model bioconvection system, but with an extended to a periodic boundary to apply the methodologies based on dynamical systems theory. In doing so, we introduce an equilibrium density profile, independently of the self-motility, which enables us to control the effects of the self-motility in the bulk region. We will show emergence of the edge state representing localization and bistability with linear stability analysis.

31.2 Setting

We consider a two-dimensional fluid model with a buoyancy force acting in the y (vertical) direction defined in a periodic rectangular domain $(x, y) \in [0, 8\pi] \times [0, 2\pi]$ with an elongation in the x (horizontal) direction. We consider the Boussinesq approximation for the velocity field, $\boldsymbol{u}(x, y, t)$, and the continuity equation for the density deviation $m(x, y, t)$ from an equilibrium vertical profile $m_0(y)$, we may write down the governing equations in the nondimensionalized form as

$$\frac{\partial \boldsymbol{u}}{\partial t} + (\boldsymbol{u} \cdot \nabla)\boldsymbol{u} = \mathrm{Pr}(\nabla^2 \boldsymbol{u} - \mathrm{Ra}m\,\hat{\boldsymbol{y}}), \tag{31.1}$$

$$\frac{\partial m}{\partial t} + (\boldsymbol{u} \cdot \nabla)m = \nabla^2 m - \mathrm{Pe}\partial_y m - u_y \partial_y m_0, \tag{31.2}$$

with the incompressibility condition, $\nabla \cdot \boldsymbol{u} = 0$. Here, the Rayleigh number Ra represents normalized number density, and the Prandtl number Pr is the ratio of momentum diffusivity to diffusivity of SPPs. Unlike a standard thermal convection system, Eq. (31.2) contains an extra term from the self-propelled speed, represented by the Péclet number Pe, and also the term from the non-uniform density gradient.

31.3 Results

We now theoretically and numerically investigate the linear stability and nonlinear steady solutions of Eqs. (31.1) and (31.2) by setting $n_0 = \sin(y)$ in the two-dimensional periodic box. To examine the impact on self-propelled speed represented by Pe, we change the value of Pe with fixing Ra (average density), Pr (diffusivity) and size of the domain.

31.3.1 Linear Stability of Trivial Solution

We first assume the following form of mode expansion for the vorticity $\omega(x, y, t)$ and the density deviation $m(x, y, t)$:

$$\omega(x, y, t) = \exp(\lambda t + ikx) \sum_{l=-M}^{M} A_l \exp(ily), \tag{31.3}$$

$$m(x, y, t) = \exp(\lambda t + ikx) \sum_{l=-M}^{M} B_l \exp(ily), \tag{31.4}$$

where M is the cutoff wavenumber for the vertical profile. By substituting the expansion Eqs. (31.3) and (31.4) into Eqs. (31.1) and (31.2), eigenvalue problem for linear stability is written as

$$\lambda A_l = Pr(-k^2 A_l - Raik B_l), \tag{31.5}$$

$$\lambda B_l = -k^2 B_l - il Pe B_l - \frac{1}{2}\left(\frac{ikA_{l+1}}{k^2 + (l+1)^2} + \frac{ikA_{l-1}}{k^2 + (l-1)^2}\right), \tag{31.6}$$

for each l. We then numerically solved the eigenvalue problem, Eqs. (31.5) and (31.6). We choose the cut-off wavenumber as $M = 32$ and then the linear stability analysis becomes a eigenvalue problem for a square matrix with $2M + 1 = 65$ dimensions. In Fig. 31.1, we plot the contour of the maximum growth rate, defined as the maximum of the real part of λ, for the parameter set of $k = 0.25$ and $Pr = 2.5$. The neutral curve is then shown as the contour with $\mathrm{Re}(\lambda) = 0$. As seen in Fig. 31.1, all the contour curves increase with Pe, which indicates that the increase of Pe stabilizes the trivial solution.

Besides, we numerically observed that the maximum eigenvalue is always a real value. Additionally, Eqs. (31.5) and (31.6) are independent of Pr at $\lambda = 0$. By these two observations, the neutral curve should be independent of Pr.

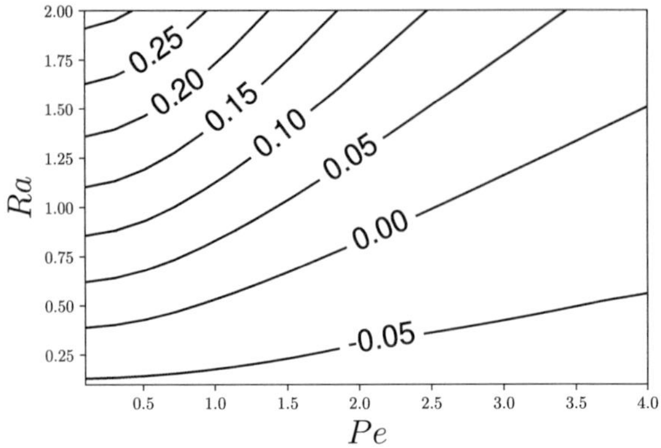

Fig. 31.1 Contour of the maximum growth rate, Re(λ), obtained by the linear stability analysis for $Pr = 2.5$

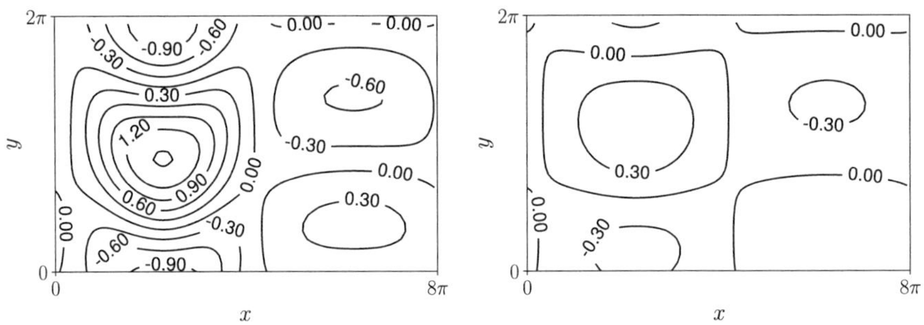

Fig. 31.2 Density deviation $m(x, y)$ of the nonlinear stationary solutions for $Ra = 0.4$, $Pr = 2.5$, and $Pe = 1$. (Left) the stable upper branch solution, $\boldsymbol{X}_{UB}$. (Right) the unstable lower branch solution, $\boldsymbol{X}_{LB}$. The unstable lower branch solution is obtained by the bisection method

31.3.2 Nonlinear Solution

We then proceed to examine the system bistability, focusing on the parameter values of $(Ra, Pr, Pe) = (0.4, 2.5, 1)$, where the trivial solution is linearly stable, for an illustrative purpose. In this parameter set, we found a stable nonlinear solution [upper branch (UB) solution, $\boldsymbol{X}_{UB}$] in addition to the trivial solution. The density deviation $m(x, y)$ for the UB solution is shown by a contour map in Fig. 31.2(Left). The density profile is spatially localized, noting the scales are different in the horizontal and vertical directions.

To evaluate the horizontal accumulation, which may be experimentally observed as a localized spot when seen from the top or bottom of a chamber, we consider vertically-averaged density, defined as, $n(x) = (1/2\pi) \int m(x, y)dy$. A thick line shown in Fig. 31.3 indicates $n(x)$ for the UB solution, which exhibits a positive peak at $x \approx 2\pi$, whereas no clear negative peak is observed.

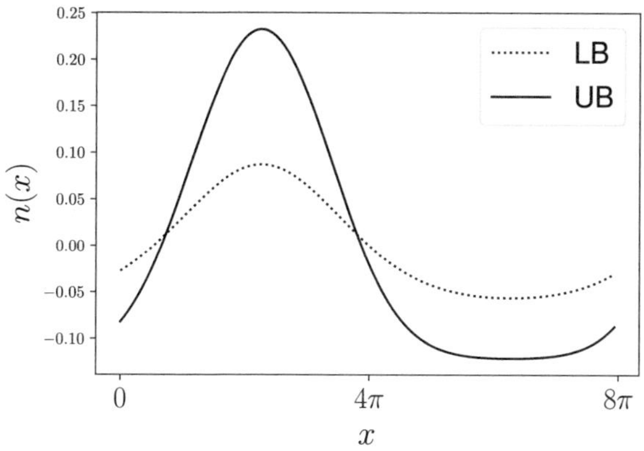

Fig. 31.3 Vertically-averaged density $n(x)$ for the upper and lower branch solutions

We then examine the bistability of the system. When two stable solutions coexist in phase space, we expect additional unstable solution exists. In this case, there must exists a basin boundary and an edge state on the basin boundary.

To obtain an unstable solution in a basin boundary, we perform the bisection iteration method used in the previous study of wall-bounded flow [10], and introduce $\boldsymbol{X}_\alpha = \alpha \boldsymbol{X}_{UB}$ with an additional parameter $\alpha \in [0, 1]$ and use this as our initial conditions. Since $\boldsymbol{X}_0$ is the trivial solution, there exists $\alpha_c \in [0, 1]$ such that $\boldsymbol{X}_{\alpha_c}$ is located on the basin boundary. In the bisection method, we seek a state with the energy norm $E = (u_x^2 + u_y^2)/2$ of $\boldsymbol{X}_{\alpha_c}$ being unchanged for a long time, and such α_c has been numerically detected (Fig. 31.4). As seen in Fig. 31.4, at the particular value of $\alpha = \alpha_c$, the energy norm E is unchanged after initial transient time duration, as shown in the red line. Once α exceeds the critical value, $\alpha > \alpha_c$, the energy norm E approaches E_0, the energy norm of the UB solution, i.e., X_{UB} is an attracting state. If $\alpha < \alpha_c$, on the other hand, E decays to zero, being consistent with the stable trivial solution. The state $\boldsymbol{X}_{\alpha_c}$ is found to be an unstable stationary solution on the basin boundary, hence, an edge state. We refer this solution as the lower blanch (LB) $\boldsymbol{X}_{LB}$ solution hereafter.

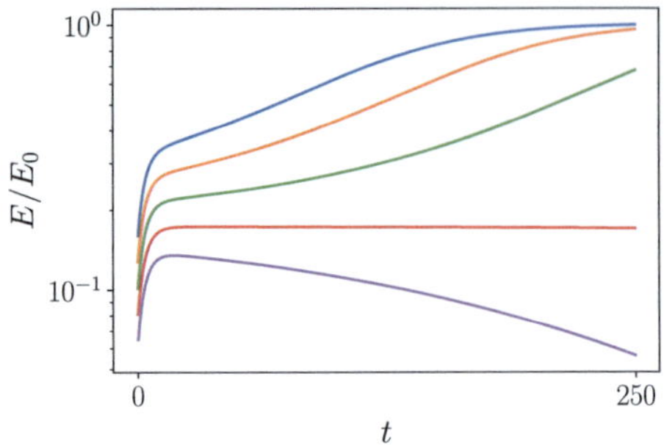

Fig. 31.4 Bisection method to obtain a state on basin boundary. Time evolution of energy norm of the velocity, normalized by the energy norm of the UB solution, E_0, for $Ra = 0.4$, $Pr = 2.5$, and $Pe = 1$. The different colors corresponds to different initial conditions, $\boldsymbol{X}_\alpha$. The red curve with stationary energy norm is the unstable lower branch (LB) solution, $\boldsymbol{X}_{LB}$, which is identified as an edge state

The density deviation of the LB solution for the parameter set, $(Ra, Pr, Pe) = (0.4, 2.5, 1)$, is shown in Fig. 31.2(Right). As seen in the UB solution (Fig. 31.2(Left)), the LB solution is also spatially localized. However, the absolute values of $m(x, y)$ are relatively smaller than those of LB solution. The dotted line shown in Fig. 31.3 indicates vertically-averaged density $n(x)$ for the LB solution, which exhibits a weakly localized peak compared with the UB solution.

31.4 Conclusions

In this study, to understand the effect of the self-propelled speed on emergence of localized, bistable flow patterns of a suspension of SPPs, we introduced a model system motivated by bioconvection in a doubly periodic box with a fixed equilibrium density profile in the vertical direction. Linear stability analysis of the model convection system clarifies that an increase of self-motility in the vertical direction results in stabilizing the trivial nonconvection state. Even when the trivial solution is linearly stable, nonlinear solution can be stable, unlike Rayleigh–Bénard convection. This localized, bistable nature is analogous to the turbulence transition in shear flows. By using the methodology based on the dynamical systems theory developed in these systems, we numerically obtained an unstable stationary solution which is identified as an edge state, whose stable manifold forms a basin boundary between two stable solutions.

Acknowledgements K.I. acknowledges the Japan Society for the Promotion of Science (JSPS) KAKENHI (Grant No. 21H05309) and the Japan Science and Technology Agency (JST), FOREST (Grant No. JPMJFR212N).

References

1. Childress, S., Levandowsky, M., Spiegel, E.A.: Pattern formation in a suspension of swimming microorganisms: equation and stability theory. J. Fluid Mech. **63**, 591 (1975)
2. Taheri, M., Bligen, E.: Bioconvection of gravitactic microorganisms in rectangular enclosures. Int. J. Heat Mass Transf. **50**, 4652 (2007)
3. Yamashita, H., Kamikubo, T., Muku, K., Suematsu, N.J., Izumi, S., Iima, M.: Emergence of a Euglena bioconvection spot controlled by non-uniform light. Front. Ecol. Evol. **11**, 1132956 (2023)
4. Joseph, D.D.: Nonlinear stability of the Boussinesq equations by the method of energy. Arch. Ration. Mech. Anal. **22**, 163 (1966)
5. Avila, M., Mellibovsky, F., Roland, N., Hof, B.: Streamwise-localized solutions at the onset of turbulence in pipe flow. Phys. Rev. Lett. **110**, 224502 (2013)
6. Duguet, Y., Schlatter, P., Henningson, D.S.: Formation of turbulent patterns near the onset of transition in plane Couette flow. J. Fluid Mech. **650**, 119 (2010)
7. Kawahara, G., Uhlmann, M., van Veen, L.: The significance of simple invariant solutions in turbulent flows. Annu. Rev. Fluid Mech. **44**, 203 (2011)
8. Hiruta, Y., Toh, S.: Subcritical laminar–turbulent transition as nonequilibrium phase transition in two-dimensional Kolmogorov flow. J. Phys. Soc. Jpn. **89**, 044402 (2020)
9. Hiruta, Y., Toh, S.: A simple thermal convection system showing subcritical transition and localized turbulences in two-dimensional periodic domains. J. Phys. Soc. Jpn. **91**, 013401 (2022)
10. Itano, T., Toh, S.: The dynamics of bursting process in wall turbulence. J. Phys. Soc. Jpn. **70**, 703 (2001)

Chapter 32
Experimental Study of Turbulent Spots in Free-Stream Turbulence Induced Boundary Layer Transition

André Weingärtner, Imge Yigili, and Jens H. M. Fransson

Abstract Predicting the behavior of turbulent spots is critical in modeling the late stage of laminar-turbulent transition in boundary layers. While these spots have been studied for over 70 years and are relatively well understood, their development under the influence of free-stream turbulence (FST) remains less explored. Previous investigations suggest that the characteristics of the incoming FST influence the spreading parameters of turbulent spots. The present study delves deeper into this phenomenon, utilizing a novel experimental setup featuring wall-mounted microphones to track individual, naturally occurring turbulent spots under varying FST conditions. The findings reveal variations in both, the growth rates and the number of generated spots.

Keywords Natural turbulent spots · Free-stream turbulence · Microphone measurements

32.1 Introduction

The late stage of the laminar-turbulent transition process of a boundary layer is associated with the birth of turbulent spots. These discrete, localized regions of turbulence are known to grow in size as they propagate downstream inside the boundary layer, merge with adjacent spots, and eventually develop into fully turbulent flow [1, 2]. Although the transitional region, in which turbulent spots are present, is often relatively short, in a turbine, it can make up as much as 50% of the chord length of an individual blade [3], making their behavior relevant for macroscopic effects such as skin friction and heat transfer.

Artificially generated spots, originating, for example, from a spark discharge or a pulsed wall-normal jet, have been investigated in detail by means of ensemble-averaging, commonly employing hot-wire anemometry. It has been found that spots

A. Weingärtner · I. Yigili · J. H. M. Fransson (✉)
FLOW, Department of Engineering Mechanics, KTH Royal Institute of Technology, Stockholm, Sweden
e-mail: jensf@kth.se

K. Kato et al. (eds.), *Proceedings of the 10th IUTAM Symposium on Laminar-Turbulent Transition*, IUTAM Bookseries 44, https://doi.org/10.1007/978-981-96-9829-5_32

feature an arrow-head shape pointing downstream and propagate slightly slower than the free-stream speed U_∞, while growing both in length and width. The tip of the arrow head, i.e. the leading edge of the spot, has a speed of around $U_{\rm LE} = 0.88\,U_\infty$, the speed of the trailing edge is about $U_{\rm TE} = 0.5\,U_\infty$, and the spot has a half-spreading angle of $\alpha = 8 - 11°$ [2].

In contrast, naturally occurring spots are harder to study due to the impossibility of ensemble-averaging; individual structures must be tracked in space and time using non-intrusive methods with a sufficiently high spatial resolution. For example, spanwise arrays of hot-film sensors were used previously, making it possible to track turbulent spots as they travel downstream [4]. In this study, the inception of arrowhead-shaped turbulent spots was observed for low levels of free-stream turbulence (FST) intensity, but at elevated levels the spots were irregular in shape and difficult to identify as discrete regions with clear boundaries. Fransson [5] investigated artificially generated turbulent spots in an asymptotic suction boundary layer with and without FST, showing a dependence of the spreading parameters on the properties of the FST, namely on the turbulence intensity Tu and integral length scale Λ_x.

In the present study, we use a multitude of wall-mounted microphones under individual pinhole orifices in a flat plate to measure wall-pressure fluctuations in the boundary layer. This makes it possible to track individual, naturally born turbulent spots in the boundary layer and assess their behavior depending on the FST conditions.

32.2 Experimental Setup

Measurements were performed in the MTL wind tunnel at the Fluid Physics Laboratory at KTH Royal Institute of Technology. The wind tunnel features a nominal FST turbulence intensity of 0.025% and a maximum flow velocity of 69 m/s for an empty test section. The experimental setup consists of a flat plate of 4.5 m in length that covers the entire 1.2 m width of the test section. The plate has an asymmetric leading edge and an adjustable trailing flap, which can together accomplish a relatively short leading-edge pressure gradient. In the test section, FST can be generated by mounting different grids upstream of the leading edge. With an installed rail system, the movement of the grid relative to the leading edge of the plate is facilitated, making it possible to tailor specific FST conditions, both for Tu and Λ_x.

The flat plate is equipped with a total of 1320 wall-mounted microphones, distributed in 23 spanwise arrays of 53 sensors each and 101 streamwise microphones. Each array consists of two staggered rows of microphones to improve the spatial resolution. The center-to-center distance between two microphones, when packed in the spanwise direction, is 4 mm, featuring a spanwise resolution of 2 mm. This results in a total width of an array of 104 mm. The two microphone rows are separated by $\sqrt{12} \approx 3.46$ mm in the streamwise direction, which is taken into account in the data processing, assuming a convective velocity of $0.7\,U_\infty$ (e.g. [2]). Each microphone is mounted under an individual pinhole with a diameter of 0.4 mm, and the arrays are installed from underneath, leaving the measurement surface of the plate free from any discontinuities. The slots where microphone arrays can be mounted are closely spaced in the upstream region on the plate with gradually increasing distance further downstream. A total of 212 microphones, i.e. four spanwise arrays, can be individually amplified and sampled simultaneously at a frequency of $F_s = 5\,\text{kHz}$ using an NI cDAQ-9189 fully equipped with eight NI9205 modules. For more information about the setup, the interested reader is referred to the PhD thesis by Weingärtner [6].

32.3 Spot Tracking Procedure

The spot tracking process is briefly described in the following enumeration. A more detailed description, including sample figures, can be found in [6].

1. Microphone signals were binarized into laminar (0) and turbulent (1) using an in-house program, resulting in a logical matrix, where turbulent spots were identified as connected regions of ones.
2. Detected spots were refined by filling laminar *holes*, smoothing the data, and extracting contours, which were then fitted to a characteristic shape. This returned the leading/trailing and lateral edges of each spot.
3. Spots were tracked across arrays in the streamwise direction using their center location, using $0.7\,U_\infty$ as an initial guess for the propagation speed, and step 2 was repeated on the identified corresponding spot on the next microphone array.
4. The evolution of each spot's edges was plotted against the downstream distance (i.e. array spacing), showing linear development consistent with constant leading and trailing-edge speeds and a constant spreading angle. Spots with less than three data points or a poor linear fit were excluded.

32.4 Results

In this paper, we present natural spot characteristics under the influence of three different FST cases. The conditions are measured at the leading edge and are shown in Table 32.1. Besides the turbulence intensity Tu and the integral length scale Λ_x, the table also shows the average spanwise spacing of the FST-induced streamwise streaks λ_z, measured by two-point spanwise correlation with two hot-wire probes. The streamwise range in which the measurements were performed is also given, and λ_z was determined at the most upstream of these locations. After the experiments by Fransson [5], it was hypothesized that this quantity together with the u_{rms}^{max} inside the boundary layer could have an effect on the spreading characteristics of turbulent spots.

Table 32.1 Turbulence properties for the investigated cases

Case	Tu (%)	Λ_x (mm)	λ_z (mm)	Measurement region (m)
1 •	1.54	13.8	18.2	1.58–2.28
2 •	2.06	14.0	14.4	0.76–1.22
3 •	3.41	13.8	11.2	0.32–0.62

For the FST conditions, it was attempted to keep the integral length scale close to constant, varying the turbulence intensity. Consistent with previous investigations, the spanwise wavelength λ_z decreases with increasing Tu, at constant Λ_x. For each of these cases, microphone data was acquired for 15 min in order to have enough passages of spots to arrive at converged statistics. The streamwise locations of the measurements were selected so that spots were just visible on the first microphone array, but not further upstream.

It could already be observed by eye that the number of detected turbulent spots varied vastly among the cases. Case 3, with the highest turbulence intensity, was found to have a large number of spots that merged quickly, making the tracking process challenging. Nonetheless, by sampling for a sufficient amount of time, it was possible to track individually occurring spots and to arrive at converged statistics. Data is presented in the form of histograms in Fig. 32.1, plotting U_{LE}, U_{TE}, and α. At first glance, it can be seen that both the trailing edge velocity and the spreading angle are close to normally distributed. The leading-edge velocity, however, is clearly skewed with a tail towards higher speeds.

The dashed lines in Fig. 32.1 represent the mean values. The mean leading edge speeds feature only a small variation with the FST properties, while being notably larger than what is commonly reported in the literature for artificial turbulent spots, i.e. without FST. In fact, a large number of spots appear to travel faster than the mean free-stream velocity. Although seemingly paradoxical, this does not infer that fluid particles are moving at this speed, but it rather represents the laminar-turbulent interface. This interface can be moving faster than the flow, as the spot grows at the same time as being convected downstream. Recall, that under FST there is a continuous forcing of turbulence on the boundary layer edge, and as the spot propagates downstream the streamwise streaks ahead of the spot become even more unstable. This makes it likely that new spots can be born at the interface between the original spot and the upstream laminar boundary layer, causing a spreading rate of turbulence larger than U_∞. It should be noted that this observation is not due to the employed measurement method. A parallel investigation [7] studies artificially generated turbulent spots using the same setup and finds results in agreement to literature data. In Fig. 32.1, mean values of the trailing-edge velocity show a clearer trend among the three FST cases. It can be seen that there is an increase with rising turbulence intensity, indicating that the streamwise growth is somewhat inhibited by the FST.

Table 32.2 Statistical parameters of turbulent spots for the presented cases

Case	U_{LE}/U_∞		U_{TE}/U_∞		α (°)		ξ	Spot rate
	μ	σ	μ	σ	μ	σ		($s^{-1}m^{-1}$)
1 •	1.02	0.12	0.53	0.04	8.8	0.86	0.14	4.6
2 •	1.01	0.15	0.57	0.06	8.0	1.14	0.11	16.9
3 •	0.99	0.17	0.58	0.07	7.5	1.40	0.09	129.5

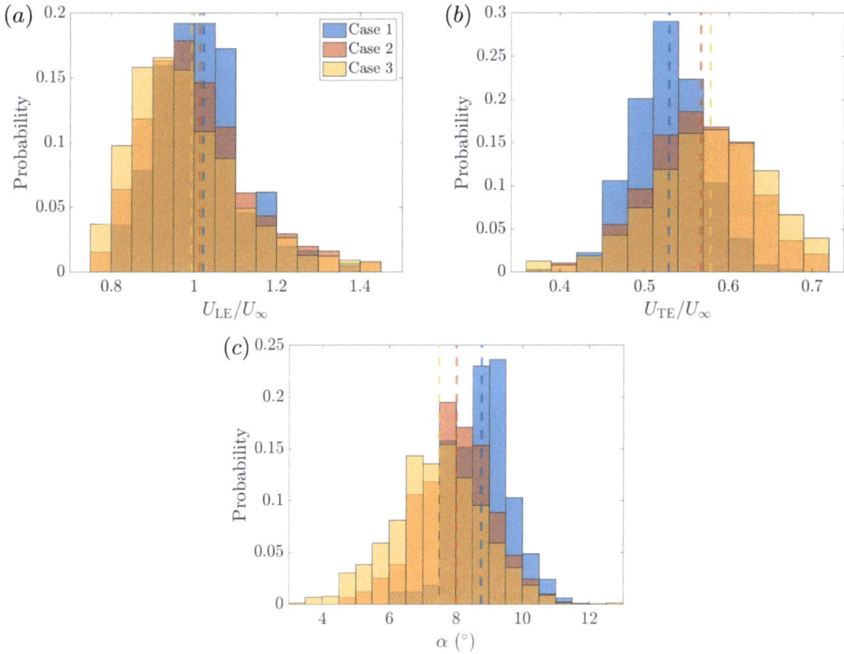

Fig. 32.1 Histograms of tracked turbulent spots for the three investigated cases. Vertical dashed lines indicate the mean values. **a**: Leading-edge velocity, **b**: trailing-edge velocity, and **c**: half-spreading angle

A similar conclusion can be drawn from the data of the spreading angle, an increase in Tu leads to a lower half-spreading angle.

Table 32.2 presents statistics of the investigated cases. Here, μ are the mean values and σ the standard deviations of the respective quantities. The mean has been discussed previously, but it is interesting to note that all measured standard deviations increase with higher turbulence intensity. From the mean quantities, the growth parameter of the turbulent spots ξ can be determined, which returns a value for the overall growth of the spot. For a triangular-shaped spot, ξ can be calculated as

$$\xi = U_\infty \left(\frac{1}{U_{TE}} - \frac{1}{U_{LE}} \right) \tan \alpha \; . \tag{32.1}$$

The expression was introduced by Gostelow et al. [8], applying the more general formulation by Emmons [1] to a triangular shape. As already indicated by the individual quantities, data in Table 32.2 show that the growth of the turbulent spots is in fact lower for cases with higher turbulence intensity.

However, it has been observed before that the length of the transition region decreases with increasing values of Tu. This means that the number of turbulent spots has to differ. Therefore, the table also shows the spot rate, which is the number of spots that are detected on the most upstream array per second, normalized with the width of the microphone array. Note that this quantity is not the spot production rate, which is also a function of the streamwise location, but instead represents the number of spots that pass this constant streamwise position. It can be observed that the number of spots rises very quickly with the FST intensity, significantly faster than the spot growth parameter decreases, therefore resulting in a shorter transition region.

Although not shown here, it could be observed that *young* turbulent spots tend to grow at a lower spreading angle and move faster in the flow compared to *mature* spots. However, this was only visible on two neighboring microphone arrays, while by the time it reached the third array, the spot had *matured* (or merged with other adjacent spots). Due to the employed method of tracking the turbulent spots, and the constraint that it has to appear on at least three arrays, these cases are not taken into account in the presented data. It appears that this type of spot was dominant in the highest turbulence intensity case, which should be taken as a caveat for the interpretation of the present data.

32.5 Conclusions

Using a novel experimental setup, the behavior of naturally born turbulent spots in FST-induced boundary layer transition was investigated. Statistical properties of natural spots are presented under the variation of FST conditions. At a higher turbulence intensity, the spots grow slightly slower, but this is outweighed by a significantly larger number of turbulent spots. Consequentially, the length of the transition zone reduces with increasing Tu, besides moving further upstream.

It is remarkable to note that each individual spot spreads linearly in space, but at quite different rates compared to other spots, implying that every spot is unique and different from another. Moreover, the variance between spots increases with the turbulence intensity. Further work will explore how a variation of the integral length scale affects the spot propagation properties and statistics thereof. In the current experiments, only the turbulence intensity of the FST was varied, while Λ_x was kept close to constant. A larger parameter variation study can reveal what the decisive FST quantity is that dictates the evolution of the spots, or if it purely depends on the amplitude and spanwise wavelength of the streaky structures.

References

1. Emmons, H.W.: The Laminar-Turbulent transition in a boundary layer - Part I. J Aeronaut. Sci. **18**(7), 490–498 (1951)
2. Schubauer, G.B., Klebanoff, P.S.: Contributions on the mechanics of boundary-layer transition. Technical report, NACA-TN-3489 (1956)
3. Mayle, R.E.: The 1991 IGTI scholar lecture: the role of Laminar-Turbulent transition in gas turbine engines. J. Turbomach. **113**(4), 509–536 (1991)
4. Anthony, R.J., Jones, T.V., LaGraff, J.E.: High frequency surface heat flux imaging of bypass transition. J. Turbomach. **127**(2), 241–250 (2004)
5. Fransson, J.H.M.: Turbulent spot evolution in spatially invariant boundary layers. Phys. Rev. E **81**, 035301 (2010)
6. Weingärtner, A.: Experimental studies of laminar boundary layers: instability, transition and control. Ph.D. Thesis, KTH Royal Institute of Technology (2024)
7. Yigili, I, Weingärtner, A., Fransson, J.H.M.: Turbulent spot measurements: wall-pressure and streamwise velocity correlation. In: IUTAM Laminar-Turbulent Transition Conference (2024)
8. Gostelow, J.P., Melwani, N., Walker, G.J.: Effects of streamwise pressure gradient on turbulent spot development. J. Turbomach. **118**(4), 737–743 (1996)

Chapter 33
Direct Numerical Simulations of K-Type Transition in a Flat-Plate Boundary Layer with Supercritical Fluids

Pietro Carlo Boldini, Benjamin Bugeat, Jurriaan W. R. Peeters, Markus Kloker, and Rene Pecnik

Abstract We investigate the controlled K-type breakdown of a flat-plate boundary-layer with highly non-ideal supercritical fluid. Direct numerical simulations are performed at a Mach number of $M_\infty = 0.2$ for one subcritical (liquid-like regime) temperature profile and one strongly-stratified transcritical (pseudo-boiling) temperature profile with slightly heated wall. In the subcritical case, the formation of aligned Λ-vortices is delayed compared to the reference ideal-gas case of Sayadi et al. (*J. Fluid Mech.*, vol. 724, 2013, pp. 480–509), with steady longitudinal modes dominating the late-transitional stage. When the wall temperature exceeds the pseudo-critical temperature, streak secondary instabilities lead to the simultaneous development of additional hairpin vortices and near-wall streaky structures near the legs of the primary aligned Λ-vortices. Nonetheless, transition to turbulence is not violent and is significantly delayed compared to the subcritical regime.

Keywords Direct numerical simulation · Laminar-to-turbulent transition · Supercritical fluids

33.1 Introduction

When a fluid at supercritical pressure is isobarically heated or cooled across the pseudo-critical (Widom) line, it undergoes a continuous phase transition characterised by strong thermophysical property variations, which can impact the transition

P. C. Boldini (✉) · J. W. R. Peeters · R. Pecnik
Process and Energy Department, Delft University of Technology, Delft, The Netherlands
e-mail: pietro.c.boldini@gmail.com

R. Pecnik
e-mail: r.pecnik@tudelft.nl

B. Bugeat
School of Engineering, University of Leicester, Leicester, UK

M. Kloker
Institute of Aerodynamics and Gas Dynamics, University of Stuttgart, Stuttgart, Germany

K. Kato et al. (eds.), *Proceedings of the 10th IUTAM Symposium on Laminar-Turbulent Transition*, IUTAM Bookseries 44, https://doi.org/10.1007/978-981-96-9829-5_33

to turbulence. Research on the hydrodynamic stability of boundary layers with fluids exhibiting non-ideal thermodynamic behaviour has recently gained interest [1]. The newly identified Mode-II instability [2] in the transcritical regime, associated with the minimum of kinematic viscosity at the Widom line, arises from the interaction between shear and baroclinic waves [3]. Non-modal stability analysis revealed that Mode II alters the streamwise invariance of optimal streamwise streaks [4]. In three-dimensional boundary-layer flows, the inviscid Mode-II instability prevailed over cross-flow instability despite flow acceleration [5]. While these studies focused on the linear regime, nonlinear interactions and breakdown to turbulence remain unexplored. As a first attempt, Boldini et al. [6] investigated the H-type breakdown of a flat-plate boundary layer under subcritical and transcritical temperature conditions. They showed that staggered Λ-vortices with inverted Λ-shape at their valleys (secondary vortex systems) lead to a violent breakdown process. Here, we examine the other main controlled transitional scenario, the K-type breakdown [7], by means of direct numerical simulation (DNS).

33.2 Methodology

The non-ideal compressible, high-order finite-difference code CUBENS is employed for the numerical simulations in this work; see Ref. [8] for details. A rectangular computational domain is selected containing the boundary layer of a flat-plate, (see Ref. [6]), with the inlet location at x_0 from the leading edge chosen based on the inlet Reynolds number of $Re_{x,0} = 10^5$. The reference length scale is the boundary-layer thickness at the domain inlet, $\delta^*_{99}(x_0) = \delta^*_{99,0}$. Non-reflecting boundary conditions with sponge zones are applied at the inlet, top, and outflow boundaries, while no-slip and fully-reflective conditions are enforced at the wall. The grid is equidistant in the streamwise and periodic spanwise directions, with grid stretching applied in the wall-normal direction.

Mimicking the ideal-gas case of Ref. [9], disturbances are introduced at the wall via a blowing/suction disturbance strip with a length of $x/\delta_{99,0} = 13$, at $Re_\delta = \sqrt{Re_x} = 381$. Here, δ is the local Blasius length scale defined as $\delta^* = (\mu^*_\infty x^* / \rho^*_\infty / u^*_\infty)^{1/2}$. The wall-normal velocity distribution is specified as

$$v(y=0) = f(x)\,[A_{2\text{-D}} \sin(\omega_0 t) + A_{3\text{-D}} \sin(\omega_0 t) \cos(\beta_0 z)]\,, \qquad (33.1)$$

where $f(x)$ is provided in Ref. [9]. The amplitudes $A_{2\text{-D}} = 2.8 \times 10^{-2}$ and $A_{3\text{-D}} = 4.0 \times 10^{-4}$, corresponding to the fundamental two-dimensional (2-D) and the oblique three-dimensional (3-D) wave pair, respectively, share the same frequency, $F_0 = \omega_0 / Re_{\delta_{99,0}} = 110 \times 10^{-6}$, as in Ref. [9]. Note that the forcing fundamental 2-D amplitude for the H-type breakdown was $A_{2\text{-D}} = 7.5 \times 10^{-3}$ [8]. The spanwise wavenumber is $\beta_0 = 2\pi/z_e$, where z_e is the spanwise computational domain. In the following, modes are denoted in a double-spectral notation $(h\omega_0, k\beta_0)$.

33.3 Cases, Base Flow and Linear Stability Analysis

The flow parameters in this study match those used for the DNS investigation of the H-type breakdown in Refs. [6, 8]. The supercritical fluid is described by the reduced Van der Waals equation of state, where reduced properties are denoted as $(\cdot)_r = (\cdot)^*/(\cdot)^*_c$, with $(\cdot)^*_c$ representing the critical point values. Transport properties are modelled using the Jossi, Stiel, and Thodos analytical relations [8]. The free-stream parameters are consistent among all cases: a supercritical pressure of $p_{r,\infty} = 1.10$, temperature of $T_{r,\infty} = 0.90$, density of $\rho_{r,\infty} = 1.805$ (in the liquid-like region), and Mach number of $M_\infty = 0.2$. Two reduced wall temperatures are considered: one in the liquid-like regime at $T_{r,w} = 0.95$, or $T^*_w/T^*_\infty = 1.056$ (subcritical case, referred to as Tw095), and the other in the vapour-like regime at $T_{r,w} = 1.10$, or $T^*_w/T^*_\infty = 1.222$ (transcritical case, referred to as Tw110). For further details on the flow parameters; see Ref. [8], while computational parameters for all cases are presented in Table 33.1. The ideal-gas K-type breakdown in Ref. [9], labelled as TadIG, is included for validation and comparison. The laminar boundary-layer profiles are discussed in Ref. [6], with the strongest density and thermophysical property variations occurring near the Widom line for case Tw110. Linear stability theory is used to analyse the linear disturbance behaviour across the entire relevant disturbance frequency range, and the stability diagrams are compared in Fig. 33.1 The spatial growth rate $-\alpha_i$ is plotted as a function of the local Reynolds number $Re_\delta = \sqrt{Re_{x,0}}/\delta_{99,0} = Re_{\delta,0}/\delta_{99,0}$ and frequency parameter F.

Within the considered DNS domain for the K-type breakdown at $F_0 = 110 \times 10^{-6}$ in Fig. 33.1c, the unstable region of mode (1, 0) is caused exclusively by Mode II. Moreover, the broad frequency bands of instability for Modes I and II are visible in the inset of Fig. 33.1c.

33.4 Direct Numerical Simulations

Figure 33.2 shows the downstream development of the maximum amplitudes (over y) of $(\rho u)'$ for cases Tw095 (Fig. 33.2a) and Tw110 (Fig. 33.2b). In the subcritical regime (Fig. 33.2a), the modal development resembles case TadIG (not shown here). However, due to the increase in wall temperature towards the Widom line, the

Table 33.1 Numerical parameters of the K-type simulations. Viscous units $(\cdot)^+$ are calculated at $\max\{C_f\}$. Case TadIG refers to the ideal-gas reference case of Ref. [9]

Case	L_x/δ_0	L_y/δ_0	L_z/δ_0	N_x	N_y	N_z	Δx^+	Δy^+_w	Δz^+	$Re_{\theta,max}$
TadIG	347	20	9.63	3000	600	150	9.3	0.61	5.2	1007
Tw095	352	20	9.63	3000	600	150	10.0	0.65	5.5	970
Tw110	570	40	9.63	8400	600	150	6.1	0.68	5.8	748

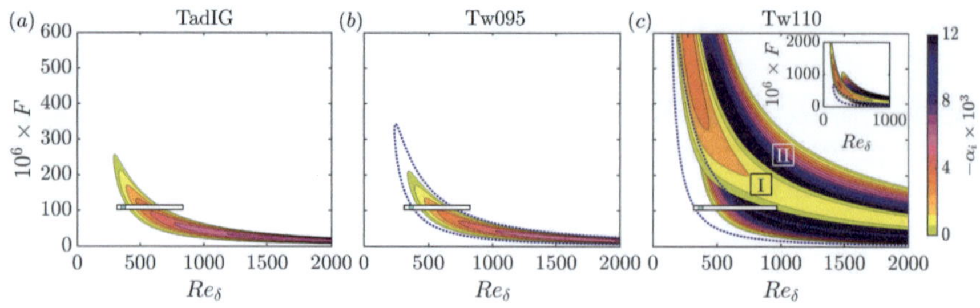

Fig. 33.1 Growth-rate ($-\alpha_i$) contours in the Re_δ–F stability diagram for 2-D disturbances: **a** TadIG, **b** Tw095, and **c** Tw110 (Modes I and II). The dotted blue lines in (**b**, **c**) represent the ideal-gas neutral stability with equal T_w^*/T_∞^*-ratio. The inset of (**c**) highlights the wide frequency band of Modes I and II. The DNS domain and perturbation strip for the fundamental breakdown, i.e. $F_0 = 110 \times 10^{-6}$, are indicated by the white and cyan bars, respectively. Re_δ is scaled with the local Blasius length scale

growth of secondary oblique disturbance waves $(1, \pm 1)$ is weakened, delaying the location of $\min\{C_f\}$. Later, all modes saturate, and from $Re_x/10^5 \approx 2.6$ onward, the nonlinearly generated longitudinal mode $(0, 1)$ reaches the highest amplitudes, surpassing the other modes over a considerable streamwise distance. In this case, the higher harmonic mode $(1, 3)$ is destabilised, whereas in case TadIG (not shown here), it is the $(1, 4)$ mode. Further downstream, in the strongly nonlinear region, the generated mean-flow distortion $(0, 0)$ reaches its largest amplitude, indicating a strong mean-flow deformation due to transition to turbulence, i.e. $\max\{C_f\}$. In contrast, in the transcritical regime (Fig. 33.2b), the nonlinear, high-amplitude primary wave initially saturates due to nonlinear effects and is surpassed by its first higher harmonic $(2, 0)$ at $Re_x/10^5 \approx 3.4$, where Λ-structures reveal at halved streamwise wavelength. Note that the higher harmonics ($(2, 0)$, $(3, 0)$, etc.) are highly unstable

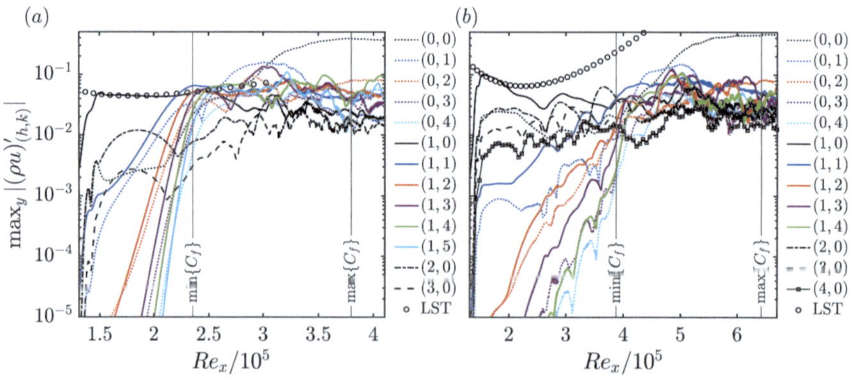

Fig. 33.2 Streamwise development of the y-maximum $(\rho u)'$ disturbance amplitudes: **a** Tw095, **b** Tw110. LST results are shown in black circles. Note that the same v-distribution has been applied to the disturbance strip in both cases

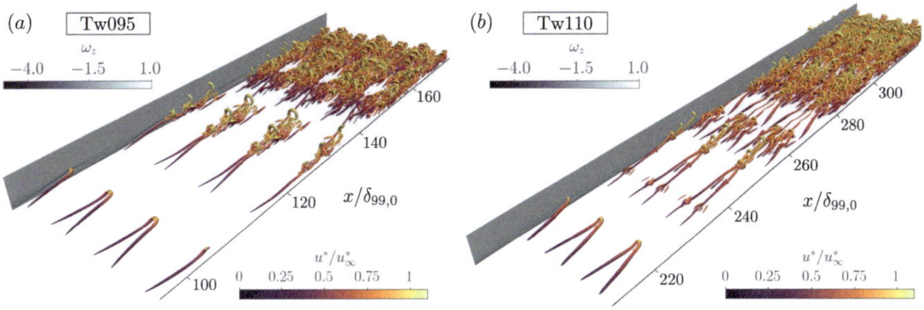

Fig. 33.3 Instantaneous isosurfaces of the Q-criterion, coloured by the streamwise velocity u: **a** Tw095 ($Q = 0.06$), **b** Tw110 ($Q = 0.016$). The side plane indicates the instantaneous spanwise vorticity ω_z. Note the different $x/\delta_{99,0}$-ranges, and that the domain is replicated three times in the spanwise direction for improved visualisation

according to LST, as shown in Fig. 33.1c, are more strongly excited at the disturbance strip than in the subcritical case, and reach a higher amplitude level than (1, 1) and (0, 1) in the early transitional stage. Once (2, 0) dominates, the secondary oblique disturbance waves experience rapid streamwise growth around the $\min\{C_f\}$ location (subharmonic resonance), together with the higher harmonic modes ((1, 2), (0, 2) (1, 3), etc.). In the transitional region, the steady longitudinal mode (0, 1) reaches its highest amplitude from $Re_x/10^5 \approx 4.3$ onwards, prior to the full transition to turbulence.

Figure 33.3 displays the instantaneous flow structures through isocontours of the Q-criterion. In the subcritical regime (Fig. 33.3a) the sequence of aligned Λ-structures, with a high-shear layer at their tips (see spanwise vorticity contours), and hairpin-shaped vortices can be clearly observed. As they propagate downstream, Λ-structures are known to elongate further [7]. This behaviour is even more pronounced here, as (0, 1) is the dominant mode in the region around $x/\delta_{99,0} = 120$. In the transcritical regime (Fig. 33.3b), the breakdown is significantly delayed compared to the subcritical regime (Fig. 33.3a). In particular, Λ-structures are aligned, and no secondary vortex structures are found at their valleys, as previously discovered in the H-type breakdown under the same flow conditions [6]. Moreover, we observe a less rapid and abrupt breakdown to turbulence than in the H-type breakdown. Interestingly, in Fig. 33.3b, around $x/\delta_{99,0} = 263$ ($Re_x/10^5 \approx 4.7$), a pair of secondary hairpin vortices are observed lifting off the sides of the main hairpin vortex, accompanied by strong vortical near-wall structures. These secondary structures originate farther upstream from the main legs of the Λ-vortex at $x/\delta_{99,0} = 238$ ($Re_x/10^5 \approx 4.35$). In this region, the streamwise-elongated vortex structure is predominantly composed of the (0, 1) mode, which has exceeded an amplitude of 10%. To highlight its structure, Fig. 33.4 a displays the two main legs of oppositely signed streamwise vorticity ω_x (symbol I: one at $z/\delta_{99,0} > z_e/2$ and the other at $z/\delta_{99,0} < z_e/2$). Additionally, two co-rotating legs of $\pm\omega_x$ (symbol II) are noticeable closer to the wall between the main vortex-pair structure, having a higher streamwise momentum ρu than the surrounding low-velocity, low-density fluid; see Fig. 33.4b. The interaction between

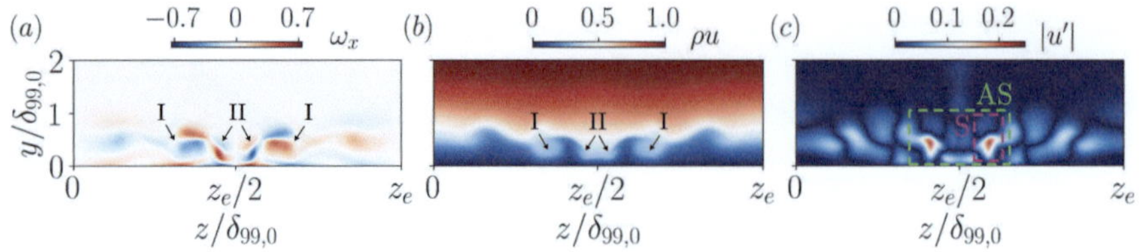

Fig. 33.4 Instantaneous isosurfaces at $x/\delta_{99,0} = 238$ ($Re_x/10^5 \approx 4.35$) in the y-z plane: **a** streamwise vorticity ω_x of counter-rotating (I) and co-rotating (II) vortex pairs, **b** streamwise momentum ρu, and **c** absolute value of streamwise velocity perturbation u' with symmetric (S) and antisymmetric (AS) mode

counter-rotating and co-rotating vortex tubes generates significant spanwise modulation, with low-speed regions, i.e. low-speed streaks, forming between the vortices of opposite vorticity signs at $z/\delta_{99,0} > z_e/2$ and $z/\delta_{99,0} < z_e/2$. Given the aforementioned nonlinear amplitude, these streaks undergo a localised secondary instability (LSI), exhibiting two different instability modes on either side of the streak or streak pair. In Fig. 33.4c, the contours of streamwise velocity perturbations over the spanwise distance are shown. At the centre of each low-speed streak, a symmetric (S) varicose mode is detected, which develops into the aforementioned symmetric hairpin vortex. Simultaneously, an antisymmetric sinuous (AS) mode is identified, causing the streak pair to meander with respect to $z/\delta_{99,0} = z_e/2$, and generating additional near-wall streamwise vortices downstream.

33.5 Conclusions

The K-type breakdown of a weakly wall-heated flat-plate boundary layer at a Mach number of 0.2 with a supercritical fluid at a reduced pressure ($p_r = p^*/p_c^*$) of 1.10 has been investigated. In the subcritical heating regime, where $T_\infty^* < T_{pc}^*$ and $T_w^* > T_{pc}^*$, breakdown to turbulence is slightly delayed compared to the reference incompressible case in Ref. [9]. Conversely, in the transcritical (pseudo-boiling) heating regime, where $T_w^* > T_{pc}^*$, large-amplitude higher harmonics dominate the initial breakdown stage, delaying both the location and strength of the fundamental resonance. Furthermore, subharmonic resonance between mode (2, 0) and (1, 1) is active just before the transition onset. Farther downstream, a competition between symmetric and antisymmetric modes of localised secondary instability emerges near the legs of the primary aligned Λ-vortices at the peak stations. Unlike the rapid and violent H-type breakdown in the transcritical wall-heating regime, the transcritical K-type breakdown is more gradual and does not exhibit strong secondary vortical structures at the spanwise valley positions. Compared to the ideal-gas regime at the same forcing amplitudes, transcritical K- and H-type breakdowns exhibit a stronger and nonlinear response to the 2-D fundamental forcing wave, driven by significantly enhanced density fluctuations in highly non-ideal flow conditions.

References

1. Robinet, J.-C., Gloerfelt, X.: Instabilities in non-ideal fluids. J. Fluid Mech. **880**, 1–4 (2019)
2. Ren, J., Marxen, O., Pecnik, R.: Boundary-layer stability of supercritical fluids in the vicinity of the Widom line. J. Fluid Mech. **871**, 831–864 (2019)
3. Bugeat, B., Boldini, P.C., Hasan, A.M., Pecnik, R.: Instability in strongly stratified plane Couette flow with application to supercritical fluids. J. Fluid Mech. **984**, A31 (2024)
4. Boldini, P.C., Bugeat, B., Peeters, J.W.R., Kloker, M., Pecnik, R.: Transient growth in diabatic boundary layers with fluids at supercritical pressure. Phys. Rev. Fluids **9**, 083901 (2024)
5. Ren, J., Kloker, M.: Instabilities in three-dimensional boundary-layer flows with a highly non-ideal fluid. J. Fluid Mech. **951**, A9 (2022)
6. Boldini, P.C., Gaspar, R., Bugeat, B., Costa, P., Peeters, J.W.R., Pecnik, R.: Direct numerical simulation of H-type transition in a flat-plate boundary layer with supercritical fluids. In: 14th International ERCOFTAC Symposium on Engineering Turbulence Modelling and Measurements (ETMM14), Barcelona, Spain (2023)
7. Rist, U., Fasel, H.: Direct numerical simulation of controlled transition in a flat-plate boundary layer. J. Fluid Mech. **298**, 211–248 (1995)
8. Boldini, P.C., Hirai, R., Costa, P., Peeters, J.W.R., Pecnik, R.: CUBENS: a GPU-accelerated high-order solver for wall-bounded flows with non-ideal fluids. Comput. Phys. Commun. **309**, 109507 (2025)
9. Sayadi, T., Hamman, C., Moin, P.: Direct numerical simulation of complete H-type and K-type transitions with implications for the dynamics of turbulent boundary layers. J. Fluid Mech. **724**, 480–509 (2013)

Chapter 34
Characteristic Fluctuations Observed in Boundary Layer Transition Around Rotating Dimpled Sphere

Shoji Sakaue, Yusuke Tominaga, Takakage Arai, Masahide Onuki, and Takahiro Sajima

Abstract In this study, we experimentally investigate the boundary layer transition on the surface of a rotating dimpled sphere using hot-wire anemometry, focusing on understanding the relationship between characteristic velocity fluctuations observed during the transition process and the dimple configuration. The results indicate that the boundary layer transition is driven by high-frequency fluctuations, which occur as the dimples pass continuously in the streamwise direction, independent of the relative orientation between the freestream and the sphere's rotation direction. Additionally, the direction of rotation affects the location of the transition point, causing it to shift in the direction of rotation.

Keywords Transition · Dimpled sphere · Hot-wire measurements · Characteristic fluctuation

34.1 Introduction

It is well known that the drag coefficient of a sphere placed in uniform flow decreases sharply within a specific range of Reynolds numbers [5]. This phenomenon occurs because the boundary layer on the surface of the sphere transitions to turbulence, causing the separation point to move downstream. As a result, the wake behind the sphere reduces in size, and pressure drag decreases. For spheres with surface roughness or dimples, the boundary layer transition occurs at a lower Reynolds number compared to that of a smooth sphere, resulting in drag reduction at lower Reynolds numbers [1]. A golf ball is a typical example of this phenomenon, and it has been confirmed that the aerodynamic forces acting on the dimpled sphere are influenced by both the arrangement and depth of the dimples [2–4]. However,

S. Sakaue (✉) · Y. Tominaga · T. Arai
Osaka Metropolitan University, Sakai, Japan
e-mail: shoji-sakaue@omu.ac.jp

M. Onuki · T. Sajima
Sumitomo Rubber Industries, Ltd., Kobe, Japan

K. Kato et al. (eds.), *Proceedings of the 10th IUTAM Symposium on Laminar-Turbulent Transition*, IUTAM Bookseries 44, https://doi.org/10.1007/978-981-96-9829-5_34

since the golf ball rotates during flight, the flow around it is more complex than that around a stationary smooth sphere. Consequently, the boundary layer transition on the rotating dimpled sphere remains not fully understood.

This research was conducted as part of a collaborative project between Sumitomo Rubber Industries, Kobe University, and Osaka Metropolitan University, with the aim of improving golf ball flight distance by clarifying the relationship between dimple configurations, rotational speed, and boundary layer transition, all of which influence the aerodynamic characteristics of the sphere (for example, see [6]). In this study, the turbulent transition of the boundary layer on the rotating dimpled sphere, particularly the characteristic disturbances occurring during the transition process, was investigated experimentally in detail using hot-wire anemometry.

34.2 Experimental Apparatus and Settings

Figure 34.1 illustrates the schematic of the experimental setup. The origin of the coordinate system is located at the center of the sphere, with the x-axis aligned along the streamwise direction, the y-axis normal to the flow, and the z-axis in the spanwise direction. The sphere was mounted in cantilever within the wind tunnel test section and could be rotated clockwise around the z-axis. The cross-sectional plane

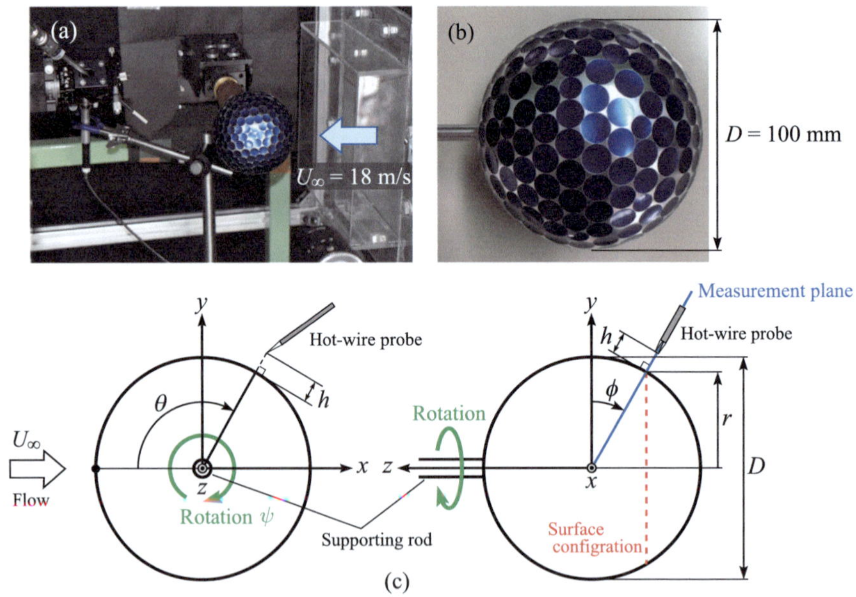

Fig. 34.1 Schematics of experimental setup: **a** over view, **b** dimpled model sphere, **c** coordinate system.

that includes the x-axis is defined as the measurement plane. The angle between the y-axis and the measurement plane is denoted as ϕ, the angle relative to the front stagnation point is θ, the phase angle of the rotating sphere is ψ, and the height from the surface is h. For the experiments, a dimpled sphere with a diameter of $D = 100$ mm was used. The Reynolds number, based on the freestream velocity U_∞ and the sphere diameter D, was $Re = 1.1 \times 10^5$, which corresponds to the typical flight conditions for a real golf ball. The spin ratio SR, defined as the ratio of the circumferential speed of sphere to the freestream velocity, was set at $SR = 0.1$. Under these conditions, the freestream velocity was $U_\infty = 18$ m/s, and the sphere's rotational speed was 344 rpm, or 5.7 rps. Measurements were conducted using a constant-temperature hot-wire anemometer with a tungsten wire probe that had a diameter of 5 μm and a sensing length of 1 mm.

34.3 Results and Discussion

First, the results for the equatorial plane of rotation $\phi = 0°$ are presented [6]. Figure 34.2 illustrates the mean velocity $\bar{u}$ and the root mean square (rms) velocity fluctuations u'_{rms} as functions of the height h on the upper surface of the dimpled

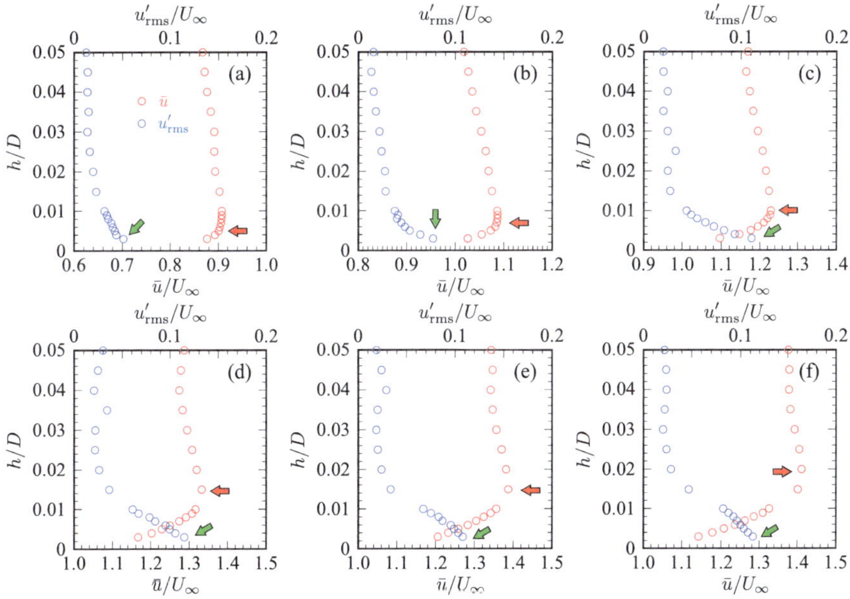

Fig. 34.2 Mean and rms velocity profiles at $\phi = 0°$, **a** $\theta = 40°$, **b** $\theta = 50°$, **c** $\theta = 60°$, **d** $\theta = 70°$, **e** $\theta = 80°$, and **f** $\theta = 90°$. The red circles indicate the mean velocity $\bar{u}$ and the blue circles show the rms velocity fluctuations u'_{rms}

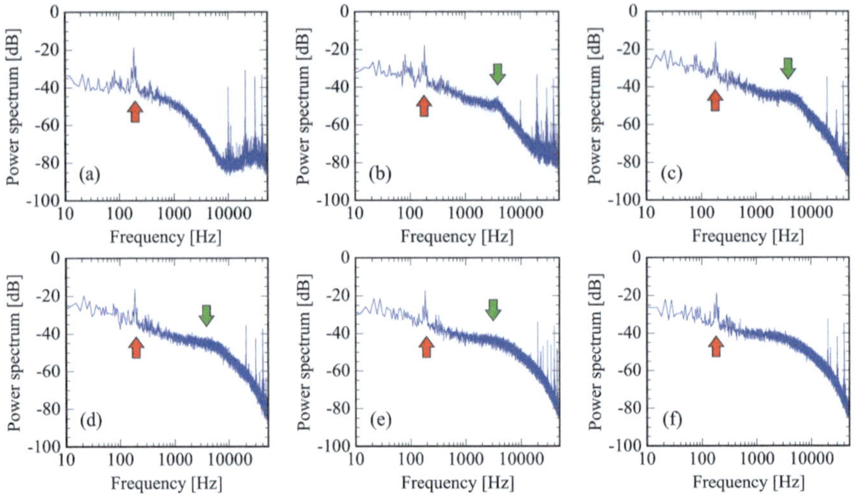

Fig. 34.3 Power spectrum distributions of velocity fluctuations at $h/D = 0.3 \times 10^{-2}$ and $\phi = 0°$, **a** $\theta = 40°$, **b** $\theta = 50°$, **c** $\theta = 60°$, **d** $\theta = 70°$, **e** $\theta = 80°$, and **f** $\theta = 90°$

sphere, where the rotation direction is aligned with the freestream direction. The red circles represent the mean velocity, while the blue circles denote the rms velocity fluctuations. At $\theta = 40°$, the boundary layer was considered too thin to enable accurate measurements of the inner boundary layer. As the measurement position moved downstream, the boundary layer thickness, as indicated by the red arrows, increased. Correspondingly, the rms value of the velocity fluctuations near the surface, marked with the green arrows, also increased. After $\theta = 60°$, the boundary layer was assumed to be in a transitional state.

To investigate the characteristics of the fluctuations leading to the boundary layer transition, spectral analysis was conducted. Figure 34.3 presents the power spectrum distributions of velocity fluctuations measured at the point closest to the sphere surface, marked with a green arrow in Fig. 34.2. As illustrated in Fig. 34.3, a peak in the power spectrum distribution was observed near 200 Hz, indicated by the red arrow. This peak is thought to result from disturbances associated with the arrangement of the dimples. Additionally, a dominant frequency band, marked by the green arrow, was confirmed around 2 kHz in Fig. 34.3b–e. This characteristic high-frequency fluctuation appeared in the region where the intensity of the velocity fluctuations began to increase, but was not observed in either upstream or downstream regions. Based on these observations, the high-frequency disturbance is considered to play a critical role in triggering the boundary layer transition.

The phenomena presented in Figs. 34.2 and 34.3 were similarly observed on the lower side of the dimpled sphere, where the direction of rotation opposed the freestream, as well as in measurement cross-sections other than the equatorial plane of rotation. Consequently, the region where the rms velocity fluctuations began to increase and a dominant high-frequency band appeared in the power spectrum

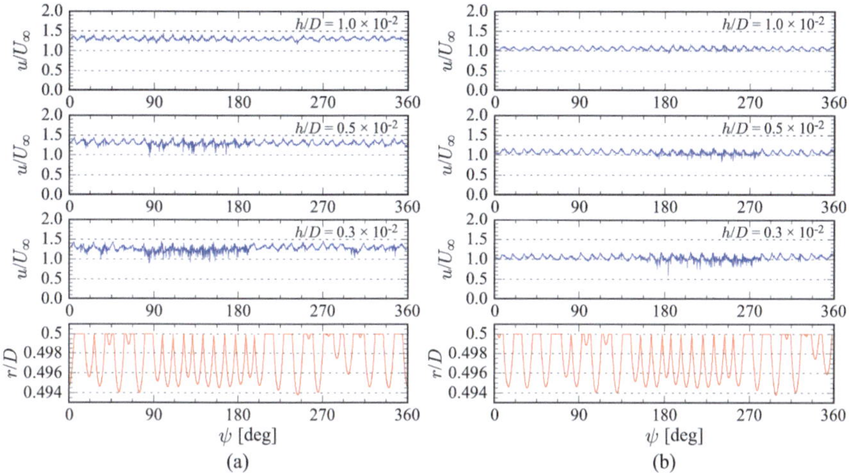

Fig. 34.4 Time-series of velocity fluctuations in transition region on the equatorial plane of rotation $\phi = 0°$ plotted against the rotation phase ψ of the sphere: **a** at $\theta = 70°$, **b** at $\theta = -55°$, and $h/D = 1.0 \times 10^{-2}$, $h/D = 0.5 \times 10^{-2}$, and $h/D = 0.3 \times 10^{-2}$ from top to bottom. The bottom row illustrates the surface configuration of the cross-section normal to z-axis, including the measurement point

distribution was identified as the transition region. To further investigate the high-frequency disturbances identified in the power spectral distribution in Fig. 34.3, a detailed examination of the velocity fluctuations within the transition region was conducted.

Figure 34.4 presents the time-series waveforms of velocity fluctuations on the equatorial plane of rotation: (a) at $\theta = 70°$, situated on the upper side of the sphere where the direction of rotation is aligned with the freestream [6], and (b) at $\theta = -55°$, on the lower side of the sphere where the direction of rotation opposes the freestream. Each graph presents results for $h/D = 1.0 \times 10^{-2}$, $h/D = 0.5 \times 10^{-2}$, and $h/D = 0.3 \times 10^{-2}$, from top to bottom. The bottom row of each figure illustrates the surface configuration of the cross-section normal to the z-axis, including the measurement point. The velocity fluctuations are presented in relation to the phase ψ throughout a single rotation of the dimpled sphere, synchronized with the passage of the dimples. At all measurement points, nearly identical waveforms were observed, irrespective of the sphere's rotation period. At $h/D = 1.0 \times 10^{-2}$, away from the surface of the sphere, velocity fluctuations with a nearly constant period were observed. The frequency of these fluctuations corresponded to that indicated by the red arrow in Fig. 34.3. In contrast, at $h/D \leq 0.5 \times 10^{-2}$, closer to the sphere surface, spike-like velocity fluctuations were observed. These fluctuations occurred when a specific arrangement of dimples passed through the measurement point, and their intensity increased as the measurement point approached the sphere surface. Thus, it is inferred that the boundary layer transition in the equatorial plane of a rotating dimpled sphere is primarily caused by high-frequency fluctuations induced

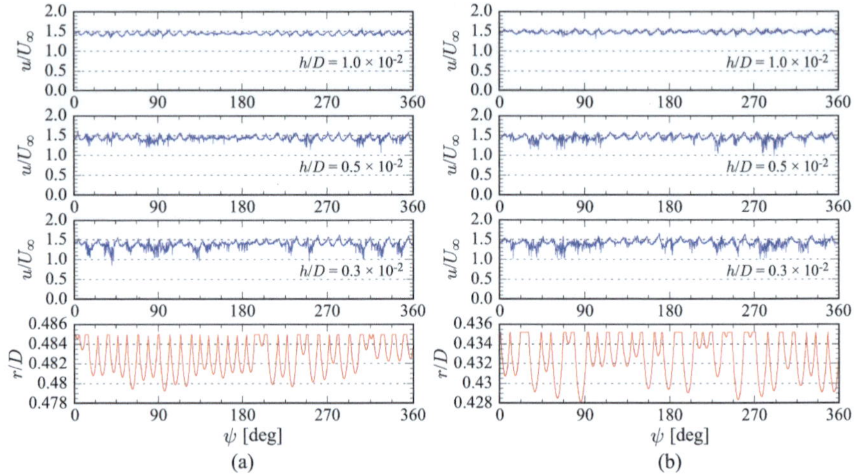

Fig. 34.5 Time-series of velocity fluctuations in transition region on other than the equatorial planes of rotation plotted against the rotation phase ψ of the sphere: **a** at $\phi = 15°$, $\theta = 70°$, **b** at $\phi = 30°$, $\theta = 80°$, and $h/D = 1.0 \times 10^{-2}$, $h/D = 0.5 \times 10^{-2}$, and $h/D = 0.3 \times 10^{-2}$ from top to bottom. The bottom row illustrates the surface configuration of the cross-section including the measurement point

by specific dimple arrangements. These high-frequency fluctuations arise as dimples continuously pass through the measurement point, independent of the relationship between the freestream and the sphere's rotational direction. They are considered to result from the instability of the locally separated flow occurring in the dimples. Furthermore, the effect of the rotational direction is considered as follows: when the flow near the surface is relatively accelerated, the transition point moves downstream, whereas when the flow is decelerated, the transition point moves upstream.

Figure 34.5 presents the time-series waveforms of velocity fluctuations during one rotation of the sphere, measured at locations other than the equatorial plane on the upper side of the sphere: (a) at $\phi = 15°, \theta = 70°$ and (b) $\phi = 30°, \theta = 80°$. The display of all figures is consistent with Fig. 34.4. At $h/D = 1.0 \times 10^{-2}$, located farther from the sphere's surface, periodic fluctuations with a nearly constant period were also observed. However, the wavenumber of these periodic fluctuations decreased as ϕ increased. This can be attributed to the decreasing circumference of the sphere with increasing ϕ, which reduces the number of dimples along the circumference. Consequently, the frequency of these fluctuations becomes lower than the peak frequency at the equatorial plane, as indicated by red arrow in Fig. 34.3. Closer to the sphere's surface, at $h/D \leq 0.5 \times 10^{-2}$, spike-like velocity fluctuations were also detected. However, no clear relationship was found between these characteristic spike-like velocity fluctuations and the specific arrangement of the dimples. In Fig. 34.5a, the dimples pass the measurement point almost continuously, however, the spike-like velocity fluctuations are not synchronized with the passage of the dimples. In contrast, Fig. 34.5b shows that these spike-like velocity fluctuations persist even during intervals when the dimples do not pass continuously. This observation suggests that the fluctuations may not be directly associated with the presence of the dimples at the measurement point.

The surface configurations in Figs. 34.4 and 34.5 correspond to cross-sections normal to the z-axis. On the equatorial plane of rotation, the streamlines passing through the measurement points are aligned with the direction of the dimple arrangement, as shown in Fig. 34.4. It is considered that the disturbances caused by the locally separated flow within the dimples propagate along the similar dimple configuration. On the other hand, in the measurement cross-sections located other than the equatorial plane, it is assumed that the flow reaches the measurement point from the stagnation point on the leading edge. The arrangement of dimples along the streamline is assumed to differ from the surface configuration shown in Fig. 34.5. Therefore, to investigate the relationship between dimple arrangements and the disturbances leading to boundary layer transition, it is necessary to extract the surface configurations along the streamlines, synchronized with the flow direction.

34.4 Conclusion

The turbulent transition of the boundary layer on the rotating dimpled sphere was measured in detail using hot-wire anemometry, and the relationship between characteristic velocity fluctuations observed during the transition process and the dimple configuration was investigated. The results indicate that the turbulent transition is induced by high-frequency fluctuations, which result from the instability of locally separated flow within the dimples, and which occur as the dimples pass continuously in the streamwise direction. Furthermore, it was found that the direction of rotation affects the location of the transition point, causing it to shift in the direction of rotation.

References

1. Achenbach, E.: The effects of surface roughness and tunnel blockage on the flow past spheres. J. Fluid Mech. **65**, 113–125 (1974)
2. Barlow, J.B., Domanski, M.J.: Lift on stationary and rotating spheres under varying flow and surface conditions. AIAA J. **46**, 1932–1936 (2008)
3. Berman, P.W., Harvey, J.K.: Golf ball aerodynamics. Aeronaut. Q. **27**, 112–122 (1976)
4. Choi, J., Jeon, W.-P., Choim, H.: Mechanism of drag reduction by dimples on a sphere. Phys. Fluids **18**, 041702 (2006)
5. Schlichting, H.: Boundary-Layer Theory, 7th ed. McGraw-Hill (1979)
6. Tominaga, Y., et al.: Comparison between experimental and numerical results on boundary layer transition around rotating dimpled sphere (in Japanese). J. Japan Soc. Fluid Mech. **40**, 398–401 (2021)

Chapter 35
Transitional Phenomena on an Elliptic Profile

Marco Costantini, Benjamin D. Dimond, Christian Klein, Stephan Sattler, and Massimo Miozzi

Abstract Intermittency and turbulent spots were experimentally studied on an elliptic profile with 16% relative thickness at two moderate chord Reynolds numbers. The measurements were performed on a two-dimensional model in the Large Water Tunnel Braunschweig. A temperature-sensitive paint enabled the global, non-intrusive, time-resolved determination of the surface temperature maps, and thus the study of the transitional phenomena. Two different adverse pressure gradients, as measured by pressure taps, were implemented by changing the profile angle-of-attack. With a weaker pressure gradient at a Reynolds number of 5.0×10^5, large variations in the transition front were observed in space and time via temperature-sensitive paint; moreover, the spatial and temporal development of an isolated turbulent spot was evaluated from the temperature-sensitive paint data. A reduction of the spanwise, long-wavelength variations of the transition front was observed for a stronger adverse pressure gradient, but turbulent spots were still identified. In this pressure gradient case, marked intermittent events and streamwise oscillations were observed in the transition region at Reynolds numbers of 2.5×10^5 and 5.0×10^5, respectively.

Keywords Elliptic profile · Boundary-layer transition · Turbulent spots · Intermittency · Temperature-sensitive paint · Underwater applications

Supplementary Information The online version contains supplementary material available at https://doi.org/10.1007/978-981-96-9829-5_35.

M. Costantini (✉) · B. D. Dimond · C. Klein
German Aerospace Center (DLR), Göttingen, Germany
e-mail: marco.costantini@dlr.de

S. Sattler
TU Braunschweig, Braunschweig, Germany

M. Miozzi
Institute of Marine Engineering (CNR-INM), Rome, Italy

K. Kato et al. (eds.), *Proceedings of the 10th IUTAM Symposium on Laminar-Turbulent Transition*, IUTAM Bookseries 44, https://doi.org/10.1007/978-981-96-9829-5_35

35.1 Introduction

Boundary-layer transition and separation on elliptic profiles have been investigated in fundamental research studies since the 1930's [1]. More lately, the impact of these phenomena on the fluid dynamic behavior of such profiles has become relevant also in practical applications, as for example on canard rotor/wing aircraft [2] and in hydrokinetic energy harvesting technologies [3], which are operated at moderate Reynolds numbers. Nevertheless, recent experimental investigations have focused on the aerodynamic characteristics of elliptic profiles via measurements of the aerodynamic coefficients, rather than on the complex evolution of transitional and separating boundary layers. The transition of the boundary layer developing over an elliptic profile with 16% relative thickness (such as that examined in [2]) is studied in the current work. The experiments were carried out in the Large Water Tunnel (GWB) of TU Braunschweig [4] on a two-dimensional model investigated at two moderate chord Reynolds numbers: $\mathrm{Re} = U_\infty c/\nu_\infty = 2.5$ and 5.0×10^5, where U_∞ and ν_∞ are the freestream velocity and kinematic viscosity, respectively, and c the profile chord length. The measurements were performed for a wide range of angles-of-attack ($-4° \leq \mathrm{AoA} \leq 30°$), thus enabling the study of the boundary-layer evolution for a variety of stability situations. The investigation of the onset of stall and of the separated-flow behavior at large angles-of-attack is reported in [5], whereas the present work focuses on the transitional phenomena occurring at relatively small angles-of-attack.

35.2 Experimental Setup

The measurements were conducted in the GWB water tunnel, a closed-circuit facility featuring a test section with a length of 6 m and a cross-section of $1 \times 1\ \mathrm{m}^2$. The experiments were performed at $U_\infty = 1.3$ and 2.6 m/s for the two examined Reynolds numbers $\mathrm{Re} = 2.5$ and 5×10^5, respectively. The investigated two-dimensional model had a chord length of $c = 0.2$ m and was mounted in the two turntables installed at the GWB test-section side walls, as shown in Fig. 35.1 (left). The model upper surface was equipped with a Temperature-Sensitive Paint (TSP) [6] for the global, non-intrusive, time-resolved measurement of the surface temperature distribution, and thus of boundary-layer transition phenomena. The TSP technique has already been successfully applied in underwater investigations of the laminar-turbulent transition of boundary layers over flat plates [7, 8], cylinders [9], isolated hydrofoils [10] and hydrofoils in the wake of a marine propeller [11]. The TSP used in this work is analogous to those used, e.g., in [9–11], and appears as a yellow coating in Fig. 35.1 (left). In the experiments, the model surface temperature was increased by an internal flow of warm water; the imposed temperature difference between model surface and flow (less than 6 K) enhanced the laminar-turbulent surface temperature differences

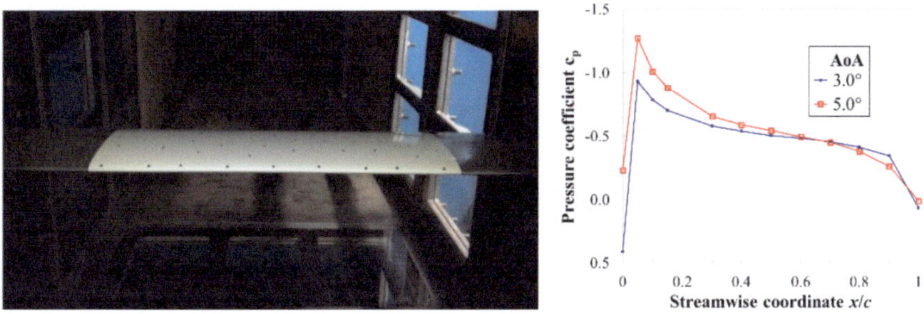

Fig. 35.1 Left: Elliptic profile model mounted in the GWB test section (view in flow direction). Right: upper surface pressure distributions measured at Re = 5.0×10^5 for AoA = 3.0° and 5.0°

[6–11], and thus enabled capturing the spatio-temporal development of the transitional phenomena via TSP. The TSP images were acquired with a high-speed camera at a frequency of 1 kHz over a mid-span region of approx. 0.2×0.3 m^2, with a spatial resolution of approx. 0.3 mm (i.e., 0.15% c). The model was also instrumented with 16 pressure taps (orifice diameter: 0.8 mm) for the measurement of the surface pressure distribution. Details on the experimental setup, on the data acquisition and on the TSP image processing are reported in [5].

35.3 Results and Discussion

The observed transitional phenomena are presented exemplarily for two different angles-of-attack: AoA = 3.0° and 5.0°. The pressure distributions measured on the model upper surface at these two incidences are shown in Fig. 35.1 (right) for the higher Reynolds number, those measured at Re = 2.5×10^5 being qualitatively similar to those at Re = 5.0×10^5. At both angles-of-attack, the streamwise pressure gradient was adverse for the whole model upper surface downstream of the leading-edge region; as expected, the adverse pressure gradient was more pronounced at the larger angle-of-attack. Instantaneous temperature distributions (snapshots) measured at Re = 5.0×10^5 for AoA = 3.0° and 5.0° are presented in Fig. 35.2 (top-left and bottom-left, respectively). In these figures, the flow is from the left; the y-axis is oriented from the model port side to the starboard side, $y = 0$ being the port end of the TSP region (the model centerline is at $y/c = 1.5$). The transition region corresponds to the region of negative temperature change in the streamwise direction. As a result of the more pronounced adverse pressure gradient, transition was shown to occur further upstream and over a shorter streamwise distance at the larger angle-of-attack. The different characteristics of the transition regions for these two cases are elucidated by the time series of the surface temperature distributions obtained via TSP and provided as Supplementary Material (Videos 1 and 3 for AoA = 3.0° and 5.0°, respectively). At AoA = 3.0°, the transition front presented a strongly

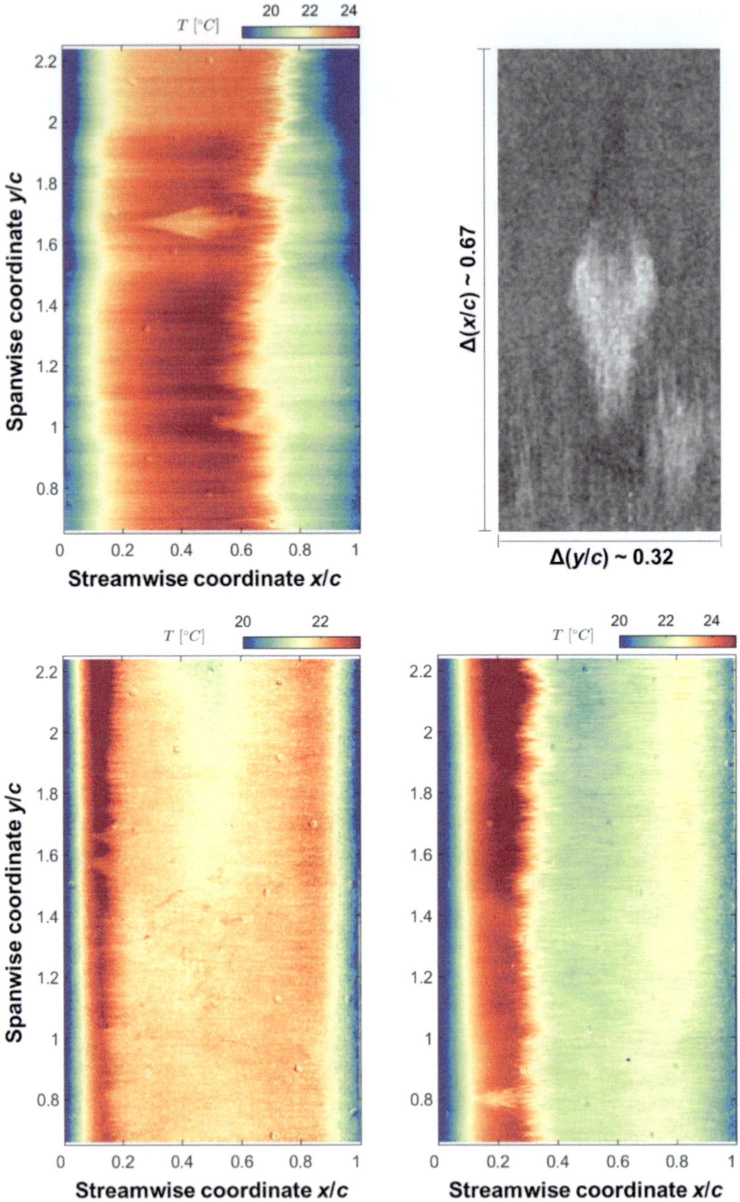

Fig. 35.2 TSP results on the model upper surface. Instantaneous surface temperature distributions (flow from the left): AoA = 3.0° and Re = 5.0×10^5 (top, left); AoA = 5.0° and Re = 5.0×10^5 (bottom, left); AoA = 5.0° and Re = 2.5×10^5 (bottom, right). Top, right (flow from the top): turbulent spot identified for the case at AoA = 3.0° and Re = 5.0×10^5 with the temporal gradient of the surface temperature

intermittent character, with marked variations in space and time and a frayed appearance, which was probably the result of a "natural" transition process [12] induced by a weakly adverse pressure gradient. With a more pronounced adverse pressure gradient (AoA = 5.0°), the transition front showed a reduced variation in the spanwise direction; in this case, transition likely occurred within a laminar separation bubble. Nevertheless, the surface temperature distributions in this case also presented a significant variation in time: one of the most striking phenomena was an oscillation of the transition front in the streamwise direction, which was possibly related to the "flapping" of the laminar separation bubble [10]. Moreover, turbulent spots were observed in both cases: their appearance in the surface temperature distributions is a wedge-shaped thermal signature [12], such as that isolated in Fig. 35.2 (top-left) for AoA = 3.0°. The analysis of the spot evolution in space and time was based on [12], and is exemplarily shown for this turbulent spot in Fig. 35.2 (top-right) and in Video 2. In these results, the flow is from the top, and the temporal gradient of the surface temperature distribution is displayed as grayscale (positive and negative temporal gradients correspond to dark and bright regions, respectively). This representation of the results enabled to clearly identify the characteristic wall signature of the turbulent spot, i.e., a downstream-pointing arrowhead shape with "wing tips" (rounded outer sides of the spot trailing-edge interface), which appears in Fig. 35.2 (top-right) and in Video 2 as a bright region. The spatially and temporally resolved temperature distributions also enabled to determine the (half) spreading angle of the turbulent spots, which was found to be approx. 12° for the examined turbulent spot. This value is in line with the expectations from the literature for adverse pressure gradients [12].

The influence of a variation in the Reynolds number on boundary-layer transition is exemplarily shown for AoA = 5.0°. A snapshot of the surface temperature distribution measured at this incidence for Re = 2.5×10^5 is displayed in Fig. 35.2 (bottom-right), while a time series is provided in Video 4. Transition was found to occur further downstream at Re = 2.5×10^5, as compared to Re = 5.0×10^5 (Fig. 35.2, bottom-left, and Video 2). This observation was the expected consequence of the Reynolds number impact, being the measured pressure distribution similar for both cases. Nevertheless, the time series of the surface temperature maps revealed differences in the spatial and temporal features of the transition region. This is further elucidated in Fig. 35.3 (left), where the maps of the kurtosis of the wall temperature on the port half of the examined surface are shown for the two Reynolds numbers. Figure 35.3 (right) shows the temporal evolution of the wall temperature at the locations marked by the white circles in the left figures.

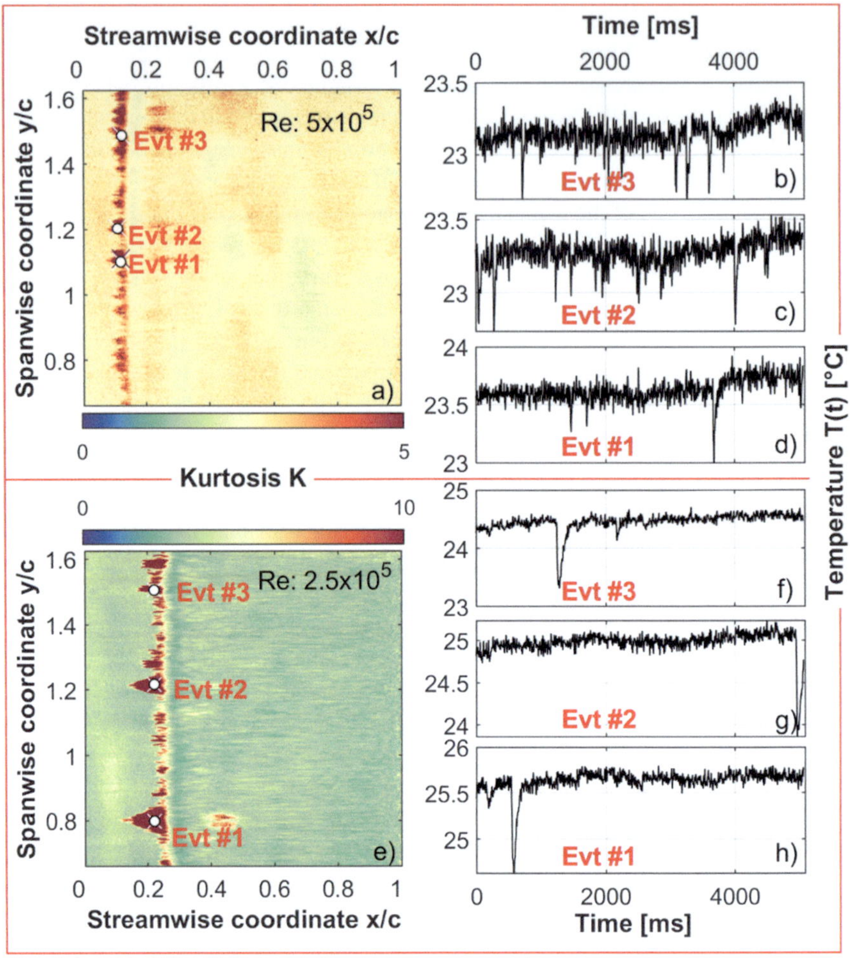

Fig. 35.3 Left: maps of the kurtosis of the surface temperature on the port half surface of the TSP area (flow from the left). Right: time series of the surface temperature at the locations indicated in the corresponding kurtosis maps. The top and bottom figure show the cases at Re = 5.0 and 2.5 × 10^5, respectively. AoA = 5.0° for both cases

At AoA = 5.0° and Re = 2.5 × 10^5, the transition front did not present long-wavelength variations in the spanwise direction (such as those observed in Fig. 35.2, top-left), while it did show short-wavelength ones. These are emphasized in Fig. 35.3 (bottom-left) in the kurtosis map, which provides a measure of events in the wall temperature probability distribution occurring far from the temporal mean. The three events marked with "Evt #1–#3 in the figure are particularly intense, being related to the passage of turbulent spots (such as that identified in Fig. 35.2, bottom-right). The time series of the wall temperature at these three locations, reported in Fig. 35.3 (bottom-right), show a pronounced drop in the surface temperature due to the spot passage and the following temperature recovery, in line with the results discussed in [12]. An analogous analysis conducted for the case at AoA = 5.0° and Re = 5.0×10^5 (Fig. 35.3, top) shows some similarities but less intense events, resulting in reduced kurtosis values and less pronounced temperature drops in the time series, as compared to the case at Re = 2.5×10^5. As discussed above, the streamwise oscillation of the transition front had a prominent impact on the temporal evolution of the surface temperature at AoA = 5.0° and Re = 5.0×10^5.

35.4 Conclusions

Transitional phenomena occurring on an elliptic profile (16% relative thickness) were experimentally investigated in the GWB water tunnel at two chord Reynolds numbers (Re = 2.5 and 5.0×10^5) and two different streamwise adverse pressure gradients (angles-of-attack AoA = 3.0° and 5.0°). Intermittency and turbulent spots were examined via TSP, which provided surface temperature distributions resolved in space and time. Strong intermittency was observed in the transition region at AoA = 3.0° and Re = 5.0×10^5; for this case, the spatio-temporal evolution of an isolated turbulent spot was exemplarily obtained from the TSP data analysis, including the determination of the spot spreading angle. At the same Reynolds number but with AoA = 5.0°, the transition front showed a weaker variation in the spanwise direction, but a marked oscillation in the streamwise direction. A Reynolds number reduction to Re = 2.5×10^5 for AoA = 5.0° led to a re-establishment of marked intermittency in the transition region.

The present results illustrate the complexity of transition scenarios even for a geometrically simple profile at moderate Reynolds numbers, for which the time-series of surface temperature maps enable to quantify the transitional phenomena. These data therefore not only support the experimental evaluation of the aero/hydrodynamic performance of elliptic profiles, but can also serve for the validation and improvement of models for the numerical prediction of boundary-layer transition.

Acknowledgements The experiments were funded and performed within the framework of the DLR *ADaMant* project. The authors acknowledge: C. Fuchs, U. Henne, T. Hermann, K. Kienass and T. Kleindienst (DLR) for the support during the preparation and conduction of the test campaign; M. Brinsa, N. Burbach, B. Eilerts, D. Fengewisch, T. Fenske, H. Fink, M. Göhler, J. Grimme, D. Grulich, E. Hasse, N. Kassler, B. Lange, V. Len, M. Moser, L. Rehbein, L. Rettberg, M. Reuter, N. Schmidt, S. Siebertz, and M. Tegeler (DLR) for the engineering, manufacturing and instrumentation of model and heating system; A. Volkmann and S. Koch (DLR) for providing the external thermostatic bath and the amplifier for the thermocouple signals, respectively; D. Durante (CNR-INM) for the preliminary numerical simulations of the elliptic profile; A. De Vincenzo (Polytechnic University of Milan) for the support to the analysis of the pressure tap data; P. Scholz (TU Braunschweig) for the advice during the preparation of the test campaign; and C. Grabe and L. Koop (DLR) for the constant support to this experimental study.

Disclosure of Interests The authors have no competing interests to declare that are relevant to the content of this article.

Supplementary Material

- **Video 1**: Re = 5.0×10^5, AoA = 3.0°, Video-1_Re-500000_AoA_3deg.mp4;
- **Video 2**: turbulent spot isolated from Video 1, Video-2_Isolated-Turbulent-Spot.mp4;
- **Video 3**: Re = 5.0×10^5, AoA = 5.0°, Video-3_Re-500000_AoA_5deg.mp4;
- **Video 4**: Re = 2.5×10^5, AoA = 5.0°, Video-4_Re-250000_AoA_5deg.mp4.

References

1. Schubauer, G.B.: Air Flow in the Boundary Layer on an Elliptic Cylinder. NACA Rep. No. 652 (1939)
2. Kwon, K., Park, S.O.: Aerodynamic characteristics of an elliptic Airfoil at Low Reynolds Number. J. Aircraft **42**(6), 1642–1644 (2005)
3. Kim, D., Strom, B., Mandre, S., Breuer, K.: Energy harvesting performance and flow structure of an oscillating hydrofoil with finite span. J. Fluids Structs. **70**, 314–326 (2017)
4. Scholz, P., Sattler, S., Wulff, D.: Der Große Wasserkanal "GWB"—Eine Versuchsanlage für zeitauflösende Messungen bei großen Reynoldszahlen. In: Proceedings DLRK 2013, Stuttgart, Germany (2013)
5. Costantini, M., Dimond, B.D., Klein, C., Sattler, S., Miozzi, M.: Underwater investigation of stall onset on an elliptic profile via temperature-sensitive paint. In: A. Dillmann, G. Heller, E. Krämer, C. Breitsamter, L. Krenkel, C. Wagner (eds.), New Results in Numerical and Experimental Fluid Mechanics XV. Notes on Numerical Fluid Mechanics and Multidisciplinary Design. 156, Springer Nature, Cham (2026). https://link.springer.com/book/9783032111142
6. Liu, T., Sullivan, J.P., Asai, K., Klein, C., Egami, Y.: Pressure and Temperature Sensitive Paints. 2nd edn. Springer Nature, Cham, Switzerland (2021)
7. Lemarechal, J., Klein, C., Henne, U., Puckert, D.K., Rist, U.: Detection of Lambda- and Omega-vortices with the temperature-sensitive paint method in the late stage of controlled laminar-turbulent transition. Exp. Fluids **60**, 91 (2019)
8. Lemarechal, J., Klein, C., Henne, U., Puckert, D.K., Rist, U.: Spatially and temporally resolved visualization of turbulent spots in Blasius boundary layer transition. In: Rösgen, T. (ed.) Proceedings ISFV 18, ETH Zurich, Switzerland (2018)
9. Capone, A., Klein, C., Di Felice, F., Beifuss, U., Miozzi, M.: Fast-response underwater TSP investigation of subcritical instabilities of a cylinder in crossflow. Exp. Fluids **56**, 196 (2015)
10. Miozzi, M., Capone, A., Costantini, M., Fratto, L., Klein, C., Di Felice, F.: Skin friction and coherent structures within a laminar separation bubble. Exp. Fluids **60**, 13 (2019)
11. Miozzi, M., Costantini, M.: Temperature and skin-friction maps on a lifting hydrofoil in a propeller wake. Meas. Sci. Technol. **32**, 114007 (2021)
12. Dimond, B.D.: Investigation of boundary layer transition at high Reynolds numbers using time-resolved temperature-sensitive paint. Ph.D. thesis, Georg-August-Universität Göttingen (2024)

Chapter 36
Demonstration of Natural Laminar Vertical Tail at Flight Reynolds Number in ETW

Naoko Tokugawa, Takahiro Ishida, Keiji Ueshima, and Keisuke Ohira

Abstract Natural laminar flow design was applied to the vertical tail of 120-seat class transport aircraft to reduce CO_2 emissions through drag reduction. While the Reynolds number based on the mean aerodynamic chord of a vertical tail is close to that of a main wing, the vertical tail's large sweep angle and symmetric airfoil makes natural laminar flow design more challenging. The design was verified experimentally at flight Reynolds number conditions in the European Transonic Windtunnel, and the transition Reynolds number based on the distance from the leading edge to the transition location, considering the effect of sweep, was the largest that has ever been observed experimentally, despite the symmetrical airfoil.

Keywords Natural laminar flow · Vertical tail · Experimental verification

36.1 Introduction

The increasing frequency and severity of large-scale weather phenomena such as heat waves, hurricanes, drought and wildfire, attributed to man-made global warming, are threatening our peaceful daily lives. 2% of the CO_2 that contributes to global warming is emitted by aircraft, so reducing aviation CO_2 emissions is urgently required to achieve the goal of net zero by 2050 [1]. Since skin friction drag accounts for about 50% of the total drag of a large transport aircraft, boundary-layer control techniques have been receiving attention [2–4]. Among these, natural laminar flow (NLF) design has the potential to significantly reduce skin friction drag, but although it has been studied for many years and considered for many aircraft [5–8], it has not been sufficiently practical so far due to issues in design, manufacture and operation: it is difficult to achieve a design that suppresses crossflow instability, which is dominant in the boundary layer on the swept wings of high subsonic aircraft, while satisfying

N. Tokugawa (✉) · T. Ishida
Aviation Technology Directorate, Japan Aerospace Exploration Agency, Tokyo, Japan
e-mail: tokugawa.naoko@jaxa.jp

K. Ueshima · K. Ohira
Analysis and Design Solution Division, Ryoyu Systems Co., Ltd., Nagoya, Japan

K. Kato et al. (eds.), *Proceedings of the 10th IUTAM Symposium on Laminar-Turbulent Transition*, IUTAM Bookseries 44, https://doi.org/10.1007/978-981-96-9829-5_36

aerodynamic or structural requirements; removing roughness and gaps, which are the source of disturbances in crossflow instability, creates manufacturing difficulties; and surface roughness due to insect adhesion needs to be addressed in day-to-day operations. However, the development of measures to address these issues is making NLF design increasingly attractive [9–12].

Considering this situation, the Japan Aerospace Exploration Agency (JAXA) has launched the "Demonstration of Laminar FIN System (DoLFINS)" research program aimed at the practical application of NLF targeting a vertical tail. Since a vertical tail has a relatively small area compared to wings and fuselage and its basic purpose is to increase lateral stability, it is less flight-critical and so an easier target for the practical application of new technology while still being worthwhile. On the other hand, NLF design for a vertical tail is more technically challenging than for a main wing for a number of reasons: it has a higher sweep angle, its Mean Aerodynamic Chord (MAC) is almost the same as that of a wing despite its smaller area, and its airfoil is symmetrical. Regarding the latter reason, with a main wing it is possible to achieve natural laminar flow by manipulating the shape of the lower surface to adjust wing thickness and/or increase flow acceleration at the leading edge, but this is not possible with a symmetrical airfoil. The DoLFINS program addressed not only the design [13] and demonstration of an NLF vertical tail, but also established manufacturing requirements for surface roughness [14] and developed a highly-washable paint.

In order to evaluate the influence of roughness on the transition location at flight Reynolds numbers of the order of 10^7, we developed a multi-fidelity turbulent transition analysis system [14] in which the flow field near the roughness is solved by direct numerical simulation (DNS). However, the entire flow field is analyzed by Reynolds-averaged Navier-Stokes (RANS), which includes a laminar region to significantly reduce the calculation cost. Furthermore, it is possible to predict the transition characteristics by performing a stability analysis based on the nonlinear parabolized stability equation (PSE) analysis. In this paper, we verify the effectiveness of NLF wing design applied to the vertical tail of a high subsonic narrow-body aircraft under flight Reynolds number conditions in wind-tunnel tests.

36.2 Design Object and Natural Laminar Flow Design

The design object is the vertical tail of the TRA2022 technical reference aircraft [15], a 120-seat narrow-body transport aircraft designed by JAXA. The vertical tail has a sweep angle of 40.18 deg and a MAC of 3.89 m.

The NLF design used the Computational Fluid Dynamics (CFD)-based inverse problem design method [13], in which an airfoil shape that achieves a target pressure distribution is iteratively designed by repeatedly comparing the pressure distribution of the current airfoil with that of the target, and then modifying the airfoil to reduce the difference. The target pressure distribution, which is the key to inverse problem design, was determined to suppress both crossflow and Tollmien-Schlichting instabilities, and features a rapid acceleration at the leading edge followed by a short flat distribution and re-acceleration, based on transition prediction. The designed two-dimensional airfoil sections at each spanwise station are then smoothly connected to construct a three-dimensional shape. The NLF design process was applied first to the full aircraft configuration, and then to a wind-tunnel test model, which has a shortened fuselage and omits the horizontal stabilizer in order to be accommodated in the wind-tunnel test section. The difference in the pressure distribution due to the difference in the shape between the vertical tail wind-tunnel model and the full configuration had a significant impact on the NLF design.

Numerical transition analysis was applied to create the target pressure distribution and to predict or analyze the transition location observed in experiments. The transition analysis was performed using JAXA's Stability And Transition Analysis System (SATAS) in-house code [16]. The boundary-layer profile was obtained by SATAS from the surface pressure distribution using a compressible 3D boundary-layer analysis method based on Iyer's formulation. The surface pressure distribution was obtained numerically by another JAXA in-house code, the unstructured flow solver FAST Aerodynamic Routines (FaSTAR) [17]. The computational grid is tetrahedron-based and unstructured, with 62 million cells. The governing equation is the 3D compressible Reynolds-averaged Navier-Stokes equations, and the Spalart-Allmaras turbulence model incorporated the quadratic constitutive relation [18]. The transition location was predicted based on a linear-stability analysis using the e^N method. For growth rate integration, the envelope method and fixed-β method can be applied. The transition location on the vertical tail optimized for the wind-tunnel model was expected to be 38% based on $N_{\text{envelope}} = 12$. It was reduced to 30% for $N_{\text{envelope}} = 11$.

36.3 Wind-Tunnel Test

The wind-tunnel model used in this test was a 20% model of the TRA2022 vertical tail and part of the fuselage. Such models are known as a "half-model" configuration, although the term is not strictly correct for a vertical tail. The coordinate axes are X in the uniform flow direction, Z in the vertical direction from the root to the tip of the vertical tail, and Y in the horizontal direction perpendicular to X and Z. It should be noted that the origin of the Z-axis is on the virtual longitudinal axis of the full aircraft configuration. The Z-coordinate is normalized based on S, which is the distance from the origin to the tip of the vertical tail. The wing-body junction, where the vertical tail joins the fuselage, is at $Z/S = 0.44$. The chord-wise coordinate is x/c, which is the distance from the leading edge x non-dimensionalized by the local chord length c. Although leading edge separation and attachment line contamination are prevented by the airfoil design itself, a spacer and a miniature strake were installed on the half-model to completely eliminate attachment line contamination. The spacer was placed between the fuselage and the wind-tunnel wall instead of a peniche. To detect the transition location, two Temperature-Sensitive Paint (TSP) [21] pods were installed at $x/c = 0.10 - 0.70$, $Z/S = 0.46 - 0.68$ and $x/c = 0.10 - 0.70$,

$Z/S = 0.73 - 0.95$. The manufacturing tolerance of the wind-tunnel model shape is extremely small, including the TSP surface after painting. The arithmetic mean roughness was 0.05 μm or less on both the bare and TSP-covered surfaces of the model. The experiments were conducted in the European Transonic Windtunnel (ETW) [19], a special facility that can achieve flight Reynolds numbers of the order of 10^7 using low-temperature nitrogen as the test gas and moderate pressurization. The ETW is a closed-circuit type of wind tunnel, and the test section is 2.00 m in height, 2.40 m in width, and 8.73 m in length. The velocity fluctuation of uniform flow is about 0.1% or less [20]. The application and measurement of the TSP to visualize the transition location were supported by the German Aerospace Center (DLR), which is a cooperative partner with the ETW. The selected "2Color-TSP" paint was composed of two dyes, ruthenium and europium, because of the wide temperature range in the ETW itself and in the JAXA 2 m×2 m transonic wind tunnel, where a subsequent use is planned. During the wind-tunnel test, a sudden total temperature step of +8 K was applied to clarify the difference in heat transfer between laminar and turbulent regions. In addition to TSP, the static-pressure distribution, wing deformation, forces, model attitude, acceleration, and internal temperature were measured. This paper presents only the results of TSP and static-pressure distribution measurements.

A total of 116 tests were performed, including confirmation of reproducibility. This paper is focused on the design point; that is, a flight speed of Mach 0.78, Reynolds number based on mean aerodynamic chord $\mathrm{Re}_{\mathrm{MAC}}$ of 23.6 million, and sideslip angle $\beta_{\mathrm{sideslip}}$ of 0 deg.

36.4 Results

As shown in Fig. 36.1a, the experimentally-measured pressure distributions are in remarkably good agreement with CFD predictions, and similar good agreement is observed for other sections and the fuselage. These results show that the model shape was manufactured with extremely good tolerance, and the flow fields agreed very well between the wind tunnel and CFD. This indicates that both the experimental method and the CFD analysis method are valid.

The transition location detected by TSP is shown in Fig. 36.1b. The contour shows the ratio of the luminance measurement I during the test to a reference luminance value I_0. Areas with a small luminance ratio, shown in light gray, correspond to laminar flow, while areas with a large luminance ratio, shown in black, correspond to turbulent flow. The boundary between the two is the transition location where the surface temperature changes rapidly. Despite great care being taken to clean the model and the airflow in the wind tunnel, some turbulent wedges occurred. However, regions of natural transition were widely observed between the wedges. The natural transition positions identified from 12 TSP images observed at design point Mach number and Reynolds numbers, including different total temperatures, were in excellent agreement. The transition location agrees with the $N_{\mathrm{envelope}} = 11$ line predicted by the envelope method as shown in Fig. 36.1b.

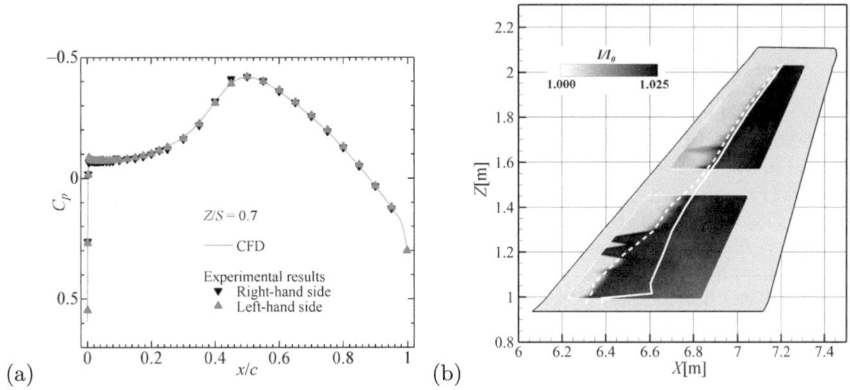

Fig. 36.1 Experimental results. **a** Surface pressure distributions at $Z/S = 0.7$. **b** TSP contour of the luminance ratio I/I_0. The dashed and solid lines indicate the predicted transition locations obtained with $N_{\text{envelope}} = 11$ and 12, respectively

It is well-known that the transition Reynolds number based on the distance from the leading edge to the transition location is estimated to decrease as sweep angle increases [9]. The maximum value of transition Reynolds number obtained in this experiment campaign was higher than the value expected for the sweep angle, and was the largest that has ever been observed experimentally, even in spite of the symmetrical airfoil.

36.5 Conclusion

Natural laminar flow design was applied to the vertical tail of a high subsonic narrow-body aircraft, and its effectiveness was verified under flight Reynolds number conditions in the ETW. The surface pressure distributions were in fair agreement with CFD. The transition Reynolds number based on the length from the leading edge to the transition location, considering the effect of the sweep angle, was the largest that has ever been observed in experiments, even despite the symmetrical airfoil.

Acknowledgements The authors would like to thank the members of DoLFINS and related colleagues in JAXA. The authors are also grateful to M. Schulz, the late M. Wright, and all members from ETW: C. Klein, U. Henne, D. Yorita from DLR, and M. Müller from Deharde. The numerical work was conducted using the computational resources of JAXA Supercomputer System Generation 3 (JSS3).

References

1. ICAO: Report on the Feasibility of a Long-Term Aspirational Goal (LTAG) for International Civil Aviation CO_2 Emission Reductions, ICAO Committee on Aviation Environmental Protection (CAEP), pp. 1–12 (2022)
2. Walsh, M.J.: Turbulent boundary layer drag-reduction using riblets, AIAA Paper 1982–0169 (1982)
3. Hirokawa, S., Ohashi, M., Eto, K., Fukagata, K., Tokugawa, N.: Turbulent friction drag reduction on clark-y airfoil by passive uniform blowing. AIAA J. **58**, 4178–4180 (2020)
4. Krishnan, K.S.G., Bertram, O., Seibel, O.: Review of hybrid Laminar flow control systems. Prog. Aero. Sci. **93**, 24–52 (2017)
5. Tani, I.: On the Design of Airfoils in Which the Transition of the Boundary Layer Is Delayed, Reports of the Aeronautical Research Institute, Tokyo Imperial University, No. 250 (1943)
6. Tokugawa, N., Kwak, D., Yoshida, K., Ueda, Y.: Transition measurement of natural laminar flow wing on supersonic experimental Airplane NEXST-1. J. Aircraft **46**, 1496–1604 (2008)
7. Ishikawa, H., Ueda, Y., Tokugawa, N.: Natural Laminar Flow Wing Design for a Low Boom Supersonic Aircraft, AIAA Paper 2017–1860 (2017)
8. Manning, V., Kroo, I.: Multidisciplinary Optimization of a Natural Laminar Flow Supersonic Aircraft, AIAA-1999-3102 (1999)
9. Lynde, M.N., Campbell, R.L., Viken, S.A.: Additional findings from the common research model natural laminar flow wind tunnel test, AIAA Paper 2019-3292 (2019)
10. Banchy, M.N., Campbell, R.L., Hiller, B.R., Bozeman, M.D.: CDISC Remote Design Method to Simulate Aircraft Interference Effects for the CATNLF Flight Test, AIAA Paper 2024-0678 (2024)
11. Larsson, R.: Analysis of the effect of flow transition over the wings of the blade demonstrator. In: Proceedings of the 33rd Congress of ICAS (2022)
12. Fujino, M., Yoshizaki, Y., Kawamura, Y.: Natural-laminar-flow airfoil development for a lightweight business jet. J. Aircraft **40**, 609–615 (2003)
13. Ishida, T., Ueshima, K., Ohira, K., Tokugawa, N.: Pressure based NLF design for vertical tail of conventional aircraft and wind tunnel test model, IUTAM Transition, p. 26 (2024)
14. Ohira, K., Ishida, T.: Development of a multifidelity analysis system for roughness-induced transition in high Re flow over infinite swept TAMU wing, IUTAM Transition, S2-1 (2024)
15. Kwak, D., Nomura, T., Tokugawa, N., Kurita, M., Murayama, M.: Introduction of research project for environmental conscious aircraft technology in JAXA. Proc. Greener Avi. 1–10 (2016)
16. Ueda, Y., Yoshida, K., Matsushima, K., Ishikawa, H.: Supersonic natural-laminar-flow wing-design concept at high-reynolds-number conditions. AIAA J. **52**, 1294–1306 (2014)
17. Hashimoto, A., Murakami, K., Aoyama, T., Ishiko, K., Hishida, M., Sakashita, M., Lahur, P.: Toward the Fastest Unstructured CFD Code 'FaSTAR', AIAA Paper 2012-1075 (2012)
18. Spalart, P.R.: Strategies for turbulence modeling and simulation. Int. J. Heat Fluid Flow **21**, 252–263 (2000)
19. ETW. https://www.etw.de/
20. You, J.H., Wright, M.C.N., Quix, H.: Recent activities on flow quality assessment at the European Transonic Windtunnel. In: Proceedings of the 33rd Congress ICAS (2022)
21. Egami, Y., Klein, C., Quest, J., Ondrus, V., Beifuss, U.: Development of new two-component temperature sensitive paint (TSP) for cryogenic testing. Mea. Sci. and Tech. **23**(11), 115301 (2012)

Chapter 37
Various Complexity and Information Measures for Analyzing the Laminar-Turbulent Transition Process in the Mixing Layer

Masashi Ichimiya and Ikuo Nakamura

Abstract Various complexity and information measures were tested to quantitatively express the laminar-turbulent transition process in the mixing layer. The digitized instantaneous velocity data of the hot-wire anemometer was used. From the complexity analysis, an approximated Kolmogorov complexity, and the normalized compression distance, NCD were tested. Next, the Shannon and permutation entropy, which consider the complexity of data sequences, were tested from the information analysis. The tested quantities increased monotonically during the transition process successfully.

Keywords Kolmogorov complexity · Shannon entropy · Permutation entropy · Mixing layer

37.1 Introduction

An important property of turbulence is that its motion is random. In the laminar-turbulent transition process in a free-shear layer, as one example from an ordered laminar flow to the random turbulent flow, the present authors recognized that there is no obvious measure that changes monotonically as the turbulence intensity in the wall-shear flows. The randomness factor based on the distribution of the power spectrum of the turbulence has been proposed [1], though it does not change monotonically [2]. To improve this situation, the characteristics of changes in the complexity of quantities during the transition process were focused. An analysis of Kolmogorov complexity [3] on the numerical data of instantaneous velocity from a hot-wire

M. Ichimiya (✉)
Tokushima University, Tokushima, Japan
e-mail: ichimiya@tokushima-u.ac.jp

I. Nakamura
Nagoya University, Nagoya, Japan

K. Kato et al. (eds.), *Proceedings of the 10th IUTAM Symposium on Laminar-Turbulent Transition*, IUTAM Bookseries 44, https://doi.org/10.1007/978-981-96-9829-5_37

anemometer was performed [4]. The present authors also performed analyses using Shannon entropy [5] and Bandt and Pompe's permutation entropy [6]. There are several examples of applying Kolmogorov complexity and Shannon and permutation entropies to turbulent and transitional flows; see our paper for details [4, 7]. In addition, see ref. [8, 9] for recent studies on these topics. However, no research has applied the permutation entropy to the turbulent transition process of single-phase flow, as in this study. For a historical overview of the definition of turbulence, which is a prerequisite for quantitative expression of the transition progress, as well as a detailed explanation of the concept of randomness, which is essential for defining turbulence, and the Kolmogorov complexity for evaluating it, see our paper [4].

37.2 Analytical Method

37.2.1 Kolmogorov Complexity

Kolmogorov complexity, K(*X*) of the object, *X* is the length of the shortest program for outputting the object. If the object has some regularity, the program to calculate the object can be shortened if one reflects the regularity in the programming. However, if there is no regularity, that is, if the object is genuinely random, there is no way to show the object by the computer by creating a program that shows the object as it is.

However, it has proven impossible to calculate K(*X*). To overcome this shortcoming, as a way to approximate K(*X*), a method to utilize a practical compression program has been proposed [10]. That is, as an approximation of K(*X*), to use a bit length, C(*X*), of the compressed data, *X*, obtained by the practical compression programs.

In this study, to examine experimental data, i.e., digitized output of hot-wire anemometers, and approximate their complexity, we used the ratio of the length of the compressed data length by the program, C, and the original data length, and designated the approximated Kolmogorov complexity, AK.

In practice, distinguishing the complexity between elements belonging to different sets is more important than considering a single complexity itself. The fundamental aspect of the distinction is the distance defined in the set. Therefore, a corresponding use of C(*X*) for an approximation of K(*X*) has been proposed as the normalized compression distance, NCD.

$$\mathrm{NCD}(X) = \frac{C(XY) - \min\{C(X), C(Y)\}}{\max\{C(X), C(Y)\}}. \tag{37.1}$$

The data *XY* can be made by simply connecting the data *Y* after the data *X*. By utilizing NCD, clustering of practical data becomes possible.

The minimum value is subtracted from each data set, then the data is normalized and converted to a 3-digit integer. Therefore, one velocity datum corresponds to an integer in the range of 000–999. Such 98,304 integer numbers (32,768 velocity data) aligned in a row make a text file with a capacity of 98,304 bytes. The data was compressed in 7z format on a Windows PC. The compression format was determined from the viewpoint that the NCD results would best satisfy the distance axioms [4].

37.2.2 Shannon Entropy

Shannon entropy represents the degree of uncertainty, and in this study, it is used to describe the laminar-turbulent transition process, for which no previous application has been reported. It is calculated as the expected value of self-information by Eq. (37.2) [5].

$$H = -\sum_{i=1}^{n} p_i \log p_i. \tag{37.2}$$

where p_i is the probability, $P(X_i)$, of each n event, X_i, given a probability distribution, P, in a discrete probability space. The logarithm is usually set to base 2. In this case, the unit [bit] is used. Also, when $p_i = 0$, we define $p_i \log p_i = 0$. In this study, the Shannon entropy is obtained by calculating the probability for the 262,144 data of the streamwise fluctuating velocity, u.

37.2.3 Permutation Entropy

The Shannon entropy does not consider the magnitude relationship based on the order of the data. For this reason, Bandt and Pompe [6] proposed permutation entropy, which considers the magnitude relationship.

When a sequence with n elements ($x_{t+1}, x_{t+2},\ldots, x_{t+n}$) is taken $T - n + 1$ times from a time series with T elements ($x_1, x_2,\ldots, x_T$) for $0 \leqq t \leqq (T - n)$, if there are no equal values among the n elements, then the number of events (permutations) of size relationships is $n!$. In this case, the probability, p_i, of each permutation is (the number of permutations in the range $0 \leqq t \leqq (T - n))/(T - n + 1)$. Then, the entropy (permutation entropy), H_p, can be calculated in the same way as the Shannon entropy. We used 262,144 sampled data points as the number of elements, T, and 8 data as n.

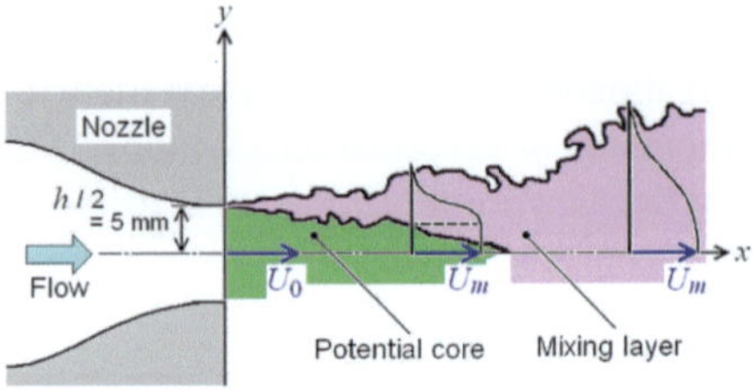

Fig. 37.1 Schematic diagram of two-dimensional mixing layer and coordinate system

37.3 Velocity Data in Mixing Layer

The flow is a mixing layer that is formed at the initial stage of a jet issuing from a rectangular nozzle into a stationary fluid [11]. The jet is ejected from a two-dimensional nozzle exit, $x = 0$, with a width of 310 mm and a height of $h = 10$ mm. Figure 37.1 shows the upper half of the flow field and the coordinate system.

The experimental conditions were a Reynolds number of 5,000 ($U_0 \fallingdotseq 7.5$ m/s) based on the velocity, U_0, at the nozzle exit and the nozzle exit height, h. An X-type hot-wire probe with a sensing area of 5 mm in diameter and 1 mm in length was used for the measurements. The output voltage of the probe was sampled 262,144 times (for approximately 52 s) at a sampling frequency of 5 kHz. The measurements were carried out in the range of vertical coordinate $y \geqq 0$ up to $x/h = 25$, sufficiently downstream from the point where the potential core disappears and the mixing layer becomes self-preserving.

37.4 Results and Discussion

37.4.1 Approximated Kolmogorov Complexity

In Fig. 37.2, the contour map of AK in the x–y plane is shown. The area between the maximum and minimum values is divided into 10 equal parts, and 9 contour lines are drawn. The fact that AK is small shows that during the compression process, any regularity is detected in the data, then the randomness becomes small. After the beginning of the transition, the regularity then disappears and the flow approaches random in accordance with the transition progress.

After AK increased to $x/h = 12$, it decreased slightly downstream. That is, the randomness presented by the data compressibility rather decreases downstream. Since AK does not necessarily monotonically change, it is not possible to measure the turbulent transition by AK alone.

When observed in the vertical, y, direction, AK is large around the nozzle edge, $y/(h/2) = 1$, where the fluctuating velocity is also large, although it is small around the centerline at $x/h = 3$ where the potential core remains. Downstream, in $x/h \geqq$

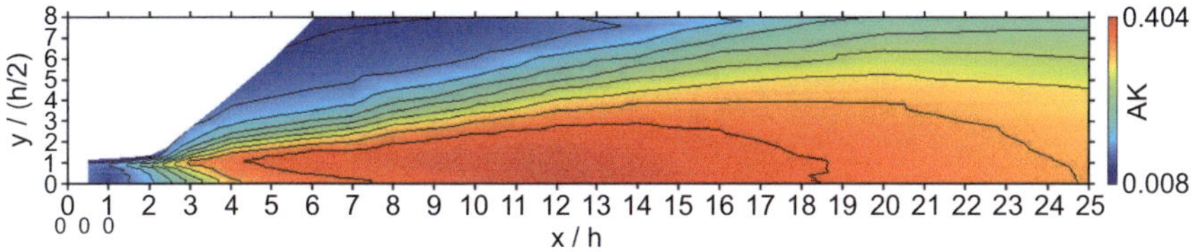

Fig. 37.2 Contour map of approximated Kolmogorov complexity in *x*–*y* plane

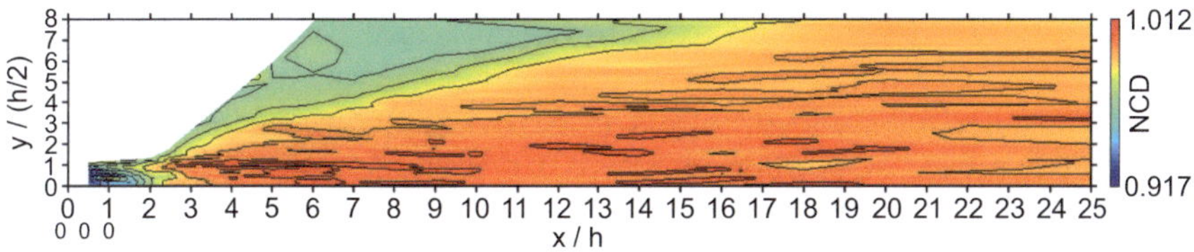

Fig. 37.3 Contour map of NCD in *x*–*y* plane

8, AK is approximately constant over the *y*-region from the centerline to the position of maximum fluctuating velocity. It decreased monotonically with increasing *y* thereafter. Hence, the randomness decreases with the decrease in fluctuating velocity.

37.4.2 Normalized Compression Distance

In Fig. 37.3 the contour map is shown. Values in the respective figures are NCD(*X*, *Y*) whose *Y* data are taken at that position. The position of the reference data, *X*, is $x/h = 0.5$, $y/(h/2) = 0$ in the laminar flow. Since the NCD value varies within a range of approximately 0.9 to 1, we consider changes in randomness from changes within that narrow range. However, since NCD at the reference position is approximately zero, the contour map of Fig. 37.3 is drawn except for around the reference position.

NCD is greater in a region of great fluctuating velocity. Unlike AK, in the streamwise variation, after monotonically increasing, the value is also substantially constant, though scatters are rather large. Hence it does not take the same value in two positions, NCD is appropriate as a measure for the laminar-turbulent transition in a mixing layer, though the increase soon ends.

37.4.3 Shannon Entropy

The contour map of Shannon entropy is shown in Fig. 37.4. For $x/h \leqq 3$, the entropy is maximum near $y/(h/2) = 1$ at the edge of the nozzle. Further inside (on the smaller $y/(h/2)$ side), the entropy is smaller due to the potential core. In other words, the

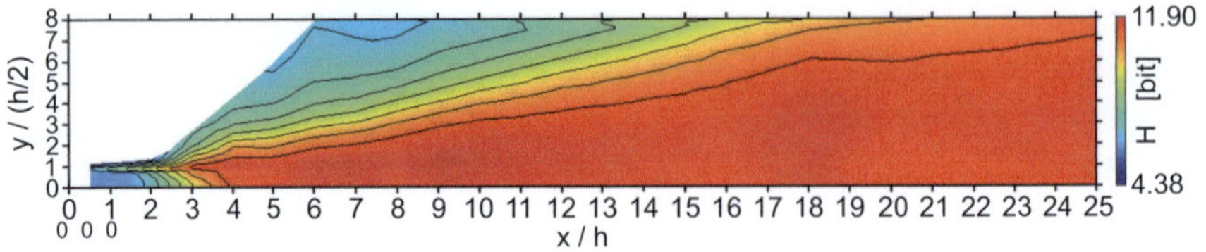

Fig. 37.4 Contour map of Shannon entropy in x–y plane

entropy is small in laminar flow. After the disappearance of the potential core, the entropy is large over a wide vertical range including the center line, $y = 0$.

The entropy increases monotonically up to $x/h = 4$, after which it becomes almost constant. Therefore, the Shannon entropy is also useful for measuring the progress of laminar-turbulent transition. The entropy also increases monotonically until it reaches a maximum, after which it does not decrease much. This is different from AK and has the same excellent properties as NCD. However, as with NCD, the maximum occurs well upstream of $x/h = 10 \sim 14$, where the transition is complete.

Here, why Shannon entropy is small in laminar flow and large in turbulent flow is considered. First, in laminar flow, the range of the velocity fluctuation is narrow and the frequency of taking equal values is high, so the number of events in the probability distribution is small. This results in a smaller Shannon entropy. On the other hand, in turbulent flow, the range of the velocity fluctuation is wide, so the number of events in the probability distribution is large. This results in a large Shannon entropy.

37.4.4 Permutation Entropy

The contours of the permutation entropy are shown in Fig. 37.5. It varies complicatedly and is distinctly different from the Shannon entropy.

The permutation entropy faithfully reflects the increase in the velocity fluctuation and the change in the fluctuation pattern to irregular.

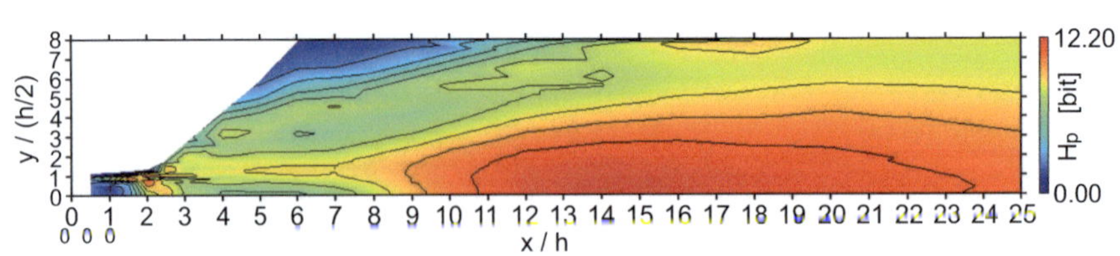

Fig. 37.5 Contour map of permutation entropy in x–y plane

37.5 Conclusions

To quantitatively evaluate the randomness of the flow, a finite discrete sequence of the fluctuating velocity during the laminar-turbulent transition process of the mixing layer was analyzed using Kolmogorov complexity, and information quantities.

AK can quantitatively display the regularity of laminar flow and the randomness of turbulent flow. The NCD varies monotonically downstream and can be a measure of transition progression. The Shannon entropy of the probability distribution of the fluctuating velocity increases monotonically and can be a measure of turbulent transition. The permutation entropy of the fluctuating velocity reflects an increase in the fluctuation velocity during the transition process of the mixing layer and a change in the pattern of irregular fluctuations.

Disclosure of Interests The authors have no competing interests to declare that are relevant to the content of this article.

References

1. Sato, H., Saito, H.: Fine-structure of energy spectra of velocity fluctuations in the transition region of a two-dimensional wake. J. Fluid Mech. **67**(3), 539–559 (1975). https://doi.org/10.1017/S0022112075000456
2. Ichimiya, M., Kamada, S., Okajima, A., Osaki, T.: Effect of local periodic disturbance on mixing layer downstream of two-dimensional jet (Spatial structure and quantitative representation of laminar-turbulent transition process). J. Fluid Sci. Tech. **8**(1), 90–105 (2013). https://doi.org/10.1299/jfst.8.90
3. Kolmogorov, A.N.: Three approaches to the quantitative definition of information. Probl. Inform. Transm. **1**(1), 1–7 (1965)
4. Ichimiya, M., Nakamura, I.: Randomness representation in turbulent flows with Kolmogorov complexity (In mixing layer). J. Fluid Sci. Tech. **8**(3), 407–422 (2013). https://doi.org/10.1299/jfst.8.407
5. Shannon, C.E.: A mathematical theory of communication. Bell Syst. Tech. J. **27**, 379–423, 623–656 (1948). https://doi.org/10.1002/j.1538-7305.1948.tb01338.x
6. Bandt, C., Pompe, B.: Permutation entropy: a natural complexity measure for time series. Phys. Rev. Lett. **88**(17), 174102 (2002). https://doi.org/10.1103/PhysRevLett.88.174102
7. Ichimiya, M., Nakamura, I.: Analysis of laminar-turbulent transition process in mixing layer with various information measures, Trans. JSME (in Japanese) **83**(845) (2017). https://doi.org/10.1299/transjsme.16-00497
8. Zhang, J., Jin, N., Ren, W.: Detecting the fluid flow transitions in oil–gas–water three-phase flows using multivariate rescaled range permutation entropy analysis. Eur. Phys. J. Plus **137**(808) (2022). https://doi.org/10.1140/epjp/s13360-022-03026-6
9. Bahamonde, A.D., Cornejo, P., Sepúlveda, H.H.: Laminar to turbulent transition in terms of information theory. Physica A **629** (2023). https://doi.org/10.1016/j.physa.2023.129190
10. Li, M., Chen, X., Li, X., Ma, B., Vitányi, P.M.B.: The similarity metric. IEEE Trans. Inform. Theory. **50**(12), 3250–3264 (2004). https://doi.org/10.1109/TIT.2004.838101
11. Ichimiya, M., Kato, T., Morimoto, T.: Effect of local periodic disturbance on mixing layer at exit of two-dimensional jet. J. Fluid Sci. Tech. **6**(6), 887–901 (2011). https://doi.org/10.1299/jfst.6.887

Chapter 38
Experimental Study of the Impact of Distributed Surface Roughness on Boundary-Layer Transition over a Flat Plate

Victoria Prieto, Félix Ducaffy, Maxime Forte, and Erwin R. Gowree

Abstract An experimental study was conducted to investigate the impact of distributed surface roughness on the linear amplification of Tollmien-Schlichting instabilities in a boundary layer without pressure gradient. Two distinct types of roughness were considered: evenly distributed and stochastic. A preliminary ΔN model for predicting the onset of transition based on average roughness heights $S_a/\overline{\delta}_1 \leq 0.16$ ($S_a \leq 67\mu$ m) is proposed. Furthermore, the effect of roughness extent in the streamwise direction on boundary layer transition is also studied, proving its impact on the linear amplification of modal instabilities.

Keywords Distributed surface roughness · Tollmien-schlichting instabilities · Over-amplification · Flat plate

38.1 Introduction

The efficiency of control strategies aimed at extending the laminar portion of the boundary layer is significantly affected by external disturbances such as turbulence, acoustics, or distributed surface roughness. According to Morkovin [10], for low turbulence levels, the transition path followed, known as natural transition, is divided into three stages: receptivity, linear amplification, and breakdown.

Among the possible disturbances affecting transition, it is well known that non-negligible surface roughness can shift the transition location upstream, as demonstrated by the experiments of von Doenhoff and Braslow [5], who used uniformly distributed, isolated roughness elements to analyze the effect on the transition, or Jeong et al. [7], who studied the effects of higher-density roughness on velocity pro-

V. Prieto (✉) · E. R. Gowree
DAEP, ISAE-SUPAERO, Toulouse, France
e-mail: victoria_maria.prieto_rueda@onera.fr

F. Ducaffy · M. Forte
DMPE, ONERA, Université Toulouse, Toulouse, France

K. Kato et al. (eds.), *Proceedings of the 10th IUTAM Symposium on Laminar-Turbulent Transition*, IUTAM Bookseries 44, https://doi.org/10.1007/978-981-96-9829-5_38

files and transition positions. On the other hand, Corke et al. showed that such shifts can be attributed to increased linear amplification of modal instabilities, referred to as over-amplification [2]. However, unlike discrete surface imperfections such as steps or gaps, which have been relatively well studied and several semi-empirical relations exist to predict transition [1, 3, 9], the effect of distributed surface roughness on the laminar-turbulent transition remains less understood. Some efforts have been made to model transition in the presence of distributed surface roughness, where Crouch [4] proposed an approach for addressing crossflow modes.

The objective of this study is to better understand the influence of different surface roughness parameters on the amplification of the Tollmien-Schlichting (TS) instabilities that are naturally present in a two-dimensional incompressible boundary layer. Lastly, the ultimate goal is to propose a semi-empirical ΔN model for transition prediction based on the over-amplification phenomenon, whose definition is presented in Sect. 38.3.1.

38.2 Experimental Set-Up

The present study was conducted in the TRIN 1 open-return subsonic wind tunnel located at the research facilities of ONERA Toulouse, which operates at ambient conditions. The dimensions of the test section are 2.5 m $\times$ 0.35 m $\times$ 0.6 m (length $\times$ height $\times$ width). It is preceded by a settling chamber and a contraction section with a ratio of 17, which results in turbulence levels varying between $4 \cdot 10^{-4}$ and $1.2 \cdot 10^{-3}$ for a frequency range of [5 Hz, 7 kHz]. In these experiments, velocities between 42 and 65 m/s were considered, which correspond to Reynolds numbers between $2.8 \cdot 10^6$ m^{-1} and $4.3 \cdot 10^6$ m^{-1}.

The model studied in this paper is a 1.2 m long aluminum flat plate with an optimized leading edge (193.8 mm long in the streamwise direction) shown in Fig. 38.1. The plate is mounted horizontally between the test section walls and contains a 300 mm x 280 mm slot located downstream of the leading edge. This allows for the placement of different types of inserts equipped with a desired surface roughness feature. The plate is equipped with 28 static pressure taps, aligned in the streamwise direction at 160 mm from the centerline of the leading edge up to a distance of 1000 mm. An adjustable trailing edge flap is used to regulate the pressure gradient; in this configuration it is set to $\beta = 0°$ in order to minimize the suction peak as shown

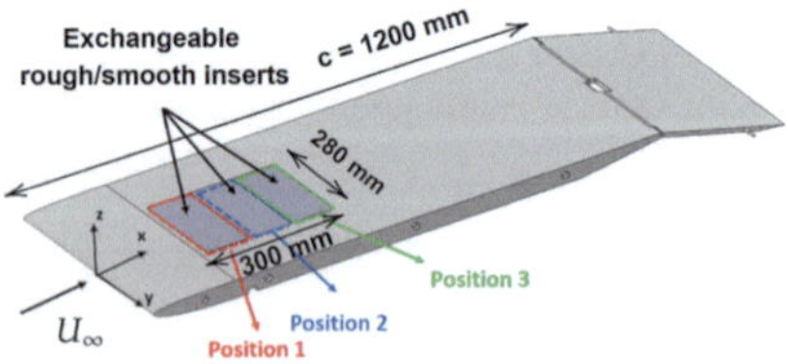

Fig. 38.1 Flat plate under study

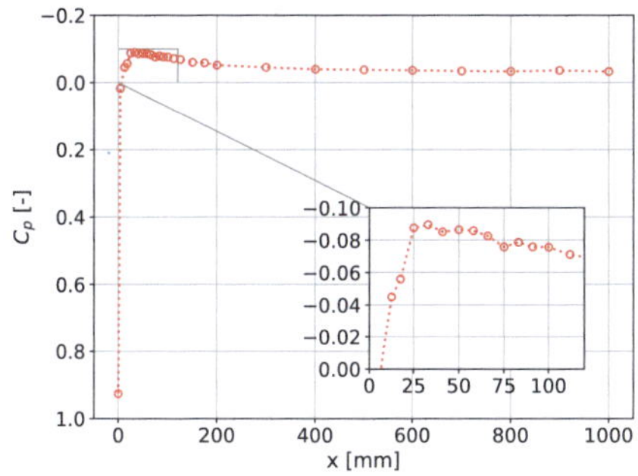

Fig. 38.2 C_p at Re $= 2.8 \cdot 10^6$ m^{-1}

in Fig. 38.2. Lastly, time-resolved velocity measurements in both the streamwise and wall-normal directions were obtained through the use of hot-wire anemometry.

In this study, the Bruker Alicona InfiniteFocus SL was used for roughness characterization, providing three-dimensional optical measurements. The roughness levels were classified based on the average roughness height, S_a, while additional parameters such as S_q (root-mean-square) and S_{sk} (skewness) were also considered. This paper studies two distinct types of distributed surface roughness in two separate experimental campaigns: evenly distributed and stochastic roughness. The first experimental campaign involved the use of SLA 3D-printed resin inserts with evenly distributed roughness patterns, including a checkerboard (see Fig. 38.3a) and flattened domes, which were studied by Ducaffy [6]. These patterns studied the influence of average roughness height, denoted as S_a, on the transition position, whose value varied from 10 to 130 μ m, first with a fixed extent of $L_R = 90$ mm, and later for different L_R values. The inserts were positioned in locations 1 to 3 of the plate for different experiments.

The second experimental campaign was conducted to investigate the impact of stochastic roughness. An insert containing sandpaper with a fixed roughness height of $S_a = 45$ μ m (see Fig. 38.3b) and variable streamwise extents (see Fig. 38.4) was used to maintain consistent experimental conditions across the experiments without

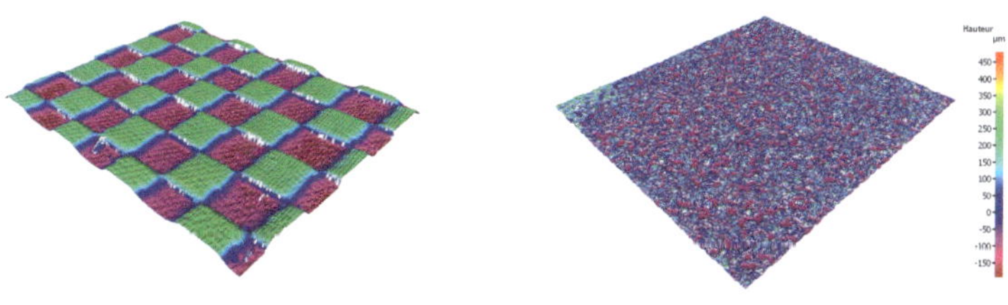

(a) Checkerboard pattern

(b) Sandpaper

Fig. 38.3 Roughness scan images

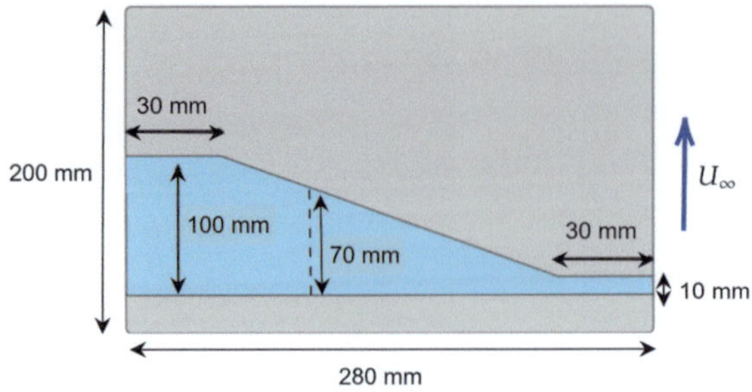

Fig. 38.4 Rough insert with variable L_R. Roughness in blue

changing the insert. The aforementioned insert was placed in both positions 2 and 3 simultaneously. The objective of this paper for this second type of insert is to examine the impact of L_R values between 10 and 70 mm on the amplification of TS instabilities.

38.3 Results and Discussion

38.3.1 The Semi-empirical Approach

One of the main objectives of this work is to propose a semi-empirical model for predicting the onset of transition due to surface roughness. This approach combines Linear Stability Theory (LST) with the e^N method, supported by experimental measurements, as shown in Fig. 38.5. The pressure distribution shown in Fig. 38.2 allows for boundary layer calculations that are later used in LST computations, which are performed with ONERA's in-house 3C3D and CASTET codes, respectively. The N-factor for the smooth baseline is obtained from these calculations. By comparing the experimental transition positions for the smooth and rough cases, the over-amplification of instabilities is quantified by a vertical shift in the N-factor, ΔN. It should be noted that the onset of transition is measured using a criterion based on the spatial gradient along the streamwise direction of the root mean square of the velocity fluctuations, u'_{rms}/U_∞, with the threshold set at $1.5 \cdot 10^{-4}$ mm^{-1} for this case.

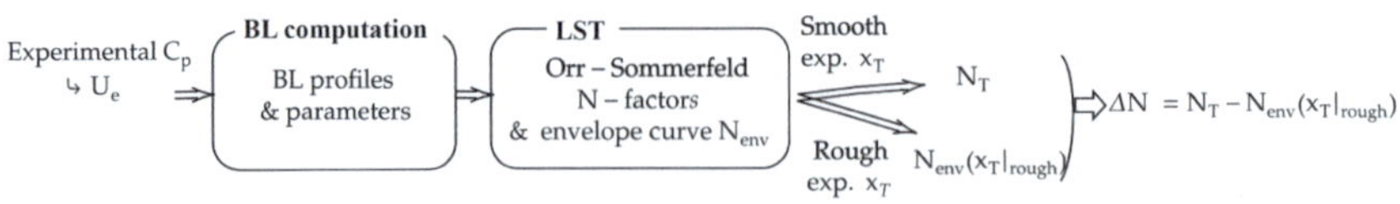

Fig. 38.5 Semi-empirical method to determine $\Delta N(\delta_1, S_a, L_R)$

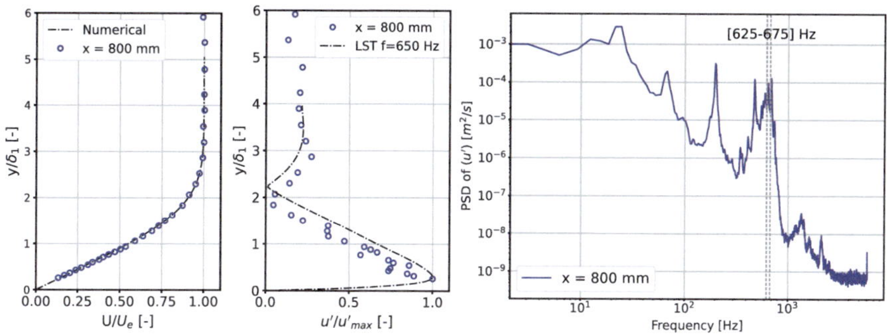

(a) Mean velocity (b) TS-amplitude (c) Spectrum at $y/\delta_1 = 0.48$

Fig. 38.6 TS profile at x = 800 mm and around 650 Hz for Re = 2.8 $\cdot 10^6$ m^{-1}

38.3.2 Validation with the Smooth Baseline Case

The validation has been conducted for a range of Reynolds numbers, but here only $Re = 2.8 \cdot 10^6$ m^{-1} is presented. From the mean velocity profiles in Fig. 38.6a , an excellent agreement between the experimental and numerical calculations shows that the boundary layer is well resolved. Figure 38.6b shows the eigenfunction and/or the fluctuating velocity profile, where the dual lobe is typical of a Tollmien-Schlichting wave sufficiently developed at x = 800 mm, and prior to the transition point located at $x_T = 947$ mm. This TS-amplitude profile is obtained by integrating the spectrum shown in Fig. 38.6c over a range of frequencies of [625, 675] Hz, which was obtained at $y/\delta_1 = 0.48$, where δ_1 is calculated from experiments and shown in Fig. 38.7. These results therefore confirm the hypothesis that the laminar-turbulent transition is driven by TS instabilities.

A confirmation of the transition being due to amplification of TS waves permits us to apply the e^N method. Figure 38.8 depicts the N-factor obtained for Re = 2.8 $\cdot$ 10^6 m^{-1}, accompanied by the growth rate of three other modes calculated from the LST. The frequency of the most amplified instability at the onset of transition is 600 Hz, which lies within the experimentally observed frequency band in Fig. 38.6c used for the determination of the TS-amplitude profile. From the experimentally observed chordwise position of the onset of transition and Fig. 38.8, the critical N-factor, N_T, is equal to about 9, which, according to Mack's criterion [8], corresponds to a turbulence intensity of $7 \cdot 10^{-4}$. In addition, this figure shows that the neutral point of the TS instability driving transition is upstream of the rough insert, confirming that the receptivity is not affected by the roughness and therefore only affects the amplification of the TS waves, which is the aim of the current study.

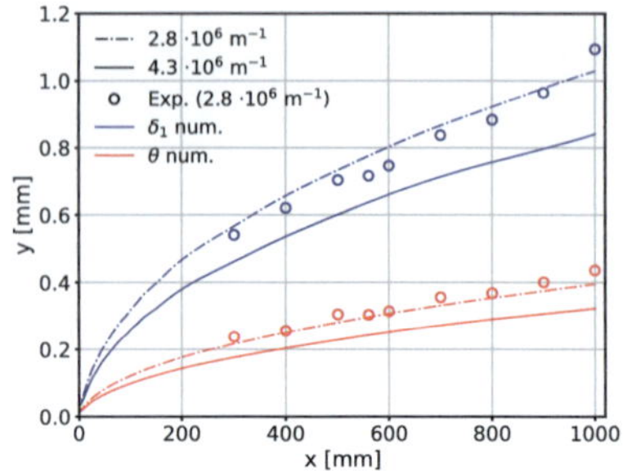

Fig. 38.7 δ_1 and θ for the lowest and highest Reynolds numbers

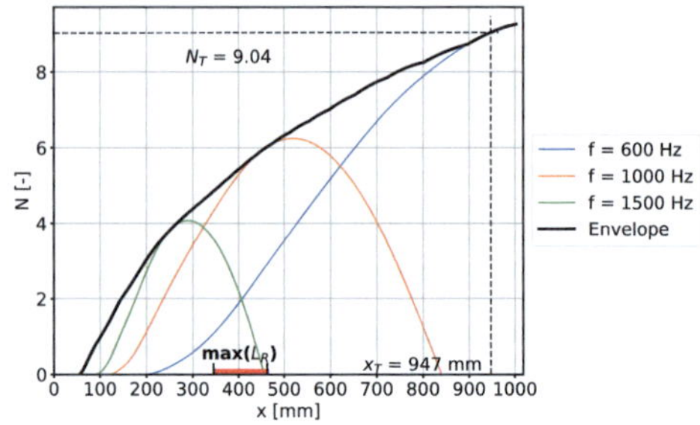

Fig. 38.8 N-factor at Re = $2.8 \cdot 10^6$ m^{-1}

38.3.3 *Impact of Surface Roughness on the Linear Amplification*

This section studies the impact of surface roughness on the amplification of TS instabilities. Figure 38.9a shows a linear correlation between ΔN and the dimensionless average roughness height $S_a/\overline{\delta}_1$ for a fixed streamwise L_R of 90 mm, with $\overline{\delta}_1$[1] calculated from the numerical δ_1 shown in Fig. 38.7. These results, which were obtained thanks to experiments conducted on uniformly distributed roughness types, led to the development of a semi-empirical relation by Ducaffy [6]:

$$\Delta N(S_a, \overline{\delta}_1) = 16.7 \frac{S_a}{\overline{\delta}_1} \tag{38.1}$$

While the model can predict transition positions for roughness within the linear amplification region, some scattering is observed, likely due to issues related to the flatness of 3D-printed inserts. The model is also limited to $S_a < 67$ µm and a reference roughness extent of $L_R = 90$ mm

To study the effect of roughness extent L_R and correct the possible errors related to the flatness of the inserts, a second experimental campaign was conducted using aluminium inserts, as shown in Fig. 38.4. In this study, different L_R values between 10 and 70 mm were studied at a fixed average roughness height, $S_a = 45$ µm, which

[1] $\overline{\delta}_1$ is calculated as the mean of δ_1 considered at the rear and front edges of the roughness.

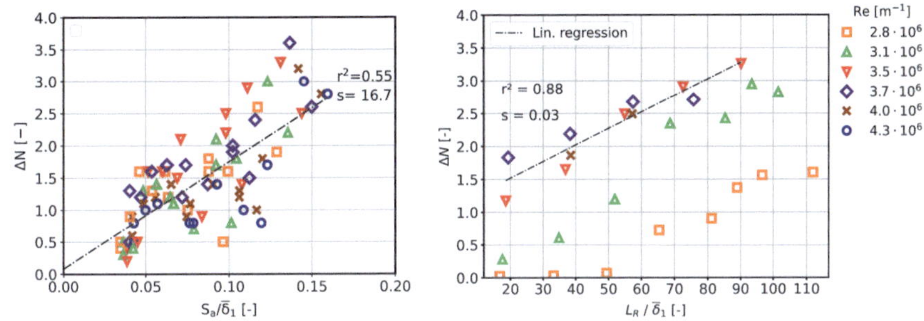

(a) Evolution with $S_a/\overline{\delta}_1$ for L_R=90mm. Evenly distributed roughness

(b) Evolution with $L_R/\overline{\delta}_1$ for S_a=45μm. Stochastic roughness

Fig. 38.9 ΔN evolution with different roughness parameters

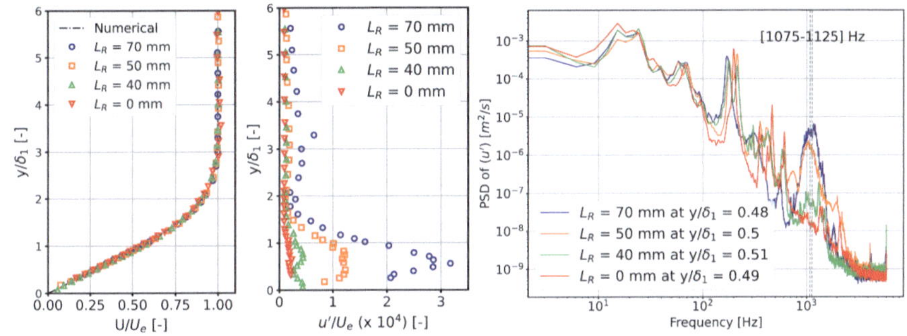

(a) Mean velocity

(b) TS-amplitude

(c) Spectra at $y/\delta_1 \approx 0.5$

Fig. 38.10 Velocity profiles varying L_R. Measured at $x \approx 400$ mm ($\Delta x = 6.2$ mm downstream of the end of roughness). $S_a = 45\ \mu$m, $\text{Re} = 2.8 \cdot 10^6 \text{m}^{-1}$. Integration frequency range $f \in$[1075, 1125] Hz

is within the range considered in the model shown in Eq. (38.1). In these experiments, first the impact of L_R on the linear amplification of instabilities was validated. Boundary-layer profiles measured 6.2 mm downstream of the rough surface (see Fig. 38.10) revealed that while roughness had no noticeable effect on mean velocity profiles, it did amplify TS-instabilities, with larger roughness lengths leading to higher amplification (see Fig. 38.10b).

Figure 38.9b shows the evolution of ΔN with $L_R/\overline{\delta}_1$ obtained from this second experimental campaign. It indicates that for Reynolds numbers $\text{Re} \geq 3.5 \cdot 10^6\ \text{m}^{-1}$, scattering is reduced, allowing for a preliminary linear correlation. However, for lower Reynolds numbers ($\text{Re} = 3.1 \cdot 10^6\ \text{m}^{-1}$ and $\text{Re} = 2.8 \cdot 10^6\ \text{m}^{-1}$), greater discrepancies between curves can be observed, with ΔN remaining zero up to $L_R/\overline{\delta}_1 = 50$ at $\text{Re} = 2.8 \cdot 10^6\ \text{m}^{-1}$. In light of these findings, a formal proposal for developing a model to accurately predict transition remains to be made, as additional results are necessary. To achieve this, further experiments with higher L_R and varying S_a values are planned, aiming to generate a more comprehensive database that will inform the development of a model incorporating the effects of both parameters.

38.4 Conclusion

The experimental study presented in this paper explored the impact of distributed surface roughness on transition of a boundary layer developing over a flat plate, focusing on two types of distributed surface roughness: evenly distributed and stochastic. It has been found that both roughness parameters under study, S_a and L_R, affect the linear amplification of natural boundary layer modal instabilities. A predictive ΔN model was fist proposed based on the linear relation found for with respect to $S_a/\overline{\delta}_1$. However, although the results obtained on the impact of L_R on the transition onset were insightful, no semi-empirical model could yet be established for this parameter. Future experiments will focus on different average roughness heights, especially lower values, and various roughness extents to develop a comprehensive ΔN model that incorporates the effect of L_R.

References

1. Beguet, S., Perraud, J., Forte, M., Brazier, J.P.: Modeling of transverse gaps effects on boundary-layer transition. J. Aircraft **54**, 794–801 (2017). https://doi.org/10.2514/1.C033647
2. Corke, T.C., Bar-Sever, A., Morkovin, M.V.: Experiments on transition enhancement by distributed roughness. Phys. Fluids **29**(10), 3199–3213 (1986). https://doi.org/10.1063/1.865838
3. Crouch, J., Kosorygin, V., Sutanto, M., Miller, G.: Characterizing surface-gap effects on boundary-layer transition dominated by tollmien–schlichting instability. Flow **2**, E8 (2022). https://doi.org/10.1017/flo.2022.1
4. Crouch, J., Ng, L., Kachanov, Y., Borodulin, V., Ivanov, A.: Influence of surface roughness and free-stream turbulence on crossflow-instability transition. In: Procedia IUTAM, vol. 14, pp. 295–302 (2015). https://doi.org/10.1016/j.piutam.2015.03.052
5. von Doenhoff, A.E., Braslow, A.L.: The effect of distributed surface roughness on laminar flow. In: Boundary Layer and Flow Control, pp. 657–681. Pergamon (1961). https://doi.org/10.1016/B978-1-4832-1323-1.50005-1
6. Ducaffy, F.: Ph.D. thesis, ISAE-SUPAERO (2022)
7. Jeong, H., Lee, S.W., Song, S.J.: Measurement of transitional surface roughness effects on flat-plate boundary layer transition. J. Fluids Eng. **141**(7), 074501 (2019)
8. Mack, L.M.: Technical Report Laminar-Turbulent Transition 22 p (SEE N78-14316 05-34), AGARD (1977)
9. Methel, J., Forte, M., Vermeersch, O., et al.: An experimental study on the effects of two-dimensional positive surface defects on the laminar–turbulent transition of a sucked boundary layer. Exp. Fluids **60**, 94 (2019). https://doi.org/10.1007/s00348-019-2741-2
10. Morkovin, M.V., Reshotko, E., Herbert, T.: Transition in open flow systems-a reassessment. Bull. Am. Phys. Soc. **39**, 1882 (1994)

Chapter 39
Model Order Reduction for Unsteady Dielectrophoretic Force-Driven Flows

Jonas Roller and Vincent Heuveline

Abstract This paper proposes a reduced-order model (ROM) for unsteady flows which are triggered by dielectrophoretic (DEP) force-driven convection using the proper orthogonal decomposition (POD). The model equations are nonlinear and time-dependent and usually computationally intensive to approximate. Our results demonstrate the feasibility of the POD approach for both the primal and dual model equations in a scenario that is associated with a real-world flow experiment.

Keywords Model order reduction · Data-driven modeling · Convection

39.1 TEHD Boussinesq Flows

A large number of applications in engineering are related to the study of heat convection between a hot and a cool wall, e.g., solar heating devices [1]. The present work involves the scenario of a dielectric fluid contained between two concentric cylinders of radii $R_1 < R_2$ that are kept at temperatures $T_1 > T_2$ and subjected to an inhomogeneous electric field. We note that the methodology in the following sections is not restricted to this particular geometry. Assuming the Boussinesq approximation of natural convection, we refer to the resulting model equations as the thermal electro-hydrodynamical (TEHD) Boussinesq equations for the fluid velocity $\mathbf{u}$, generalized pressure p, temperature θ and electrical potential Φ. They are given in [3] by

J. Roller (✉) · V. Heuveline
Engineering Mathematics and Computing Lab, Heidelberg University, Im Neuenheimer Feld 205, 69120 Heidelberg, Germany
e-mail: jonas.roller@uni-heidelberg.de
URL: https://emcl.iwr.uni-heidelberg.de/

V. Heuveline
e-mail: vincent.heuveline@uni-heidelberg.de

K. Kato et al. (eds.), *Proceedings of the 10th IUTAM Symposium on Laminar-Turbulent Transition*, IUTAM Bookseries 44, https://doi.org/10.1007/978-981-96-9829-5_39

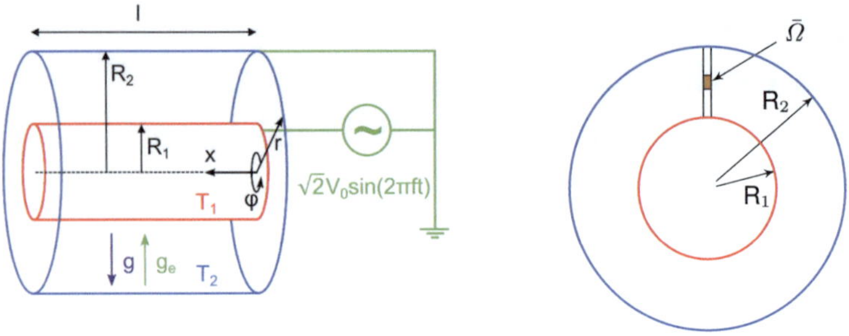

Fig. 39.1 Left: Schematic representation of the considered TEHD experiment. The inner cylinder is heated to T_1, the outer cylinder is kept at the cooler temperature T_2 and the RMS voltage V_0 is applied at the inner cylinder. Right: Computational domain Ω. The marked segment is the domain $\bar{\Omega}$ of the goal functional j introduced in Sect. 39.4

$$\begin{aligned} \partial_t \mathbf{u} + (\mathbf{u}\cdot\nabla)\mathbf{u} - \nu\Delta\mathbf{u} + \rho_r^{-1}\nabla p &= \alpha_e|\nabla\Phi|^2\nabla\theta - \alpha_g(\theta-\theta_r)\mathbf{g}, \\ \nabla\cdot\mathbf{u} &= 0, \\ \partial_t\theta + (\mathbf{u}\cdot\nabla)\theta - \kappa\Delta\theta &= 0, \\ -\nabla\cdot(\epsilon_0\epsilon_r(1-\gamma(\theta-\theta_r))\nabla\Phi) &= 0, \end{aligned} \tag{39.1}$$

in cylindrical coordinates (ϕ, r, x), with the dielectrophoretic (DEP) force $\alpha_e|\nabla\Phi|^2$ $\nabla\theta$, Earth's gravity $\mathbf{g}$ and the fluid parameters viscosity ν, thermal diffusivity κ and density ρ_r; buoyancy parameters α_e, α_g; vacuum and relative permittivity ε_0, ε_r; thermoelectric parameter γ and reference temperature θ_r. Figure 39.1 shows the geometry. In this study, we assume the flow to be dominated by the radial and azimuthal component. Therefore, our computational domain is chosen to be the radial-azimuthal plane $\Omega = \{(\phi, r) : \phi \in [0, 2\pi], r \in [R_1, R_2]\}$, which is also visualized in Fig. 39.1. We impose the following boundary conditions:

$$\begin{aligned} \mathbf{u} &= 0 && \text{at } r = R_1, r = R_2, \\ \theta = T_1, \quad \Phi &= V_0 && \text{at } r = R_1, \\ \theta = T_2, \quad \Phi &= 0 && \text{at } r = R_2. \end{aligned}$$

We also consider the *dual problem* of (39.1), which is needed to investigate sensitivity of initial conditions to perturbations and also has applications in optimal control. Assuming sufficient regularity of the dual solution, the strong formulation of the dual TEHD Boussinesq equations [3] reads as follows:

$$
\begin{aligned}
-\partial_t \hat{\mathbf{u}} + \hat{\mathbf{u}} \cdot (\nabla \mathbf{u})^T - \mathbf{u} \cdot \nabla \hat{\mathbf{u}} - (\nabla \cdot \mathbf{u})\hat{\mathbf{u}} - \nu \Delta \hat{\mathbf{u}} - \nabla \hat{p} + \hat{\theta} \nabla \theta = 0, \\
-\nabla \cdot \hat{\mathbf{u}} = 0, \\
-\partial_t \theta - \mathbf{u} \cdot \nabla \hat{\theta} - (\nabla \cdot \mathbf{u})\hat{\theta} - \kappa \Delta \hat{\theta} + \alpha_g \mathbf{g} \cdot \hat{\mathbf{u}} \\
+\alpha_e \nabla \cdot (|\nabla \Phi|^2 \hat{\mathbf{u}}) - \epsilon_0 \epsilon_r \gamma (\nabla \Phi \cdot \nabla \hat{\Phi}) = 0, \\
-\nabla \cdot (\epsilon_0 \epsilon_r (1 - \gamma(\theta - \theta_r)) \nabla \hat{\Phi}) + 2\alpha_e \nabla \cdot ((\nabla \theta \cdot \hat{\mathbf{u}}) \nabla \Phi) = 0,
\end{aligned}
\tag{39.2}
$$

with final conditions at $t = t_{\text{final}}$

$$
\begin{aligned}
\hat{\mathbf{u}}(t_{\text{final}}) &= -\partial_{\mathbf{u}} j(\mathbf{u}, p, \theta, \Phi), \\
\hat{\theta}(t_{\text{final}}) &= -\partial_{\theta} j(\mathbf{u}, p, \theta, \Phi),
\end{aligned}
\tag{39.3}
$$

for a suitable *goal functional* j and boundary conditions

$$
\hat{\mathbf{u}} = \hat{\theta} = \hat{\Phi} = 0 \quad \text{at } r = R_1, r = R_2. \tag{39.4}
$$

39.2 High-Order Approximation

To approximate solutions of the high-order models (39.1) and (39.2), we first apply a semi-discretization in time. The time stepping schemes we consider here are the two-step backward differentiation formula (BDF2) scheme [4] for the primal problem (39.1) and a continuous Galerkin-Petrov method known as cGP(1) [7] for the dual problem (39.2). Our specific version of the BDF2 scheme uses a linearized form of the convection terms at the ith time step:

$$
(\mathbf{u}^{(i)} \cdot \nabla)\mathbf{u}^{(i)} \approx ((2\mathbf{u}^{(i-1)} - \mathbf{u}^{(i-2)}) \cdot \nabla)\mathbf{u}^{(i)}. \tag{39.5}
$$

However, the time discretization does not lead to a fully linear system since we treat the DEP term semi-implicitly (implicit Φ, explicit θ)—see [4] for details. The nonlinear system is addressed by an inexact Newton-Raphson method. Each linear subproblem is discretized by the finite element method (FEM) on a mesh of quadrilaterals. We use the Taylor-Hood spaces $\mathbb{Q}_2/\mathbb{Q}_1$ for the velocity-pressure tuple $(\mathbf{u}, p)$, the Lagrange finite element space $\mathbb{Q}_2$ for θ and $\mathbb{Q}_1$ for Φ and identical spaces for the dual problem. Arising linear systems of equations are solved with the flexible generalized minimal residual method (FGMRES) method [8] that is preconditioned by a block-wise incomplete LU factorization.

39.3 Reduced-Order Modeling with POD

We utilize a snapshot-based *proper orthogonal decomposition* (POD) [2] to find the best approximation of the high-order finite element spaces by an r-dimensional subspace. The problem statement is:

Find $\{\psi_i\}_{i=1}^r \subset \mathbb{R}^m$ which solve

$$\min_{\tilde{\psi}_1,\dots,\tilde{\psi}_r \in \mathbb{R}^m} \sum_{j=1}^{n} \beta_j \left\| \boldsymbol{\alpha}_{h,j} - \sum_{i=1}^{r} \left(\boldsymbol{\alpha}_{h,j}, \tilde{\psi}_i \right)_{M_h} \tilde{\psi}_i \right\|_{M_h}^2 \qquad (39.6)$$

$$\text{s.t. } (\tilde{\psi}_i, \tilde{\psi}_j)_{M_h} = \delta_{i,j} \quad \forall 1 \le i, j \le r,$$

with the finite element *mass matrix*

$$(M_h)_{i,j} = \int_\Omega \phi_i \phi_j dx, \quad 1 \le i, j \le m, \qquad (39.7)$$

for finite element basis functions ϕ_i, ϕ_j and

$$(\mathbf{v}, \mathbf{v})_{M_h} = \mathbf{v}^T M_h \mathbf{v} \qquad (39.8)$$

for $\mathbf{v} \in \mathbb{R}^m$. Here, β_j is a time integration weight corresponding to the trapezoidal rule and the FEM solution at the jth time step is represented by the coefficient vector $\boldsymbol{\alpha}_{h,j} \in \mathbb{R}^m$. It can be shown that for $Y := (\boldsymbol{\alpha}_{h,1} \dots \boldsymbol{\alpha}_{h,n})$, $D := \text{diag}(\boldsymbol{\beta})$, the optimum is reached by setting

$$\psi_i = Y D^{\frac{1}{2}} \boldsymbol{\Phi}_i / \sqrt{\lambda_i}, \quad i = 1, \dots, r, \qquad (39.9)$$

where $(\lambda_i, \boldsymbol{\Phi}_i)$ is the ith eigenpair of the *correlation matrix*

$$C := D^{\frac{1}{2}} Y^T M_h Y D^{\frac{1}{2}}. \qquad (39.10)$$

39.3.1 Galerkin Reduced-Order Model

To formulate our reduced-order models (ROM) for (39.1) and (39.2), we transform the respective system of differential equations to its weak form. Let V be a suitable Hilbert space. The continuous problem:

Find $u \in V$ s.t.

$$(\partial_t u, v) + a(u, v) = f(v) \quad \forall v \in V, \qquad (39.11)$$

can be approximated by the *Galerkin ROM* (G-ROM) weak formulation:

Find $u_r = \sum_{i=1}^r \alpha_{r,i}(t)\psi_i(\mathbf{x}) \in V_r$ s.t.

$$(\partial_t u_r, v_r) + a(u_r, v_r) = f(v_r) \quad \forall v_r \in V_r, \tag{39.12}$$

with $V_r := \text{span}(\{\psi_i\}_{i=1}^r) \subset V$ and $\psi_i = \sum_{j=1}^m (\psi_i)_j \phi_j$ for $1 \leq i \leq r$. In general, a is a nonlinear operator, so after time discretization of (39.12), the G-ROM can be recast as a nonlinear system with r unknowns. It is desirable for $r \ll m$ to be the case, which can be achieved if the eigenvalues λ_i decay rapidly. For our TEHD Boussinesq problems (39.1) and (39.2), we compute two separate bases $\{\psi_{i,primal}\}_{i=1}^r, \{\psi_{i,dual}\}_{i=1}^r$ from primal and dual snapshots of each solution variable and employ each basis in the respective ROM. We thus use the same number of POD modes for all eight solution variables. The POD is only applied to the fluctuations of the FEM solution, as in [2]. For the pressure variable, we adapted a residual-based ROM derived from stabilization techniques [2]. The ROM time stepping scheme is chosen to be identical to that of the respective high-order model. Nonlinearities in the ROMs are treated by Picard iteration and the r-dimensional linear systems in each iteration step are solved by LU factorization. All models are implemented within the open-source finite element library HiFlow^3 [5].

39.4 Numerical Experiments

As our reference problem, we consider the circular, horizontal-gravity annular flow described in Sect. 39.1 and visualized in Fig. 39.2 which was recently investigated both numerically and experimentally [6]. We investigate the suitability of the G-ROM for the unsteady regime that is observed for sufficiently high V_0 and limit the investigation to one particular case where $R_1 = 5$ mm, $R_2 = 10$ mm, $\Delta T = T_1 - T_2 = 9$ K and $V_p = \sqrt{2}V_0 = 18$ kV, and the FEM solution is computed on the time interval [0, 20 s], starting from a steady, purely conductive initial state. The goal functional for the dual problem

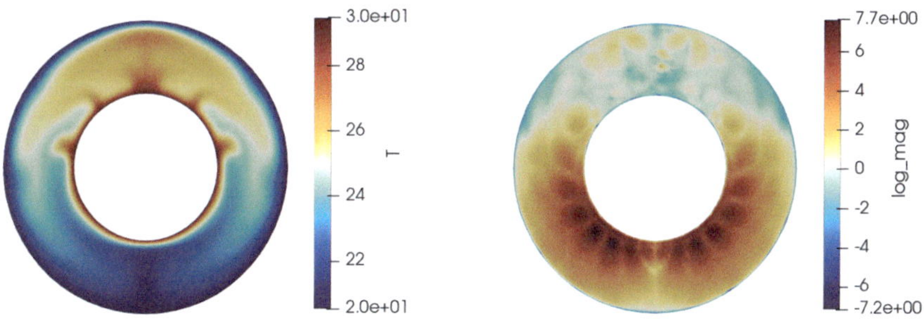

Fig. 39.2 Left: Unsteady G-ROM temperature solution θ for the reference scenario $\Delta T = 9$ K, $V_p = 18$ kV, shown at the final time $t = 20$s with POD rank $r = 24$. Right: Logarithmized G-ROM dual solution magnitude $\log(|\hat{u}| + |\hat{p}| + |\hat{\theta}| + |\hat{\Phi}|)$ for the goal functional j_{vort} at the initial time $t = 10$s with POD rank $r = 24$

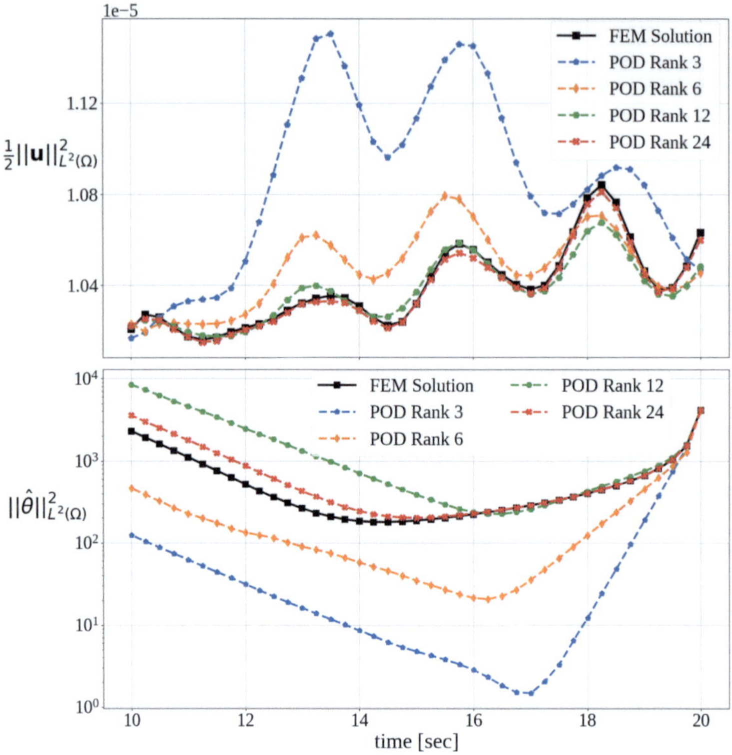

Fig. 39.3 Top: Convergence of the ROM kinetic energy to the FEM kinetic energy in the time interval [10, 20 s]. Bottom: Convergence of the squared L^2 norm of the ROM dual temperature over the same time interval

$$j = j_{vort}(\mathbf{u}) := \int_{\bar{\Omega}} \frac{1}{r}(\partial_r(ru_\phi) - \partial_\phi u_r)^2 \mathrm{d}(r, \phi), \tag{39.13}$$

is defined on $\bar{\Omega} = \{(\phi, r) : \phi = 0, r \in [7, 8\text{ mm}]\}$, i.e., squared axial vorticity in the mid-gap position at the top of the annulus (see Fig. 39.1, right). j is expected to quantify the instability in this scenario. We apply the POD and the G-ROMs to the shortened time interval [10s, 20s] to ensure that the flow under consideration is unsteady. The eigenvalues of the correlation matrices for all primal and dual solution variables decay rapidly for this problem, indicating that a small value of r could be sufficient. Our approach is further validated by Fig. 39.3 which shows the convergence of the primal kinetic energy and squared L^2 dual temperature norm with regard to r.

39.5 Conclusion

We have developed a POD-Galerkin reduced-order model combined with a residual-based pressure ROM to approximate the FEM solution of the TEHD Boussinesq equations. Since our results demonstrate a high accuracy of the ROM results with orders of magnitude fewer degrees of freedom, we can expect a major speed-up of computations and the possibility to explore DEP-driven phenomena that are not feasible to compute with FEM. We have also demonstrated that our ROM approach can replicate a physically relevant scenario. It is our intention to apply the same methodology to fully three-dimensional flow and optimal control.

References

1. Close, D.J.: The performance of solar water heaters with natural circulation. Sol. Energy **6**(1), 33–40 (1962). https://doi.org/10.1016/0038-092X(62)90096-8
2. Caiazzo, A., et al.: A numerical investigation of velocity-pressure reduced order models for incompressible flows. J. Comput. Phys. **259**, 598–616 (2014). https://doi.org/10.1016/j.jcp.2013.12.004
3. Gerstner, P.: Analysis and numerical approximation of dielectrophoretic force driven flow problems. Ph.D. Thesis. Heidelberg University (2020)
4. Gerstner, P., Heuveline, V.: Finite element approximation of dielectrophoretic force driven flow problems. ESAIM: Math. Model. Numer. Anal. **57**(3), 1691–1729 (2023). https://doi.org/10.1051/m2an/2023031
5. Gawlok, S., et al.: Hiflow3–technical report on release 2.0. Preprint Series of the Engineering Mathematics and Computing Lab, vol. 06 (2017)
6. Hamede, M.H., et al. Dielectrophoretic force-enhanced thermal convection within a horizontal cylindrical annulus. Phys. Fluids **36**(12) (2024). https://doi.org/10.1063/5.0234772
7. Schieweck, F.: A-stable discontinuous Galerkin-Petrov time discretization of higher order. J. Numer. Math. **18**(1), 25–57 (2010). https://doi.org/10.1515/jnum.2010.002
8. Saad, Y.: Iterative methods for sparse linear systems. Society for Industrial and Applied Mathematics (2003)

Chapter 40
Non-linear Interactions Between Stability Modes in a Compressible Subsonic Open Cavity

Marlon Sproesser Mathias, Felipe Oliveira Aguirre, Victor Barcelos Victorino, and Marcello Augusto Faraco de Medeiros

Abstract We study the global stability of the open cavity flow. We simulate and compute the linear stability of cases that contain both Rossiter and centrifugal modes and check whether the simulation and the linear instability results match, both in terms of the dominant modes and also regarding their frequencies. The main finding of this work is that the presence of centrifugal modes makes the mixing layer thinner, favouring higher-order Rossiter modes.

Keywords Linear stability theory · Open cavity flow

40.1 Introduction

Open cavities are a canonical geometry that can be used to model various real-world scenarios, such as small gaps between panels of an aerodynamic surface, the opening between a wing's main element and hyper-lifting devices, the space between consecutive train wagons, and so on. They can be described in terms of their length (L) and depth (D), which are non-dimensionalized by the incoming boundary layer displacement thickness at the cavity leading edge (δ_c^*). Besides the geometry, the flow is governed by other parameters, namely the Reynolds and Mach numbers (Re and Ma).

We focus on the short cavity case, where Rossiter modes dominate the 2D instability which is a well-researched case [7]. However, despite being dominated by Rossiter modes, such flows often also contain unstable centrifugal modes, due to Görtler instabilities in the recirculation area [5]. These three-dimensional modes often present much lower frequencies and amplification rates than Rossiter modes.

M. S. Mathias (✉)
Escola Politécnica, Universidade de São Paulo, São Paulo, Brazil
e-mail: marlon.mathias@usp.br

F. O. Aguirre · V. B. Victorino · M. A. Faraco de Medeiros
Escola de Engenharia de São Carlos, Universidade de São Paulo, São Carlos, Brazil

K. Kato et al. (eds.), *Proceedings of the 10th IUTAM Symposium on Laminar-Turbulent Transition*, IUTAM Bookseries 44, https://doi.org/10.1007/978-981-96-9829-5_40

Nonetheless, in this paper, we present evidence that both modes interact and modify each other's behavior.

Rossiter modes are two-dimensional and are characterized by a cycle between vortices forming in the mixing layer, impinging on the trailing edge and causing an acoustic feedback that triggers new vortices [6]. These modes are highly dependent on the Mach number [3]. Rossiter's empirical equation and its variants in general agree with experimental and numerical results. Nonetheless, it does not indicate which modes exist in the flow, nor which are their limit-cycle amplitudes. In contrast to Rossiter modes, centrifugal modes are not uniform in the spanwise direction. They are caused by Görtler instabilities in the recirculation region and may be either stationary or oscillating [2].

Our goal is to check the validity of the Linear Stability Analysis for the open cavity flow. For this, we must understand how both types of mode interact. We compute the global stability and simulate cases that contain both Rossiter and centrifugal modes and check whether results match both in terms of the dominant modes and also regarding their frequencies.

40.2 Methods

40.2.1 Flow Simulation and Stability Analysis

We used an in-house Direct Numerical Solver (DNS) for our study. It uses structured meshes and sixth-order compact spectral-like finite differences for the spatial derivatives. To perform the bi-global analysis, we employed a time-stepping approach, based on the work of [1]. Further details on our implementation and its validation are given in [4]. Both codes are publicly available and open-source.[1]

40.2.2 Parametric Space

Five cases were defined, in increasing order of instability (Table 40.1), here referred to by the initials of the unstable modes. Case `R` has only one marginally unstable mode, Rossiter 1. Similarly, in case `C`, only the second centrifugal mode is unstable. Case `RC1` has both Rossiter 1 and centrifugal 2 as marginally unstable modes. These three cases are designed to be very close to critical stability, potentially complying more closely with the linear stability predictions. Case `RC2` has a considerably higher Reynolds number, destabilizing both Rossiter and centrifugal modes, and presenting stronger non-linear effects. Finally, `RC3` maintains the Reynolds number from the previous case but contains a longer cavity with the goal of presenting higher-order

[1] https://github.com/marlonsmathias/GATT_DNS.

Table 40.1 Cases defined to investigate the interaction of Rossiter and centrifugal modes. The mode numbers depict decreasing magnitude of instability

Case	Re_{δ^*}	Ma	L/δ^*	D/δ^*	Rossiter	Centrifugal
R	100	0.5	20	10	Unstable (1)	Stable
C	150	0.3	20	10	Stable	Unstable (2)
RC1	150	0.5	20	10	Unstable (1, 2)	Unstable (2)
RC2	500	0.5	20	10	Unstable (2, 1)	Unstable (1, 2, 3)
RC3	500	0.5	40	10	Unstable (3, 4, 2, 5, 1, 6)	Unstable (1, 2, 3)

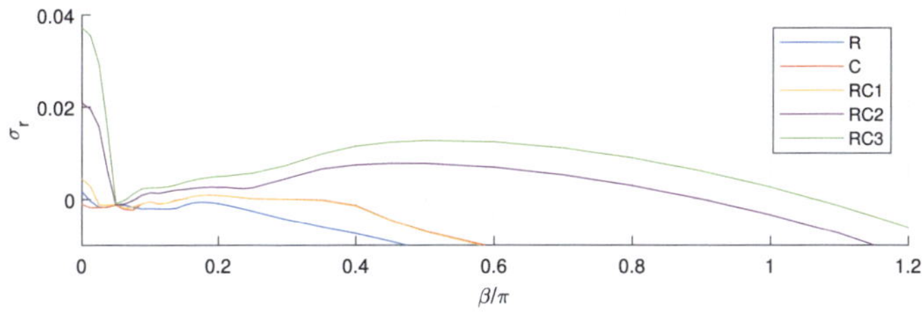

Fig. 40.1 Maximum amplification rate for each case as a function of β

Rossiter modes. The most unstable mode in this case is the third Rossiter mode, but all modes up to the sixth are unstable.

40.3 Linear Results

Figure 40.1 shows the maximum amplification rate (σ_r) of the global modes for each of our cases as a function of the span-wise wave number β. At $\beta = 0$, we see the two-dimensional Rossiter modes, which stabilize as β increases. Further increasing β causes all cases (except R) to display unstable centrifugal modes.

Cases R and RC1 only present a narrow range of β with unstable centrifugal modes, with a very low amplification rate. On the other hand, cases RC2 and RC3 have a much broader range of unstable β, reaching higher amplification rates. Nonetheless, even the most unstable centrifugal mode found still has considerably lower amplification rates when compared to Rossiter modes.

40.4 Non-linear Results

In this section, we focus on the results obtained from the DNS. The initial condition selected for each case was the base flow plus some small (usually 10^{-5}) disturbance. The cases were simulated until a limit cycle was reached and recorded for a number of cycles. Most results shown here are divided into two groups: marginally unstable cases (R, C and RC1) and more unstable cases (RC2 and RC3). This division is done for two reasons: (i) There is a considerable gap in the amplitude of the limit cycle between both groups, thus dividing them eases the visualization; (ii) The second group presents a much broader spectrum when compared to the first, which indicates a difference in the underlying physics.

Unless otherwise specified, measurements were made three-quarters along the cavity opening, at the height of the flat plate. For the 3D cases, an arbitrary span-wise position was chosen. The span-wise length chosen was $11.42\delta^*$, which corresponds to $\beta = 0.175\pi$, wide enough for the most unstable centrifugal mode of each case, and for multiples of the most unstable modes of cases RC2 and RC3.

40.4.1 Time Series

In this section, we compare the time series obtained from all cases in both 2D and 3D simulations. Figure 40.2 shows the results for both groups of cases. Two variables are shown in the figures: stream-wise and span-wise velocities. The stream-wise velocity at this position was found to be a good indicator for both Rossiter and centrifugal modes, which can be verified by looking at the eigenfunctions of each mode and observing their values at this position. The span-wise velocity is used as an indicator for three-dimensional effects; naturally, it is null for all 2D cases.

Case R, which does not present three-dimensional instabilities has remained very similar in both 2D and 3D simulations. The slight difference in the time to reach the limit cycle is due to the nature of the initial disturbances added to the base flow and is not relevant to this work. Case C has remained steady in the 2D simulation, and reached a low-frequency cycle in the 3D simulation.

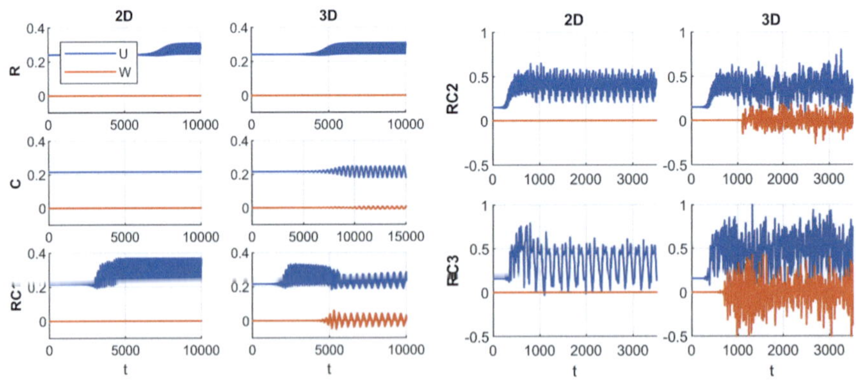

Fig. 40.2 Time series of stream-wise (U) and span-wise (W) velocities at 3/4 of the cavity opening for all cases with 2D and 3D simulations

`RC1` has an interesting behaviour in the 3D simulation. Initially, the Rossiter modes have grown much faster than the centrifugal modes, matching the linear stability predictions, which can be evidenced by the curve for the span-wise velocity taking much longer to leave its initial value than the one for stream-wise velocity. After $t = 5000$, the centrifugal modes finally reached larger amplitudes and became visible in the span-wise velocity curve, which caused the behavior of the stream-wise velocity to change. Before this event, both 2D and 3D simulations presented similar behaviors, which changed after the 3D effects grew. This is evidence that the three-dimensional modes, despite their lower frequencies and growth rates, indeed affect the two-dimensional modes. Cases `RC2` and `RC3` present much more chaotic behaviors, especially in the 3D simulations.

The curves for stream-wise velocity (blue in Fig. 40.2) reveal that, besides the oscillations, a second effect has happened, the deformation of the mean flow. Notice how the mean velocity in the limit cycle has increased when compared to the base flow. This increase was more noticeable for the more unstable cases. In case `RC1`, the mean flow after the 3D effects appeared (orange in the figure) is closer to the base flow than the mean flow before.

40.4.2 Temporal Evolution of the Spectra

All previous figures shown in this section have only considered the flow after it was fully developed, and disregarded possible changes in the spectrum with time. To analyze this facet, we have used a wavelet transform to obtain the change in the dominant frequencies as the flow evolves. Figure 40.3 shows the results for all cases. Note that, for the sake of visualization, the magnitudes shown in the images are normalized along each time step.

`R`, `C` and `RC1` settle for the frequencies seen in the fully developed flows after the initial transient. `RC2`, on the other hand, brings a more complex figure, clearly indicating Rossiter 2 as the dominant mode, while the centrifugal modes are not visible anymore. Other peaks, far from either type of mode, appear and disappear as time passes. Finally, case `RC3` does not settle for a single Rossiter mode. Most of the time R2 dominates, but it is occasionally replaced by R1 or R3. If the same analysis was done for the 2D version of `RC3`, the first Rossiter mode would clearly dominate the flow.

40.5 Conclusion

By simulating and computing the linear stability of five increasingly unstable cases, we observed both Rossiter and centrifugal modes individually and jointly. The presence of Rossiter modes causes the mixing layer to become thicker, which in turn, favours lower-order modes, which was already reported previously in the literature.

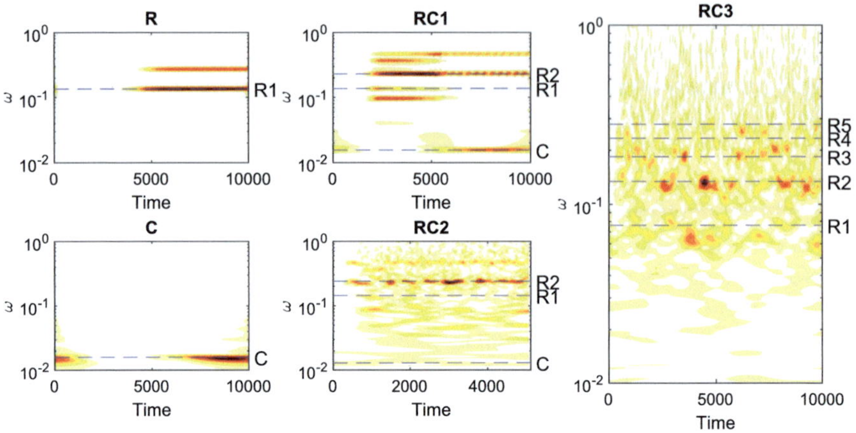

Fig. 40.3 Spectra of the stream-wise velocity at 3/4 of the cavity opening for fully-developed flows of all cases as the simulation evolves

The main finding of this work is that centrifugal modes have the opposite effect, i.e., their presence makes the mixing layer thinner, favouring higher-order Rossiter modes. This effect was clearly seen in case `RC1`, where initially only the Rossiter modes appeared, having the first mode as dominant, as predicted by the linear stability analysis. Later in the simulation, centrifugal modes appeared and made the mixing layer thinner, leading to the second Rossiter mode to dominate, contrary to the linear prediction.

A similar, yet opposite, effect is seen in case `RC3`, where the third Rossiter mode is predicted to dominate, as per the linear analysis. However, in the simulation, lower-order modes dominate the flow, as the mixing layer is considerably thicker in the mean flow when compared to the base flow. The thinning effect of the centrifugal modes was not enough to balance the thickening caused by Rossiter modes in our most unstable cases.

Acknowledgements The authors would like to thank the São Paulo Research Foundation (FAPESP/Brazil), for grants 2018/04584-0, 2017/23622-8 and 2019/15366-7; the National Council for Scientific and Technological Development (CNPq/Brazil) for grants 134722/2016-7, 307956/2019-9 and 441981/2023-2; the US Air Force Office of Scientific Research (AFOSR) for grants FA9550-18-1-0112 and FA9550-23-1-0030, managed by Dr. Geoff Andersen from SOARD and Dr. Roger Greenwood, respectively; the University of Liverpool for the access to the Barkla cluster, provided by Prof. Vassilios Theofilis; the Center for Mathematical Sciences Applied to Industry (CeMEAI) funded by FAPESP/Brazil, grant 2013/07375-0, for access to the Euler cluster; and the Center for Artificial Intelligence (C4AI-USP), with support by FAPESP/Brazil under grant number 2019/07665-4 and by the IBM Corporation, for access to the Arandu cluster.

References

1. Chiba, S.: Global stability analysis of incompressible viscous flow. J. Jpn. Soc. Comput. Fluid Dyn. **7**(1), 20–48 (1998)
2. de Vicente, J., Basley, J., Meseguer-Garrido, F., Soria, J., Theofilis, V.: Three-dimensional instabilities over a rectangular open cavity: from linear stability analysis to experimentation. J. Fluid Mech. **748**, 189–220 (2014)
3. Mathias, M.S., Medeiros, M.A.F.: The effect of incoming boundary layer thickness and Mach number on linear and nonlinear Rossiter modes in open cavity flows. Theor. Comput. Fluid Dyn. **35**, 495–513 (2021)
4. Mathias, M.S., Medeiros, M.A.F.: Optimal computational parameters for maximum accuracy and minimum cost of Arnoldi-based time-stepping methods for flow global stability analysis. Theor. Comput. Fluid Dyn. **36**(6), 1013–1036 (2022)
5. Picella, F., Loiseau, J.C., Lusseyran, F., Robinet, J.C., Cherubini, S., Pastur, L.: Successive bifurcations in a fully three-dimensional open cavity flow. J. Fluid Mech. **844**, 855–877 (2018)
6. Rossiter, J.E.: Wind-tunnel experiments on the flow over rectangular cavities at subsonic and transonic speeds. Technical Report Aeronautical Research Council Reports and Memoranda 3438, Ministry of Aviation, London (1964)
7. Yamouni, S., Sipp, D., Jacquin, L.: Interaction between feedback aeroacoustic and acoustic resonance mechanisms in a cavity flow: a global stability analysis. J. Fluid Mech. **717**, 134–165 (2013)

Chapter 41
Boundary Layer Bypass Transition Due to Gaps

Felipe Oliveira Aguirre, Victor Barcelos Victorino, Marlon Sproesser Mathias, Oliver T. Schmidt, and Marcello A. F. Medeiros

Abstract This study investigates the effects of two-dimensional gap induced flow structures on boundary layer transition. Through direct numerical simulations (DNS) and spectral proper orthogonal decomposition (SPOD), we analyze the impact of Rossiter and centrifugal modes on boundary layer transition. The research explores the interactions between these modes and how they contribute to transition mechanisms. Results reveal that while two-dimensional Rossiter modes create memory effects within turbulent flow regions, three-dimensional centrifugal modes add complexity by three-dimensionalizing the flow within the gap. Comparisons with linear stability theory (LST) confirm the presence of these modes, and their interaction hints at scenarios where combined instabilities might drive boundary layer transition. This work enhances the understanding of bypass transition processes caused by surface imperfections, contributing to flow control strategies in aerodynamics.

Keywords Boundary layer · Transition · Hydrodynamic instability · DNS · Gap · SPOD

41.1 Introduction

Many authors investigate the impact that surface imperfections have on T-S growth [5]. Perraud et al. [10] compared the variation of the factor N (ΔN) computed by T-S

F. Oliveira Aguirre · V. Barcelos Victorino · M. A. F. Medeiros (✉)
School of Engineering of São Carlos, University of São Paulo, São Paulo, Brazil
e-mail: marcello@sc.usp.br

F. Oliveira Aguirre
e-mail: felipe.aguirre@usp.br

M. Sproesser Mathias
Escola Politécnica, Universidade de São Paulo, São Paulo, Brazil

O. T. Schmidt
Department of Mechanical and Aerospace Engineering, University of California San Diego, La Jolla San Diego, CA, USA

K. Kato et al. (eds.), *Proceedings of the 10th IUTAM Symposium on Laminar-Turbulent Transition*, IUTAM Bookseries 44, https://doi.org/10.1007/978-981-96-9829-5_41

growth in a boundary layer with and without a gap. Crouch et al. [3] performed experimental ΔN correlations for cavity influence on transition with different sizes. Beguet et al. [1] revealed the limits for bypass conditions according to the gap length and depth normalized by the displacement thickness, $L/\delta^* \geq 18$ and $D/\delta^* \geq 2$, respectively. Earlier investigations have shown the presence of instabilities related to the flow over a cavity, as reported by [11], and [12]. The latter empirically proposed a physical mechanism to explain the acoustic radiation marked by a distinct frequency. The collision of vortices shed from the cavity leading edge into the cavity trailing edge generates acoustic waves that destabilize the mixing layer at the leading edge, completing the feedback cycle. Mathias and Medeiros [8] studied the dependency of the Mach number and the displacement thickness on the gap stability, concluding that the incoming boundary layer plays a role in selecting which Rossiter mode will occur, whereas the Mach number increases the acoustic energy transfer and the temporal growth rate up to a limit. Furthermore, the cavity flow presents three-dimensional oscillatory modes associated with centrifugal instability. Brès and Colonius [2] numerically investigated the stability of three-dimensional perturbations in two-dimensional flow over a cavity. Their results indicated the presence of unstable modes with wavelengths about the cavity depth that differ from each other in the frequency range. According to them, this instability arose due to a non-acoustic nature attributed to the centrifugal mechanism. They also found a minimal dependence of this mode on the Mach number, while the Reynolds number exerted influence on the instability and hierarchy of the dominant modes. References [4, 9] investigated the centrifugal instability in flow regimes without the Rossiter mode. The authors constructed the neutral stability diagram and amplification curves for different span-wise wave-number and Reynolds numbers. Victorino et al. [14] numerically investigated the interaction between Rossiter and centrifugal modes, where it's possible to see the boundary layer transition caused by the presence of both modes.

This paper presents a numerical study to examine the influence of the structures inside the gap on boundary layer transition.

41.2 Methodology

We use direct numerical simulation (DNS) and a time-stepping algorithm based on the Arnoldi method to approximate the eigenvalues and eigenfunction for the modal analysis. Both codes run with in-house developed scripts. The domain is represented by a Cartesian, structured, and non-uniform mesh. Buffer zones dissipate reflexive oscillatory motion near open boundaries by reducing the accuracy order of the spatial derivatives approximation. In the case of globally unstable flows, selective frequency damping (SFD) can be used inside the domain. Such an artifice is necessary to obtain a time-invariant solution or base flow. For details about the implementation, accuracy, validation, and method input parameters, further reading is available in [7].

As the flow over the gap has significant non-parallel effects, the linear stability analysis is performed using the global stability method. The eigenvalues and asso-

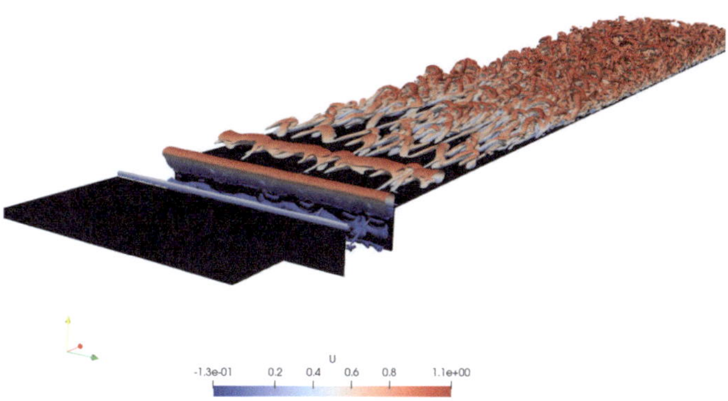

Fig. 41.1 Vortex structures plotted with Instantaneous iso-surfaces of the λ2-criterion, colored by the streamwise velocity component

ciated eigenfunction of the Jacobian matrix are approximated using a time-stepping method based on the Arnoldi method. In summary, this method adds a small perturbation to a previously obtained base flow and recursively runs the DNS solver for a defined period. After each iteration, a new orthogonal disturbance is generated for the next step. After a sufficient number of iterations, the complex eigenvalues and eigenfunction converge. Additional reading is available in [6] for more information on implementation and validation. Both codes are available on GitHub.[1]

41.3 Results

Considering a three-dimensional simulation with $Ma = 0.5$, $Re_{\delta^*} = 734$, $D = 6.11\delta^*$ and $L/D = 2$, using the SPOD [13] initially for the first 20 k_z (that is, first 20 β values, where β is the spanwise wavenumber), it will be possible to identify the modes present in the three-dimensional simulations carried out by the research group. We looked for the structures near the frequency of the 2D rossiter mode obtained previously from LST [14] (Rossiter mode frequency $f \approx 0.21$, Centrifugal mode frequency $f \approx 0.015$). Figure 41.1 shows these structures, mainly the three-dimensional features inside and close to the gap.

Initially, before identifying the modes using the SPOD, it's necessary to understand how the energy is distributed in the different span-wise wave numbers. With this in mind, in Fig. 41.2, on the left axis, we have the histogram with the PSD for each wave number present in the flow. In addition, on the right axis, we have the cumulative PSD indicating the percentage of energy contained as we increase the wave number. We see that the cumulative exceeds 99%, with 40 representatives k_z,

[1] https://github.com/marlonsmathias/GATT_DNS.

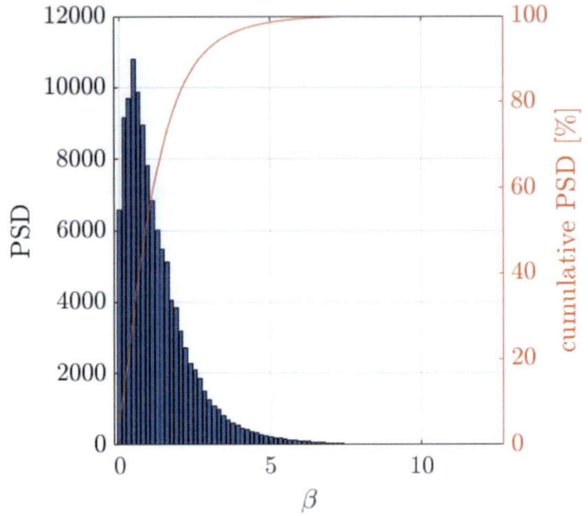

Fig. 41.2 Histogram of the PSD for each wave number in span-wise and cumulative PSD over the wave numbers

that is β up to approximately 5.5. The first 20 k_z represent a total of 75% for the cumulative energy.

The first twenty wave numbers were analyzed using the SPOD. For this first test, we kept weights only inside the gap to understand the correlation between the modes inside and the structures outside. In Fig. 41.3, analyzing the first modes, it was possible to identify a peak in the frequency of the Rossiter mode (around 0.2), the first (around 0.4), and the second harmonics (around 0.6). Additionally, some low-frequency energy in the range 0.01 to 0.05, related to centrifugal modes, are also present for $k_z > 10$.

Figure 41.4 shows the SPOD eigenfunction for a frequency of the Rossiter mode, and Fig. 41.5 shows the SPOD eigenfunction for a frequency in Centrifugal mode range. Both eigenfunctions are very similar to those expected for a Rossiter and Centrifugal from LST. For the Rossiter mode, the eigenfunction also suggests the existence of a memory in the region where the flow has fully developed turbulence.

So far it has been possible to identify modes with frequencies close to those of the predicted modes using linear stability theory. Using SPOD, we can better understand the presence and possible interaction between two- and three-dimensional unstable modes. We know that the interaction between these modes can cause the boundary layer transition process, but the process here is not entirely clear.

41.4 Conclusion

This work aimed a checking how the structures inside the cavity evolve in the flow. To do this, we applied post-processing techniques on the flow obtained through

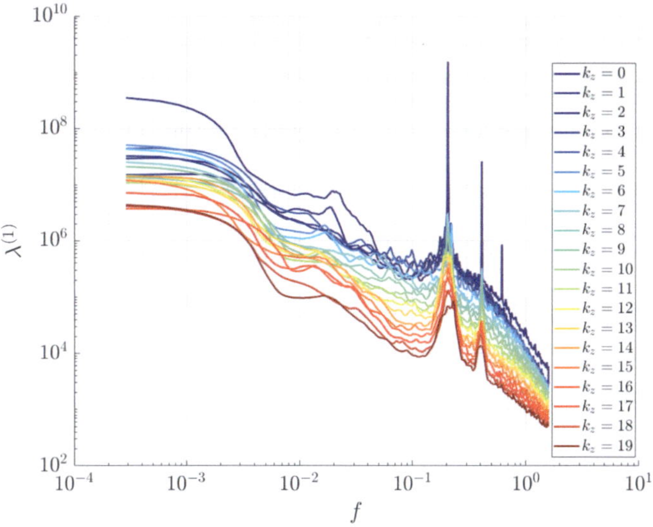

Fig. 41.3 SPOD spectrum for the first twenty wave numbers in the span-wise direction

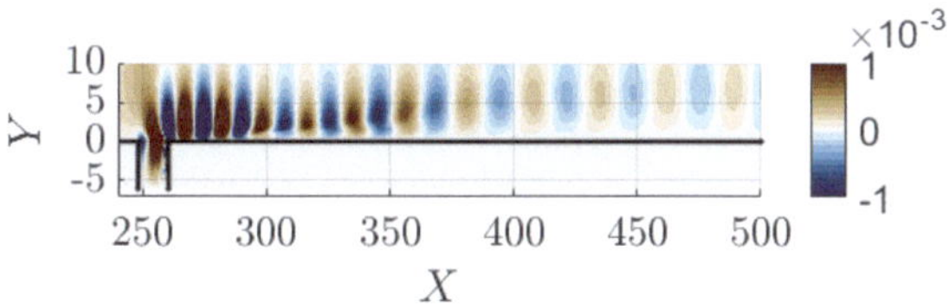

Fig. 41.4 Displays the SPOD eigenfunction of the mode with the Rossiter frequency using the V velocity component

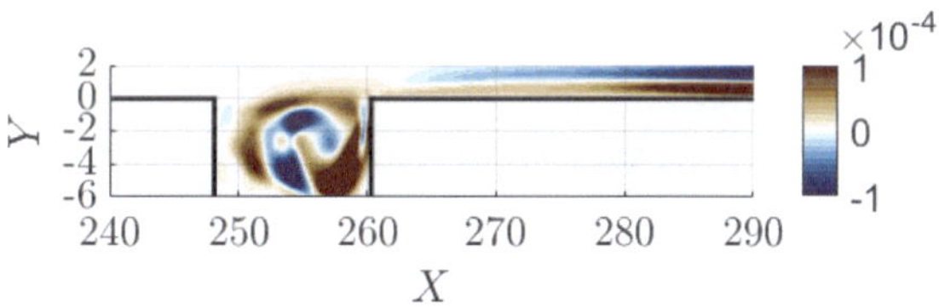

Fig. 41.5 Displays the V velocity component for the SPOD eigenfunction for the frequency close to the Centrifugal III mode

DNS to identify these modes. For this case study, LST predicted the existence of two-dimensional modes within the gap, the Rossiter modes, and three-dimensional modes, the centrifugal modes. With the Spectral Proper Orthogonal Decomposition (SPOD), it was possible to analyze the flow and relate SPOD modes to the modes found in LST. Using SPOD, we identified the predicted Rossiter mode and found its first and second harmonic structures in the simulation. The hypothesis that there are scenarios where the combination of centrifugal modes and Rossiter modes generate boundary layer transitions is still valid; however, the exact mechanism of this process has not yet been defined.

Acknowledgements We would like to thank the Coordenação de Aperfeiçoamento de Pessoal de Nível Superior (CAPES)—Programa de Excelência Acadêmica (PROEX)—Brasil for Financial Support. M.A.F.M. is sponsored by CNPq/Brazil (grant no. 307956/2019-9), sponsored by FAPESP (grant no. 2019/15336-7) and the US Air Force Office of Scientific Research (AFOSR) for grant FA9550-18-1-0112 and FA9550-23-1-0030, managed by Dr. Daniel Montes and Dr. Roger Greenwood;. We thank the Center for Mathematical Sciences Applied to Industry (CeMEAI) funded by São Paulo Research Foundation (FAPESP/Brazil), grant 2013/07375-0, for access to the Euler cluster, led by Prof. José Alberto Cuminato. This work used resources of the "Centro Nacional de Processamento de Alto Desempenho em São Paulo (CENAPAD-SP)." O.T.S. thankfully acknowledges support by AFOSR grant no. FA9550-22-1-0541 with Dr. Gregg Abate as the Program Officer.

References

1. Beguet, S., Perraud, J., Forte, M., Brazier, J.-Ph.: Modeling of transverse gaps effects on boundary-layer transition. J. Aircr. **54**(2), 794–801 (2017)
2. Brès, G.A., Colonius, T.: Three-dimensional instabilities in compressible flow over open cavities. J. Fluid Mech. **599**, 309–339 (2008)
3. Crouch, J.D., Kosorygin, V.S., Sutanto, M.I.: Modeling gap effects on transition dominated by Tollmien-Schlichting instability. In: AIAA AVIATION Forum June 15–19 (2020)
4. de Vicente, J., Basley, J., Meseguer-Garrido, F., Soria, J., Theofilis, V.: Three-dimensional instabilities over a rectangular open cavity: from linear stability analysis to experimentation. J. Fluid Mech. **748**, 189–220 (2014)
5. Himeno, F.H.T., Carvalho, A.E.B., Medeiros, M.A.F.: Analyses of Reynolds and Mach number effects on Tollmien–Schlichting wave–bump interaction in subsonic flows. J. Fluid Mech. **998**, A11 (2024)
6. Mathias, M.S., de Medeiros, M.A.F.: Optimal computational parameters for maximum accuracy and minimum cost of Arnoldi-based time-stepping methods for flow global stability analysis. Theor. Comput. Fluid Dyn. **36**, 1013–1036 (2022)
7. Mathias, M.S., Medeiros, M.A.F.: Direct numerical simulation of a compressible flow and matrix-free analysis of its instabilities over an open cavity. J. Aerosp. Technol. Manag. **10**, 1–13 (2018)
8. Mathias, M.S., Medeiros, M.A.F.: The effect of incoming boundary layer thickness and Mach number on linear and nonlinear Rossiter modes in open cavity flows. Theor. Comput. Fluid Dyn. **35**, 495–513 (2021)
9. Meseguer-Garrido, F., De Vicente, J., Valero, E., Theofilis, V.: On linear instability mechanisms in incompressible open cavity flow. J. Fluid Mech. **752**, 219–236 (2014)
10. Perraud, J., Arnal, D., Séraudie, A., Tran, D.: Laminar-turbulent transition on aerodynamic surfaces with imperfections. In: RTO-AVT-111 Symposium, Oct 2004, pp. 4–7 (2004)
11. Plumblee, H.E., Gibson, J.S., Lassiter, L.W.: Theoretical and Experimental Investigation of The Acoustic Response of Cavities In An Aerodynamic Flow. Technical Report, (USAF Report WADD-TR-61-75):1–167 (1962)
12. Rossiter, J.E.: Wind-tunnel Experiments on the Flow Over Rectangular Cavities at Subsonic and Transonic Speeds. Report, Ministry of Aviation; Royal Aircraft Establishment; RAE Farnborough (1964)
13. Schmidt, O.T., Colonius, T.: Guide to spectral proper orthogonal decomposition. AIAA J. **58**(3), 1023–1033 (2020)
14. Victorino, V.B., Aguirre, F.O., Medeiros, M.A.: Gap induced boundary layer transition. In: AIAA AVIATION 2023 Forum (2023)

Chapter 42
Application of the One-Way Navier-Stokes Equations to Model Disturbance Evolution in Laminar Separation Bubbles

Elliot J. Badcock and Shahid Mughal

Abstract We examine the effectiveness of the OWNS-R model in analysing the stability of convective disturbances in boundary-layer flows exhibiting laminar separation bubbles (LSBs). The LSBs are generated through the introduction of surface indentations over a nominally flat surface. Results highlight the effectiveness of the one-way stabilised marching procedure in scenarios without or with small LSB regions. However, they also reveal the limitations of current OWNS methodologies when applied to surface indentations that feature relatively larger LSBs. To address this shortcoming, we propose a novel modal one-way approach, M-OWNS, which enables numerically stable marching through relatively larger LSB regions.

Keywords Instability modelling · OWNS · Modal-OWNS · PSE · Boundary-layers

42.1 Introduction

The one-way Navier-Stokes (OWNS) model [5, 6] is a spatial-marching technique used to investigate the stability of slowly-varying laminar boundary layer flows. OWNS complements the parabolised stability equations (PSE) and the fully elliptic linear harmonic Navier-Stokes equations (LHNS) [4]. We examine the viability of the OWNS-Recursion (OWNS-R) [9] method in marching through regions of reverse flow, and evaluate its physics fidelity relative to PSE.

OWNS and PSE leverage the slow streamwise variation of boundary layers, decoupling upstream and downstream dynamics to capture only downstream disturbance evolution. OWNS-R achieves a stable parabolic march by recursively filtering out all upstream propagating waves in the discretised disturbance field at every streamwise point. This filtering is computationally costlier than a PSE march and requires

E. J. Badcock (✉) · S. Mughal
Department of Mathematics, Imperial College, London, UK
e-mail: ejb321@ic.ac.uk

K. Kato et al. (eds.), *Proceedings of the 10th IUTAM Symposium on Laminar-Turbulent Transition*, IUTAM Bookseries 44, https://doi.org/10.1007/978-981-96-9829-5_42

specialist knowledge of the operator's spectrum, which is baseflow specific, to effectively place *so-called* recursion parameters that drive the one-way recursion. With correctly determined parameter, the recursion is sufficient to achieve a stable marching procedure without additional parabolisation. Alternatively, the LHNS framework, while more computationally expensive, offers advantages: it does not require slowly varying baseflow, captures both convective and upstream effects, and is the most *physically rigorous* approach.

We examine LSB flows devised by [8], with flow reversal achieved by surface indentations in a Blasius boundary-layer. DNS [8], PSE, and LHNS analyses [1, 2] reveal that LSBs above a critical size trigger rapid disturbance amplification. While LHNS provides optimal analysis, its computational cost limits practical applications. OWNS, not suffering from PSE's minimum step size restriction, should correctly capture the full downstream evolution; however, our analysis reveals challenges when applied to large LSBs. The fine grid resolution required can introduce growing numerical instabilities in the solution (see Fig. 42.4c). We hypothesize that stronger reversed flows transform convective instabilities into absolute instabilities at the critical length scales necessary for accurate OWNS marching. To address this limitation, we propose modal-OWNS (M-OWNS). M-OWNS applies the one-way recursive filter of OWNS-R to disturbances with a modal or PSE type ansatz. This enables stable marching through prominent LSBs at step sizes large enough to avoid small-scale "numerical" absolute instabilities while maintaining the streamwise pressure derivative and smaller step sizes than PSE restrictions. A more detailed paper on M-OWNS and its viability to modelling modal and non-modal disturbance evolution is in preparation.

42.2 Numerics and Results

42.2.1 Indentation Coordinate System and Baseflow

To account for the curvature of the indentation, we transform the Navier-Stokes equations from the usual Cartesian coordinate system (x_1, x_2, x_3) into a indentation-specific coordinate system (x, y, z) via the transformation

$$x = x_1, \quad y = x_2 - H(x_1), \quad z = x_3, \tag{42.1}$$

where H is the indentation profile [8]. Here, x corresponds to the streamwise direction, y the wall-normal direction, and z the homogenous spanwise direction. The LHNS equations are derived by decomposing the state vector $\hat{q}$, into a time-independent baseflow Q and a time-harmonic, spanwise periodic disturbance q,

$$\hat{q}\,(x, y, z, t) = Q\,(x, y) + q(x, y)\exp\left(-i\omega t + i\beta z\right) + c.c., \tag{42.2}$$

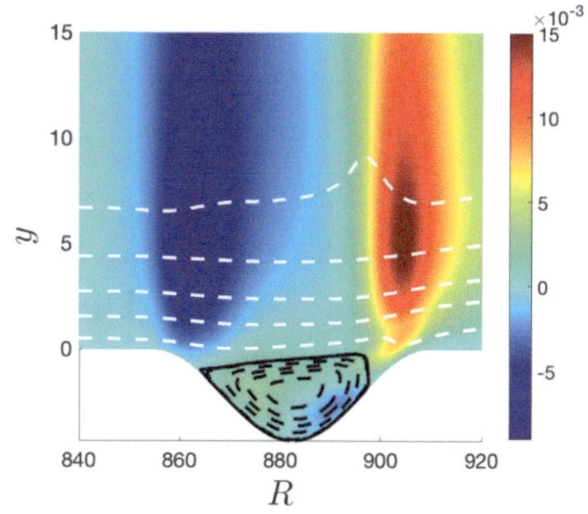

Fig. 42.1 Wall normal velocity for the $w = 81$ mm, $h = 2.17$ mm case. Positive streamwise velocity contours are mapped in white, and negative values in black. Note the LSB occupies, nearly, the entire indentation region [8]

where $c.c.$ stands for complex conjugate, and subsequently linearising with respect to q. The frequency ω and spanwise wavenumber β are real constants and remain fixed throughout all calculations.

We aim to reproduce previous results by [1, 2] with the OWNS method. The simulations, corroborated by experiment, were conducted at a freestream Reynolds number of $R_\infty = 1.2 \times 10^6$ and freestream velocity $U_\infty = 18\,\text{ms}^{-1}$. The boundary layer baseflow was calculated using a spectral/hp element discretisation implemented in the Nektar++ package (see [8]). The indentation is defined by a cosine function with variable height, h, and width, w. We focus on two different geometries that characterise these flows: $w = 162$ mm, $h = 1.62$ mm (wide and shallow, lacking a LSB) and $w = 81$ mm, $h = 2.17$ mm (thin and deep, featuring a prominent LSB). All OWNS, M-OWNS and PSE calculations are 2D in nature, and performed at a frequency of 172 Hz, with $n_y = 301$, 4th order accurate finite difference points in the wall-normal direction and a 3-point 2nd order accurate upwinded discretisation in the streamwise x direction (Fig. 42.1).

42.2.2 OWNS-R

To parabolise the system, we employ the OWNS-R method [9]. We express the LHNS equations in the following form

$$\mathbf{A}\frac{\partial q}{\partial x} = \mathbf{B}q, \tag{42.3}$$

where $\mathbf{A}$ and $\mathbf{B}$ are discretised differential operators in y. The OWNS-R method at each streamwise location proceeds as follows:

1. Solve Eq. 42.3 for the state vector q.
2. Calculate the wavenumber α that corresponds to q, via the usual kinetic energy norm.
3. Remove the influence of upstream propagating waves on q by calculating the action of $\mathbf{P}_{N_\beta}$ on q, where

$$\mathbf{P}_{N_\beta} = \left(\mathbb{I} + c \prod_{k=1}^{N_\beta} \left(\mathbf{B} - i\beta_k^+ \mathbf{A}\right) \left(\mathbf{B} - i\beta_k^- \mathbf{A}\right)^{-1} \right)^{-1}, \tag{42.4}$$

and $(\beta_k^+, \beta_j^-)_{j=1,\ldots,N_\beta}$ are complex recursion parameters, while c is a tuning parameter. This multiplication is achieved via a recursive procedure [9].

The accurate determination of recursion parameters is essential for correct parabolisation of the system. However, these parameters are flow-dependent and can be challenging to determine. A review on their placement for subsonic inviscid flow is provided by [5]. The proper determination of these parameters for boundary layer flows with and without indentations will be addressed in future works. For all OWNS and M-OWNS calculations here, we use $N_\beta = 20$ and identical recursion parameters between methods.

42.2.3 Modal-OWNS

For M-OWNS, a modal disturbance of the form,

$$q(x, y) = \tilde{q}(x, y) \exp i\Theta(x) \tag{42.5}$$

is assumed. The phase function Θ is such that $\Theta_x \approx \alpha$, where α is the wavenumber corresponding to q. This ensures that the wavelength of $\tilde{q}$ significantly exceeds wavelength of q, thus requiring fewer streamwise points to resolve the disturbance [3]. The wavenumber is approximated via

$$\alpha = \alpha_{\text{old}} - i \frac{\int_0^\infty \tilde{q}^* \frac{\partial \tilde{q}}{\partial x} dy}{\int_0^\infty \tilde{q}^* \tilde{q} dy}, \tag{42.6}$$

where * denotes the complex conjugate. This equation is updated iteratively until convergence in α is reached. For any disturbance with a modal ansatz, [7] find that minimizing the quantity $\Delta x\,(\Theta_x - \alpha)$, where Δx is the streamwise step size, is sufficient to yield an accurate PSE march. However, in a PSE analysis we are limited by finite Δx and thus this quantity cannot always be minimised. In contrast, M-OWNS is not constrained by this restriction, enabling the resolution of both modal and non-modal disturbance evolution at finer spatial scales.

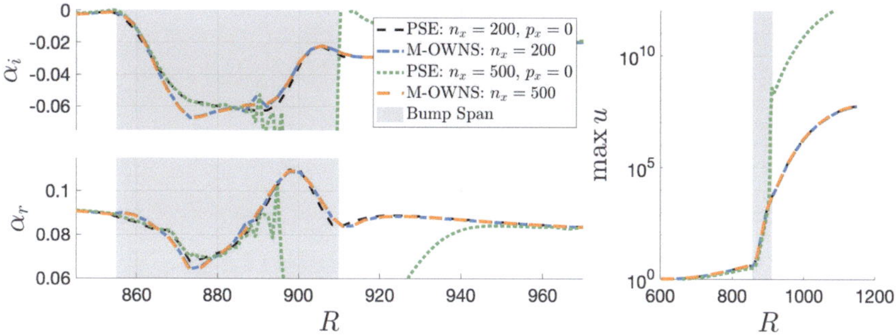

Fig. 42.2 Wavenumbers, growth rates and maximum streamwise velocity amplitude compared with OWNS, M-OWNS and PSE at streamwise positions

42.2.4 Indentation: $w = 162\,mm$, $h = 1.62\,mm$

The indentation is shallow and wide enough to prevent the formation of an LSB. Computations are performed for $500 \leq R \leq 1200$ with the 4 different models: PSE and M-OWNS at a coarse streamwise resolution of $n_x = 200$ points, and M-OWNS & OWNS results with $n_x = 2000$ points (greater than PSE step size limit). The coarse resolution results correspond to approximately 10 points per wavelength, while the fine resolution results correspond to approximately 100 points. The OWNS model only resolves with the fine resolution, unlike the new M-OWNS model, which can do both. Figure 42.2 shows the wavenumbers, growth rates, and maximum streamwise disturbance values compared for each of the methods, showing good agreement with each other and with LHNS [2].

42.2.5 Indentation: $w = 81\,mm$, $h = 2.17\,mm$

The deep and thin indentation forms a large LSB, and OWNS calculations fail upon reaching the bump. This occurs either due to insufficient streamwise resolution to capture the disturbance, or the possible occurrence of numerical instabilities that form at length scales larger than required marching step size of the original OWNS model, see Fig. 42.4c [1].

M-OWNS demonstrates a critical advantage by resolving the full disturbance field at a much coarser streamwise resolution. Figure 42.3 compares results of four approaches: PSE & M-OWNS at $n_x = 200$ and $n_x = 500$. The coarse resolution PSE remains numerically stable, while the finer PSE computation diverges. In the PSE calculation at $n_x = 500$ we turned off the adaptive iteration as the march fails as soon as we reach the LSB, see Fig. 42.4b. Our M-OWNS computation remains stable for both discretisations, with the disturbance amplitude results consistent and in agreement with LHNS of [2, pp. 69–72] (see Fig. 42.3). Some slight differences

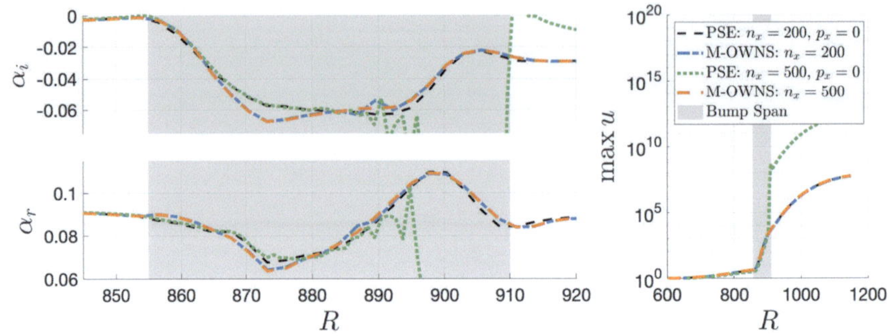

Fig. 42.3 Wavenumbers, growth rates and maximum streamwise velocity amplitudes compared for M-OWNS and PSE; $w = 81$ mm, $h = 2.17$ mm case. Notice the failure of the PSE to converge at the finer resolution

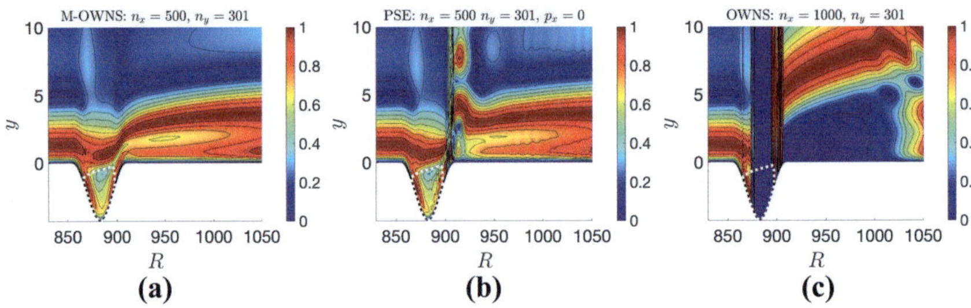

Fig. 42.4 Streamwise velocity fields that have been normalised by the maximum value at each streamwise location. **a** M-OWNS, **b** PSE **c** The OWNS method fails to converge in the region of the LSB

arise in Fig. 42.4a when compared to the corresponding result of [2], however we speculate that this may be due to the upstream effects captured with LHNS which M-OWNS restricts. The wavenumbers and growth rates of M-OWNS differ from PSE; this is attributed to the PSE's omission of the streamwise pressure derivative—notably retained in M-OWNS.

Conclusion

The OWNS model exhibits numerical instabilities when applied to boundary-layer flows with large laminar separation bubbles induced by surface indentations. To address this limitation, we proposed a novel modal OWNS approach, M-OWNS, that enables stable marching through more prominent separation regions. M-OWNS demonstrates superior robustness compared to both the original OWNS and PSE methods, converging at streamwise resolutions where the other approaches fail. This work highlights the potential of M-OWNS as an efficient and reliable technique for modelling instability development in complex boundary-layer flows, paving the way for future advancements in transition prediction and flow control applications.

Acknowledgements Funding by the United Kingdom's Defence, Science and Technology Laboratory, under contract DSTLX1000152974 is acknowledged.

References

1. Appel, T., Mughal, S., Ashworth, R.: Global stability analysis of a boundary layer with surface indentations. In: AIAA Aviation 2019 Forum. American Institute of Aeronautics and Astronautics, Dallas, Texas (2019). https://doi.org/10.2514/6.2019-3537
2. Cooke, E.: Modelling the Effect of Step and Roughness Features on Swept Wing Boundary Layer Instabilities. PhD thesis, Imperial College London (2020). https://doi.org/10.25560/83744
3. Guo, Y., Malik, M., Chang, C.-L.: A solution adaptive approach for computation of linear waves. In: 13th Computational Fluid Dynamics Conference. American Institute of Aeronautics and Astronautics, Snowmass Village, CO, U.S.A. (1997). https://doi.org/10.2514/6.1997-2072
4. Raposo, H., Mughal, S., Ashworth, R.: Acoustic receptivity and transition modeling of Tollmien-Schlichting disturbances induced by distributed surface roughness. Phys. Fluids **30**, 044105 (2018). https://doi.org/10.1063/1.5024909
5. Towne, A., Colonius, T.: One-way spatial integration of hyperbolic equations. J. Comput. Phys. **300**, 844–861 (2015). https://doi.org/10.1016/j.jcp.2015.08.015
6. Towne, A.: Advancements in Jet Turbulence and Noise Modeling: Accurate One-Way Solutions and Empirical Evaluation of the Nonlinear Forcing of Wavepackets. PhD thesis, California Institute of Technology (2016). https://doi.org/10.7907/Z99884XJ
7. Towne, A., Rigas, G., Colonius, T.: A critical assessment of the parabolized stability equations. Theor. Comput. Fluid Dyn. **33**(3–4), 359–382 (2019). https://doi.org/10.1007/s00162-019-00498-8
8. Xu, H., Mughal, S., Gowree, E.R., Atkin, C.J., Sherwin, S.J.: Destabilisation and modification of Tollmien-Schlichting disturbances by a three-dimensional surface indentation. J. Fluid Mech. **819**, 592–620 (2017). https://doi.org/10.1017/jfm.2017.193
9. Zhu, M., Towne, A.: Recursive one-way Navier-Stokes equations with PSE-like cost. J. Comput. Phys. **473**, 111744 (2023). https://doi.org/10.1016/j.jcp.2022.111744

Chapter 43
The Dual Presence of Very Large Structures in the Post-Transition State of Two-Dimensional Channel Flow

Ryo Takai, Rin Sasaki, Sattaya Yimprasert, Kentaro Kato, and Masaharu Matsubara

Abstract Two types of large-scale structures are known to exist in turbulent channel flow, with different spanwise scales that alternate for a certain Reynolds number. The aim of this study is to determine the Reynolds number at which the two types of large-scale structures alternate by means of a two-point velocity correlation experiment in the spanwise direction using a hot-wire anemometer and to investigate these structures by means of visualisation experiments. It was found that as the Reynolds number increased, the dominant structures switched from wide to narrow structures around 3500. In the flow visualisation, it was observed that the wide large-scale structures are clustered together with long streak structures.

Keywords Turbulence · Transition · Visualisation · Velocity correlation · Streak

43.1 Introduction

In two-dimensional turbulent channel flow, the existence of large-scale structures with a streamwise dimension of more than 10 times the channel width has been reported. Measurements by a hot-wire rake at very high Reynolds numbers ($Re = 144,000$) by Monty et al. [1] showed low velocity regions with streamwise and spanwise dimensions of about 10 and 0.5 times the channel width, respectively, indicating the presence of large-scale structures. The Reynolds number Re is defined as $Re = U_b d/\nu$, where U_b is the bulk (mean) velocity, d the channel width and ν the kinematic viscosity. Seki et al. [2] gradually increased the Reynolds number from the transient state of the channel flow and observed that even in the fully turbulent state ($Re \geq 2600$), a low frequency peak corresponding to about 25 times the channel width appeared in the velocity fluctuation spectrum. Matsubara et al. [3] performed

R. Takai · R. Sasaki · K. Kato · M. Matsubara (✉)
Department of Mechanical Systems Engineering, Shinshu University, Nagano, Japan
e-mail: mmatsu@shinshu-u.ac.jp

S. Yimprasert
Drive Center, Naresuan University, Phitsanulok, Thailand

K. Kato et al. (eds.), *Proceedings of the 10th IUTAM Symposium on Laminar-Turbulent Transition*, IUTAM Bookseries 44, https://doi.org/10.1007/978-981-96-9829-5_43

hot-wire measurements at higher Reynolds numbers than in the experiments of Seki et al. and showed that very large structures exist up to $Re = 4000$. Furthermore, for Reynolds numbers up to $Re = 7000$ Tanada et al. [4] confirmed the existence of large-scale structures with a streamwise length of more than 10 times the channel width, through two-point velocity correlation measurements, using two spanwise separated hot wires. They also showed that for $Re = 3300$ or less, the spanwise width of the large-scale structures is several times the channel width, while much narrower structures dominate above $Re = 3300$. In a water channel flow visualisation experiment Yimprasert et al. [5] observed characteristic oblique clusters filled with wavy streaks at $Re = 2700$, where the flow is considered to be fully turbulent. The clusters are 10 and 5 times the channel width in the streamwise and spanwise directions, respectively, close to those observed by Tanada et al. [4].

It has been assumed that after transition to a fully developed turbulent flow the properties change only gradually with changes in Reynolds number. Therefore the abrupt alternation at a certain Reynolds number between the two types of large-scale structures with very different spanwise scales is a surprising phenomenon that challenges this conventional wisdom. To understand this alternation of large-scale structures, it is essential to determine their dimensions and the Reynolds number at which they alternate. The purpose of this study is to determine the alternating Reynolds number and the dimensions of the large structures by measuring two-point velocity correlations in the spanwise direction using two hot wires and by flow visualisation.

43.2 Experimental Setups

Two channel-flow facilities were used in the present work, one air channel for two-point hot-wire velocity correlation measurement and one water channel for flow visualization using reflective flakes.

In the air facility a blower provides the air through a diffuser filled with an air filter sponge, five wire meshes and a two-dimensional nozzle with a contraction ratio of 16:1 before entering the channel section that has a width of $d = 25$ mm, a total length of 17.35 m and a height of 1.0 m. The channel section consists of two flat plates with side walls that are dumbbell shaped. The bell shape of the side walls is designed to have the same hydraulic diameter as the channel width, and helps to suppress secondary flows near the side walls. To promote transition to turbulence, a trip wire is placed 440 mm downstream of the nozzle outlet.

The coordinate system is defined with the origin at the centre of one of the flat plates, with the streamwise direction x, wall-normal direction y and spanwise direction z. The measurements were taken 300 mm upstream of the channel outlet with two constant-temperature hot-wire anemometers that measure the streamwise velocity. The sensors were made of platinum wires with a diameter of 2.5 μm and a length of 1 mm. The sampling frequency was 20 kHz and the sampling time for the mean velocity at each y position was 60 s, and 600 s for the two-point velocity correlation experiment at each z position.

The total length of the water facility is approximately 8.4 m, with the channel section having a width of $d = 7.2$ mm and a spanwise height of $h = 580$ mm. Water is circulated in the system and pumped through an orifice plate flowmeter before entering a settling box with five wire meshes after which a two-dimensional nozzle with a contraction ratio of 25:1 leads to the channel section with a length of 7320 mm. The channel flow is disturbed by a trip wire 220 mm downstream of the nozzle outlet, is accelerated by a spanwise contraction, resulting in the flow being fully developed into turbulence before entering the test section, which has a total length of 4650 mm, for a detailed description see Ref. [5].

A small amount (1.5 g) of pearlescent pigment was added to the 170 litres of water used in the facility. The reflective plate-shaped particles have a diameter of about 7 μm, are aligned with the local flow shear and reflect incoming light to reveal flow structures. The flow visualisation was performed 3500 mm downstream of the test-section inlet. Two halogen lamps were used as light sources and the video imaging area was 162×122 mm^2 (approximately $23d \times 17d$), with a resolution of 512×384 pixels.

43.3 Experimental Results

The mean velocity profiles $U(y)$ and the wall-normal distributions of the streamwise velocity fluctuations u_{rms} were measured in the air facility. The distance between the hot wire and the wall was estimated from the positional relationship between the image of the hot wire reflected from the mirrored wall and the real image simultaneously captured by a camera. For the mean velocity U shown in Fig. 43.1a, the

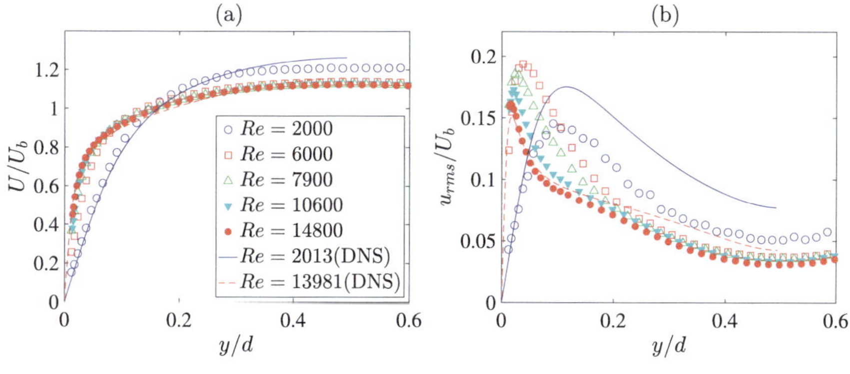

Fig. 43.1 Profiles of **a** mean streamwise velocity and **b** rms of streamwise velocity fluctuations

velocity gradient near the wall becomes steeper as Re increases, while the velocity distribution near the channel centre becomes flatter, indicating the general trend of turbulent channel flows with increasing Re. The bulk velocity U_b is evaluated by integrating the mean velocity profile using curve fitting. In the velocity fluctuation distributions in Fig. 43.1b, the peak position moves towards the wall with increasing Re, with the peak value increasing from $Re = 2000$ to $Re = 6000$ and then decreasing, indicating the general nature of wall shear turbulence. Both the mean velocity and velocity variation distributions at $Re = 2000$ and $Re = 13700$ are in close agreement with the results of direct numerical simulations (DNS) [6].

Two-point velocity correlations R were evaluated from velocity fluctuations measured by two spanwise separated hot wires. Figure 43.2 shows maps of the normalised cross-correlation R obtained for each frequency component. From the complex spectra $U_1(z_1, f)$ and $U_2(z_2, f)$ obtained by Fourier transforming the velocity time series $u_1(z_1, t)$ and $u_2(z_2, t)$ measured by the two hot wires, the cross spectrum $U_1 \times U_2^*$ was inverse Fourier transformed to obtain the two-point velocity correlation R from its real part, where z_1 and z_2 are the spanwise positions of the hot wires, t is time, f is frequency, * denotes complex conjugate. The vertical axis shows the dimensionless streamwise wavelength $\lambda = U_b / f d$ estimated from the spectral frequency f. The two hot wires were placed at the centre of the channel, and the spanwise distance $\Delta Z = |z_1 - z_2| / d$ between the hot wires was varied exponentially from 0.2 to 8 with 60 measurement points. These maps were smoothed using a simple moving average.

In the maps for $Re = 3000$, 3200 and 3400 there is a strong negative correlation in the $2 < \Delta z < 8$ region around $\lambda = 30$, indicating the presence of large structures

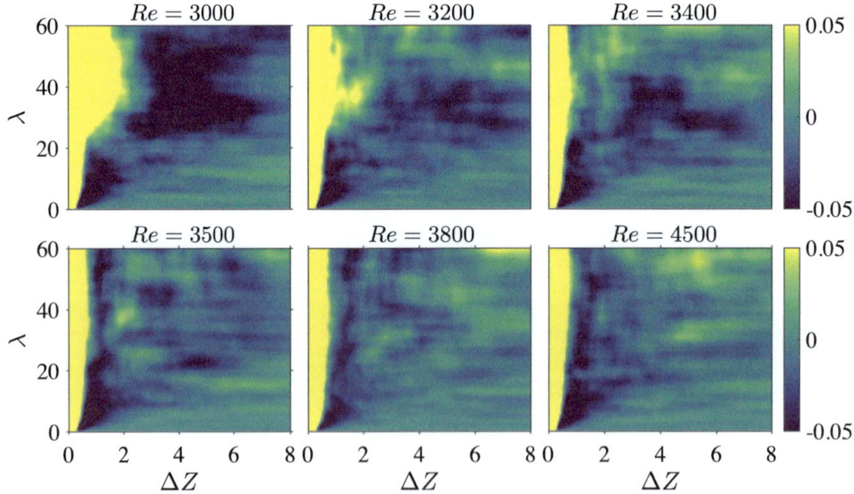

Fig. 43.2 Cross correlation of the velocity fluctuation R at the channel centre

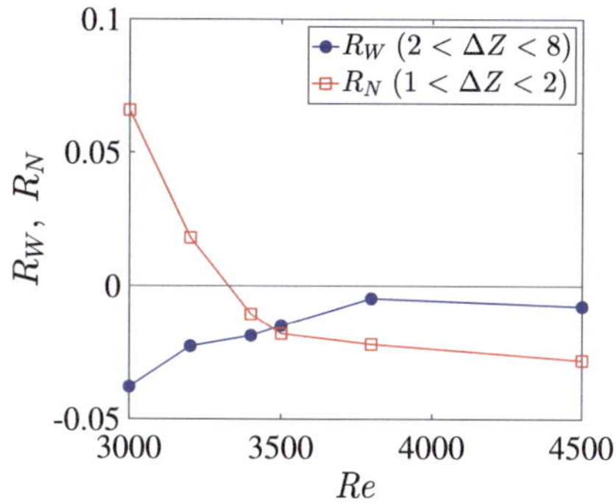

Fig. 43.3 Variation with Reynolds number of the cross-correlation coefficient averaged over $20 < \lambda < 45$ for small ($1 < \Delta z < 2$) and large ($2 < \Delta z < 8$) spanwise scales

with a streamwise length 10 times greater than the channel width d. These streamwise and spanwise scales are very close to those of the turbulent patches of transitional channel flow observed by Yimprasert et al. [5]. On the other hand, there is another strong negative correlation in the region $\Delta z < 2$, with relatively short wavelengths in the flow direction up to $Re = 3200$. In terms of their dimensions, this can be attributed to vortex structures generated close to the wall. For $Re \geq 3400$, the negative region for $1 < \Delta z < 2$ extends to even larger wavelengths, while the negative region for $2 < \Delta z < 8$ tends to disappear.

This change in scales is more clearly shown in Fig. 43.3 by averaging the correlation coefficients R_W ($20 < \lambda < 45$, $2 < \Delta z < 8$), corresponding to the wide large-scale structures, and R_N ($20 < \lambda < 45$, $1 < \Delta z < 2$), corresponding to the narrow large-scale structures. R_W shows a negative correlation at $Re = 3000$ with $R_W = -0.038$, which increases with increasing Re and becomes almost constant around a value of 0 at $Re \geq 3800$. In contrast to R_W, R_N decreases as Re increases, crossing the value 0 between $Re = 3200$ and 3400, and asymptotically reaching about $R_N = -0.03$ for $Re \geq 3500$. The intersection of these correlations is around $Re = 3500$, indicating that the dominant large-scale structures change from wide to narrow around this Re.

Figure 43.4 shows the visualisation images of the water channel flow and their moving average processed images. The visualised image is normalised by the standard deviation of the brightness of each image, after subtracting the time-averaged image of 1800 frames clipped from the video. The range in the moving average is $10d \times 3d$ in the streamwise and spanwise directions respectively. In the visualisation image at $Re = 1800$, bright regions with long streaks and dark regions with fine vortex structures are distributed in a checkerboard pattern. The fine eddy structures in the dark regions are irregularly positioned, indicating that these regions are the regular turbulent part of the flow. The long streaks in the light region are clustered in the spanwise direction and do not have fine turbulent structures, indicating that this is the non-turbulent part. The hot-wire measurements of Seki et al. [2] also show an interruption in the velocity waveform at $Re = 1800$, and this visualisation

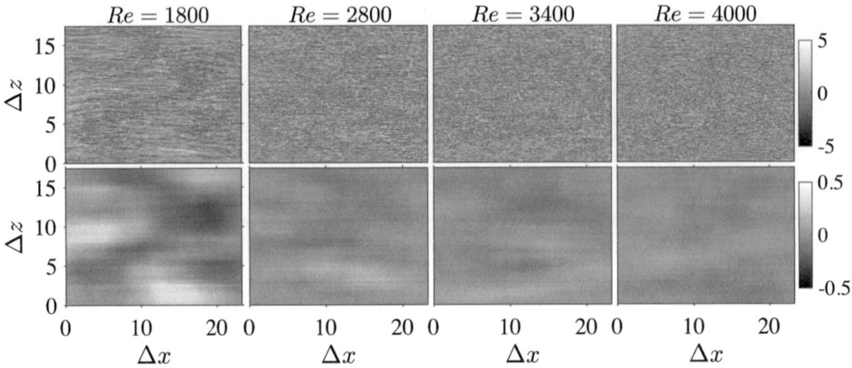

Fig. 43.4 Visualization images (upper row) and moving-average filtered visualization of instantaneous images (lower row). The flow direction is from left to right

experiment also confirms that $Re = 1800$ is a transitional flow. Comparing the visualisation image at $Re = 1800$ with the moving average image, it can be seen that the non-turbulent regions still correspond to bright areas after applying the moving average.

Visualisation images at $Re = 2800$ and $Re = 3400$, which are considered to be fully turbulent flows in the experiments of Seki et al. [2] show that the regions of regular streaks without fine structure have disappeared and the whole area is filled with irregular structures. The turbulent structures in the dark regions have a relatively short streamwise scale of less than $1d$, while the bright regions are filled with long and irregular streaks, which are densely packed together making the whole region brighter. The moving average results for $Re = 2800$ and $Re = 3400$ show that both have a checkerboard pattern of bright and dark regions, similar to the transitional flow at $Re = 1800$. The long irregular streaks are also observed at $Re = 4000$, although the contrast between bright and dark areas is weaker, as can be seen from the moving averages.

The checkerboard pattern in the moving average at $Re \leq 3400$ is more than $10d$ in the streamwise direction and about $5d$ in the spanwise direction. These dimensions are also consistent with the correlation results from hot-wire measurements, and the decrease in contrast of the moving average with increasing Re corresponds to a change in R_W, and strongly suggests that the regions of dense long irregular streaks are the wide large-scale structures. Although not shown here, observation of the visualisation photographs as Re is gradually increased suggests that the regions of long irregular streaks can be interpreted as the streaks in the non-turbulent regions at the transitional Re, which oscillate irregularly in the spanwise direction after the transition. It is inferred that while the entire flow after $Re = 2600$ is filled with irregular flow structures, the patchy large-scale structures seen at the transition Re also exist after the transition, but are no longer maintained above approximately $Re = 3500$.

43.4 Conclusions and Discussion

The measurement of the spanwise two-point cross-correlation coefficients confirms that the scales of wide large-scale structures in two-dimensional turbulent channel flow are approximately $15d \times 5d$ in the streamwise and spanwise directions, respectively, while the scales of the narrow large-scale structures are approximately $15d \times 1d$. Direct inspection of the visualisation images as well as the results from the processed images suggest the existence of regions where the wide large-scale structures are densely clustered with long irregular streaks and that the dominant wide large-scale structures are replaced by the narrow ones around $Re = 3500$.

References

1. Monty, J.P., Stewart, J.A., Williams, R.C., Chong, M.S.: Large-scale features in turbulent pipe and channel flows. J. Fluid Mech. **889**, 147–156 (2007). https://doi.org/10.1017/S002211200700777X
2. Seki, D., Matsubara, M.: Experimental investigation of relaminarizing and transitional channel flows. Phys. Fluids **24**, 24102 (2012). https://doi.org/10.1063/1.4772065
3. Matsubara, M., Horii, S., Sagawa, Y., Takahashi, Y., Saito, D.: Very-large-scale fluctuations in turbulent channel flow at low Reynolds number. Int. J. Heat Fluid Flow **62**B, 593–597 (2016). https://doi.org/10.1016/j.ijheatfluidflow.2016.11.009
4. Tanada, Y., Matsubara, M.: Alternation phenomenon of very large scale feature in turbulent channel flow. Annual Meeting Japan Soc. Fluid Mech. 184, 4 pages (2019). (in Japanese)
5. Yimprasert, S., Kvick, M., Alfredsson, P.H., Matsubara, M.: Flow visualization and skin friction determination in transitional channel flow. Exp. Fluid, **62**, 31 (2021). https://doi.org/10.1007/s00348-020-03102-6
6. Kawamura Lab, Tokyo University of Science, "Turbulence Database", published October 2018, accessed January 18, 2024, https://www.rs.tus.ac.jp/t2lab/db/

Author Index

K. Kato et al. (eds.), *Proceedings of the 10th IUTAM Symposium on Laminar-Turbulent Transition*, IUTAM Bookseries 44, https://doi.org/10.1007/978-981-96-9829-5

MIX
Papier aus verantwortungsvollen Quellen
Paper from responsible sources
FSC® C105338

If you have any concerns about our products,
you can contact us on
ProductSafety@springernature.com

In case Publisher is established outside the EU,
the EU authorized representative is:
Springer Nature Customer Service Center GmbH
Europaplatz 3, 69115 Heidelberg, Germany

Printed by Libri Plureos GmbH
in Hamburg, Germany